Palladium	Pd	46	106.4	12.02	
Phosphorus	P	15	30.9738	1.88–2.69	X_2
Platinum	Pt	78	195.09	21.45	Found in blends with other platinum series metals
Potassium	K	19	39.102	0.85	Sylvite (KCl) and Carnallite (KCl-MgCl$_2$-6H$_2$O)
Praseodymium	Pr	59	140.907	6.769	Trace quantities in Monazite
Radon	Rn	86	222.02	n/a	n/a
Rhenium	Re	75	186.2	21.04	Trace quantities in Molybdenite
Rhodium	Rh	45	102.905	12.41	Found in blends with platinum and other platinum series metals
Rubidium	Rb	37	84.57	1.532	No commercial minerals but is found with cesium oxides in biotite
Ruthenium	Ru	44	101.07	12.45	Found in blends with platinum and other platinum series metals
Samarium	Sm	62	150.35	7.536	Trace quantities in Monazite
Scandium	Sc	21	44.956	2.99	Thortveitite (30% Sc$_2$O$_3$)
Selenium	Se	34	78.96	4.28–4.79	Eucairite (CuAgSe) and Clausthalite (PbSe)
Silicon	Si	14	28.086	2.33	Numerous silicates (containing SiO$_4^-$ ions)
Silver	Ag	47	107.868	10.5	Argentite (Ag$_2$S)
Sodium	Na	11	22.9898	0.967	Halite (NaCl)
Strontium	Sr	38	87.62	2.60	Celestite (SrSO$_4$) and Strontianite (SrCO$_3$)
Sulfur	S	16	32.064	1.94–2.14	Iron Pyrite (FeS$_2$), Chalcopyrite (CuFeS$_2$), and Galena (PbS)
Tantalum	Ta	73	180.948	16.654	Tantalite ((Fe,Mn)Ta$_2$O$_6$)
Tellurium	Te	52	127.60	6.24	Alatite (PbTe), Coloradoite (HgTe), and Rickardite (Cu$_4$Te$_3$)
Terbium	Tb	65	158.924	8.23	Trace quantities in Monazite
Thallium	Tl	81	204.37	11.85	Trace amounts in sulfide ores
Thorium	Th	90	232.038	11.72	Trace quantities in Monazite
Thulium	Tm	69	168.934	9.318	Trace quantities in Monazite
Tin	Sn	50	118.69	5.77α or 7.29β	Cassiterite (SnO$_2$)
Titanium	Ti	22	47.90	4.507	Arizonite (Fe$_2$Ti$_3$O$_9$), Perovskite (CaTiO$_3$), and Rutile (TiO$_2$)
Tungsten	W	74	183.85	19.3	Wolframite ((Fe,Mn)WO$_4$)
Uranium	U	92	238.03	18.97	Pitchblende (mix of UO$_2$ and U$_3$O$_8$)
Vanadium	V	23	50.942	6.11	Small quantities of V$_2$O$_3$ are found in Magnetite iron ores
Xenon	Xe	54	131.30	n/a	n/a
Ytterbium	Yb	70	173.04	6.972	Trace quantities in Monazite
Yttrium	Y	39	88.905	4.472	Trace quantities in Monazite
Zinc	Zn	30	65.37	7.133	Sphalerite (ZnS) and Smithsonite (ZnCO$_3$)
Zirconium	Zr	40	91.22	6.506	Zircon (67.2% ZrO$_2$, 32.8% SiO$_2$)

Explanations for Unusual Chemical Symbols

Element	Symbol	Basis for Symbol
Antimony	Sb	From the Latin *stibium* which means mark.
Copper	Cu	From the Latin *cuprum* which was a simplified form of the original name *cyprium*, which was given because of the active mining of the metal in Cyprus.
Gold	Au	From the Latin *aurum*.
Iron	Fe	From the Latin *ferrum*.
Lead	Pb	From the Latin *plumbum nigrum*. The word *plumbum* means soft metal.
Mercury	Hg	From the Latin *hydragyrum*, an adaptation of a Greek word meaning liquid silver.
Potassium	K	From the Latin *Kalium* meaning alkali.
Silver	Ag	From the Latin *argentum*, meaning shining white.
Sodium	Na	From the Latin *natrium*, which came from the Egyptian word *natron* and referred to a sodium carbonate salt.
Tin	Sn	From the Latin *stannum*.
Tungsten	W	From Wolfram, the name coined by Johann Wallerius in 1747 after the tungsten ore wolframite. The name Wolfram remains in common use in Germany and Sweden.

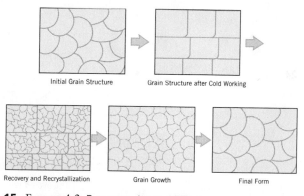

14. FIGURE 4-1 Loading–Unloading *(page 108)*

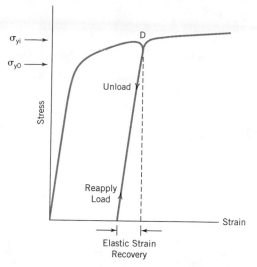

Initial Grain Structure   Grain Structure after Cold Working

Recovery and Recrystallization   Grain Growth   Final Form

15. FIGURE 4-2 Recovery *(page 109)*

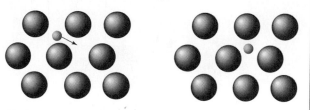

16. FIGURE 4-6 Interstitial Diffusion *(page 115)*

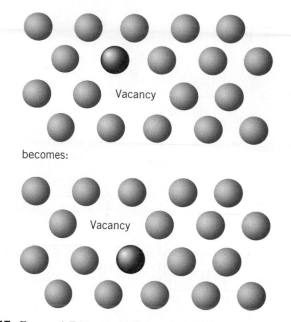

becomes:

17. FIGURE 4-7 Vacancy Diffusion *(page 115)*

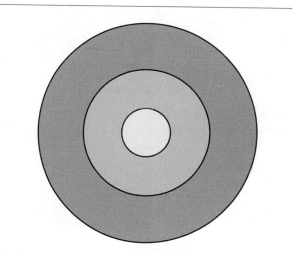

18. FIGURE 4-9 Segregation *(page 116)*

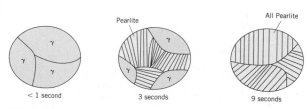

19. FIGURE 4-22 Phase Transformation and Grain Growth
(page 129)

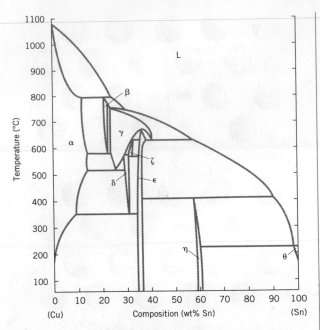

20. FIGURE 4-28 Electrochemical Cell *(page 134)*

21. Polymer Movement Near the Glass Transition Temperature *(page 160)*

Initiation
$$HO—OH + Light \longrightarrow 2HO\cdot$$

Free Radical Reacts with First Monomer

Propagation (Repeats Until Termination)

Propagation (Repeats Until Termination)

Primary Termination

Mutual Termination

22. FIGURE 5-22 Addition Polymerize *(page 162)*

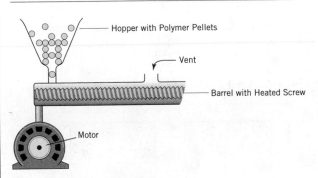

Terephthalic Acid   Ethylene Glycol

Yields

Polyethylene Terephthalate (PET) and Water

23. FIGURE 5-25 Condensation *(page 165)*

Hopper with Polymer Pellets

Vent

Barrel with Heated Screw

Motor

24. FIGURE 5-34 Extruder *(page 178)*

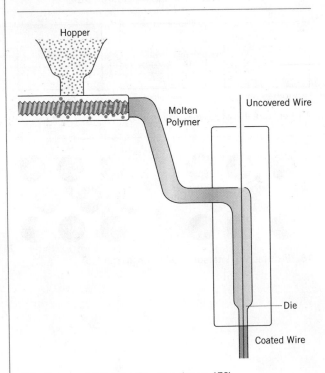

Hopper

Molten Polymer

Uncovered Wire

Die

Coated Wire

25. FIGURE 5-35 Wire Coating *(page 178)*

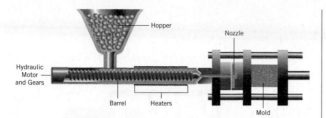

26. FIGURE 5-39 Injection Molding *(page 180)*

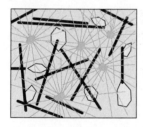

27. FIGURE 6-10 Sintering *(page 200)*

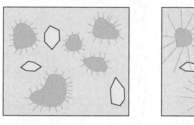
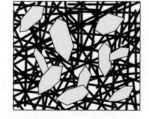

28. FIGURE 6-17 Cement Formation *(page 208)*

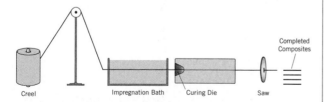

29. FIGURE 7-2 Pultrusion *(page 234)*

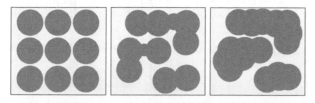

30. FIGURE 7-3 Wet Winding *(page 234)*

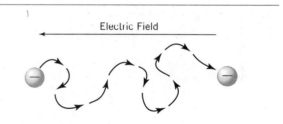

31. FIGURE 7-5 Prepregging *(page 235)*

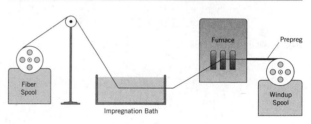

32. FIGURE 8-3 Electric Field *(page 251)*

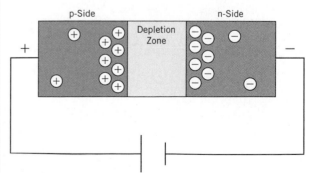

33. FIGURE 8-10 Forward Bias *(page 258)*

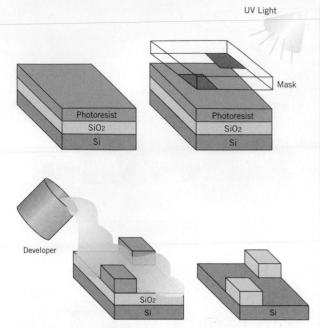

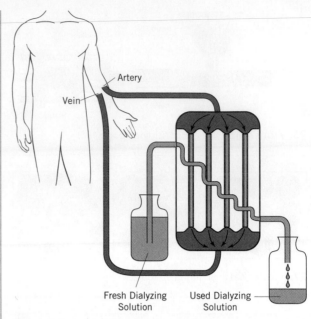

Essentials of Modern Materials Science and Engineering

1

Introduction

CONTENTS

Learning Objectives

By the end of this chapter, a student should be able to:

- Explain why you should study materials science.

- Evaluate desirable properties for specific applications.

- Explain the use and limitations of evaluation heuristics such as Ashby-style charts.

- Describe the role of economics in materials selection.

- Explain the significance of the four quantum numbers.

- Distinguish between primary and secondary bonding.

- Explain the differences between ionic, metallic, and covalent bonding and determine which type will be present given any two atoms (if their electronegativities are known).

- Explain the physical basis for dipole forces, hydrogen bonding, and Van Der Waals forces.

- Analyze the sustainability of materials and the impact of green engineering on decision making.

- Describe the fundamental properties of the major classes of materials.

Why Study Materials Science?

1.1 OVERVIEW OF MATERIALS SCIENCE

The goal of all materials science is to empower scientists and engineers to make informed choices about the design, selection, and use of materials for specific applications. Four fundamental tenets guide the study of materials science:

1. The principles governing the behavior of materials are grounded in science and are understandable.

2. The properties of a given material are determined by its structure. Processing can alter that structure in specific and predictable ways.

3. The properties of all materials change over time with use and exposure to environmental conditions.

4. When selecting a material for a specific application, sufficient and appropriate testing must be performed to ensure that the material will remain suitable for its intended application throughout the reasonable life of the product.

A materials scientist or engineer must:

- Understand the properties associated with the various classes of materials.

- Know why these properties exist and how they may be altered to make a material more suitable for a given application.

- Be able to measure important properties of materials and how those properties will impact performance.

- Evaluate the economic considerations that ultimately govern most materials issues.

- Consider the long-term effects of using a material on the environment.

What Issues Impact Materials Selection and Design?

If you are about to replace a system of copper piping, how do you decide whether to replace it with copper again, stainless steel, PVC, or something else entirely? What will you do with the copper piping that you are removing? Can it be reused elsewhere in the plant? Can it be sold to a recycling center or other plant? Will it have to be landfilled? The best answers to these questions depend on a blend of the inherent physical and chemical properties of the material. Ultimately, the decisions will be governed by the decision maker's knowledge of these properties and by economic factors.

To make an informed decision in the design or selection of a material, first you must know what properties are important to the specific application while recognizing that the list of desired properties may become longer and more complicated as product needs evolve. For example, prior to 1919, most automobiles had no windshields, which left drivers vulnerable to rain, splashed mud, and objects flung up from the road. When selecting a material to develop windshields, car designers likely put together a list of desired properties that resembled this one:

1. It must be transparent. Obviously, a windshield that cannot be seen through would be of little practical value.
2. It must be impervious to water. Otherwise, the car could not be driven in the rain.
3. It must be tough enough to resist breaking from minor impacts (small pieces of gravel, hailstones, etc.).
4. It must be inexpensive enough to not alter the price of the car substantially.
5. It must withstand various temperatures, from a few degrees below zero in the winter (more in North Dakota) to 100° Farenheit in the summer.

Most people would have little difficulty generating this list and likely would be able to identify a simple answer: glass. By 1929, nearly 90 percent of new automobiles were surrounded by glass. Unfortunately, the list of questions above was hardly complete, and early windshields were fraught with problems. If you have ever tried to drop a coin into a glass in the bottom of a fish bowl, you know that glass refracts light, distorting the apparent position of objects. This problem was so significant for drivers that some windshields stopped below eye level so that the drivers could look over the glass; others came in two parts, so drivers could open the top piece to see out. The 1920s Franklin shown in Figure 1-1 has such a split windshield and no side panels of glass at all.

In 1928, Pittsburgh Plate Glass (now PPG Industries) developed the **Pittsburgh process** that made the windshields cheaper and dramatically reduced the distortion. This process resulted from groups of materials scientists and engineers applying their knowledge of refraction and the structure of glass to design a new product that would reduce or eliminate the earlier problem.

| *Pittsburgh Process* |
Glassmaking process developed in 1928 to reduce both cost and distortion.

If everyone drove safely at all times, simple glass might have been a good enough answer. Regrettably, glass tends to shatter into sharp shards upon impact. During accidents, drivers were often cut by these shards or, even worse, were ejected through the windshield. Engineers resolved this problem by creating glass laminates, in which layers of film were placed between thin sheets of glass as shown in Figure 1-2. This so-called safety glass reduced both cuts and passenger ejections. By 1966, safety glass was required on all cars manufactured in the United States.

Scientists and engineers continued to grapple with improving the toughness and quality of the glass, but carmakers provided them with new lists of desired

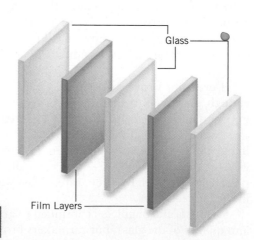

Glass

Film Layers

properties. For example, car designers requested curved panels of glass for windshields and side panels to improve both the aerodynamics and the visual appeal of cars. Until 1934, glass was produced in flat sheets. Curved windshields first appeared in 1934, but it was not until the late 1950s that a process was developed to make affordable curved side windows, which were smaller and less uniform in shape than windshields. Modern improvements include tempered safety glass that breaks in smooth fragments to reduce injury as well as tinting that allows drivers to see out while reducing glare and providing some privacy, as on the car shown in Figure 1-3.

For each improvement to occur, scientists and engineers needed to understand what characteristics of glass resulted in desirable and undesirable properties, and how to alter the structure of the material to improve its suitability for the product. Developing this understanding is the point of materials science. Find the need, select the appropriate material, and use your knowledge of that material to alter its properties to suit the needs of the specific application, which will be different from those of other applications and may change over time.

Let us consider another more modern example. The space shuttle, shown in Figure 1-4, uses an elaborate thermal protection system to shield the astronauts inside from the heat of reentry into Earth's atmosphere. When we considered the windshield materials, we noted that they must maintain their properties across a temperature range of a little more than 100°F. During reentry the space shuttle attains speeds as high as 17,000 miles per hour. The external materials on the shuttle pass rapidly from temperatures in space of close to absolute zero to as high as 3000°F.

Highly advanced black silica-based ceramic tiles cover most of the underside of the shuttle and are supplemented with insulating blankets and a second layer of white ceramic tiles. These materials provide excellent insulation and are lightweight but cannot handle the entire protection process alone. The nose of the shuttle and the leading edges of the wings experience the greatest heating during reentry. For these areas, highly specialized carbon-carbon composites are used because of their unique ability to conduct in one direction and insulate

in another. These composites shield the cabin from the heat while conducting it away from the leading edges of the shuttle.

This thermal protection system took years to develop, and scientists continuously try to improve on it. Some researchers have examined the use of specialized metallic tiles, but their greater weight has limited their application. Other scientists continue to examine combinations of advanced ceramics to improve on the existing system.

Whether dealing with something as common as windshields or as exotic as the space shuttle, the role of materials scientists and engineers is fundamentally the same. They examine the desired properties for an application, select the best available material, and use their knowledge of the structure and processing of materials to make improvements as needed. The specific challenges vary by application. Weight is a more important factor than cost for a space shuttle, which must escape Earth's gravitational pull, but less so for an automobile, which must remain inexpensive enough for most people to afford. Disposal and/or recycling are less important issues for space shuttles because there are few of them; these factors become important when considering the millions of cars currently on the road.

The range of available materials is enormous, and it is impractical to perform a detailed analysis of every possible material for every application. Instead, engineers and scientists apply their knowledge of the classes of

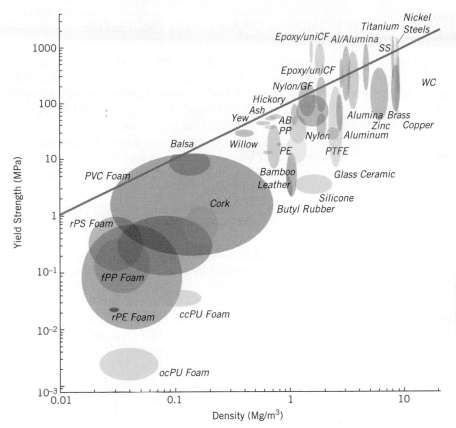

FIGURE 1-5 Ashby-Style Chart Relating Density to Yield Strength for Various Classes of Materials

From M. Ashby and K. Johnson, Materials and Design: The Art and Science of Material Selection in Product Design. Copyright © 2002 by Elsevier Butterworth-Heinemann. Reprinted by permission of Elsevier Butterworth-Heinemann.

materials along with simple guidelines, or heuristics, to help narrow the search for the best materials for specific applications. *Ashby-style charts,* such as the one shown in Figure 1-5, provide a quick and simple means of looking at how different classes of materials tend to perform in terms of specific properties. As a quick heuristic to target a search for the appropriate material, these charts may prove invaluable, but they have limitations. Ashby-style charts give no insight as to why a specific class of material outperforms another in a specific area, nor do they provide guidance on how to select between the broad range of materials within a given category or suggestions on how to optimize the performance of a specific material. These charts are useful tools to target a search but cannot replace the judgment of an engineer or a scientist trained in materials science.

Much of the rest of this book focuses on:

- The classes of materials from which a candidate may be selected for an application
- Explanations of the properties that influence the behavior of these materials and how to measure them
- Examination of the structures in materials that control these properties
- Examination of the processing strategies that can alter these structures and properties

| *Ashby-Style Charts* |
Heuristics used to provide a quick and simple means of looking at how different classes of materials tend to perform in terms of specific properties.

1.3 IMPACT OF BONDING OF MATERIAL PROPERTIES

| *Bohr Model* |
Classical representation of atomic structure in which electrons orbit the positively charged nucleus in distinct energy levels.

| *Orbitals* |
Discrete energy levels in which electrons rotate around the nucleus of an atom.

| *Quantum Mechanics* |
The science governing the behavior of extremely small particles such as electrons.

| *Quantum Numbers* |
Four numbers used to classify individual electrons based on their energy, cloud shape, orientation of cloud, and spin.

| *Principal Quantum Number* |
Number that describes the major shell in which an electron is located.

| *Second Quantum Number* |
Number that describes the shape of the electron cloud.

| *Third Quantum Number* |
Number representing the orientation of the electron cloud.

Ultimately, the properties of materials are determined by the types of atoms present, their relative orientation, and the nature of the bonding between them. A review of basic chemical principles is necessary to discuss the role of bonding. The ***Bohr model***, shown in Figure 1-6, depicts an atom with a positively charged nucleus in the center and electrons orbiting in distinct energy levels. Although it is convenient to picture the electrons as particles, they possess properties of both particles and waves. Thus, it is more convenient to think of an electron as an "electron cloud" in which the electron will be present in different parts of the cloud at different times.

The packing of these energy levels, or ***orbitals***, with electrons follows very specific rules governed by a science called ***quantum mechanics***. These rules, first presented by Erwin Schroedinger in the 1920s, allow the energy of a given electron, the shape of its electron cloud, the cloud's orientation in space, and the spin of the electron to be characterized by four numbers called ***quantum numbers***.

The ***principal quantum number*** (n) determines the energy of the electron. The innermost orbital has a primary quantum number of 1, the next 2, and so on. Figure 1-6 shows these orbitals schematically. Most commonly, letters are assigned to represent the individual orbitals. In such cases, K corresponds to $n = 1$, L to $n = 2$, and so on.

The ***second quantum number*** (λ) determines the general shape of the electron cloud. Some of the energy levels in the Bohr model split into sublevels with slightly different energies and quite different shapes. In the nth level, there are n-sublevels possible. As Figure 1-7 shows, there is only one orbital possible for the shell that corresponds to $n = 1$, but two for $n = 2$, and three for $n = 3$. The first sublevel ($\lambda = 0$) is called an s-sublevel; the second ($\lambda = 1$) is called a p-sublevel. Figure 1-7 summarizes the nomenclature and energy levels.

The ***third quantum number*** (m_λ) indicates how the electron cloud is oriented in space. m_λ can have any integer value (including 0) from $-\lambda$ to $+\lambda$. As such, an s-sublevel can have only $m_\lambda = 0$, while a p-sublevel could have values

Electrons revolve around the nucleus in discrete energy levels. For an electron to change energy levels, it must either gain or lose specific amounts of energy.

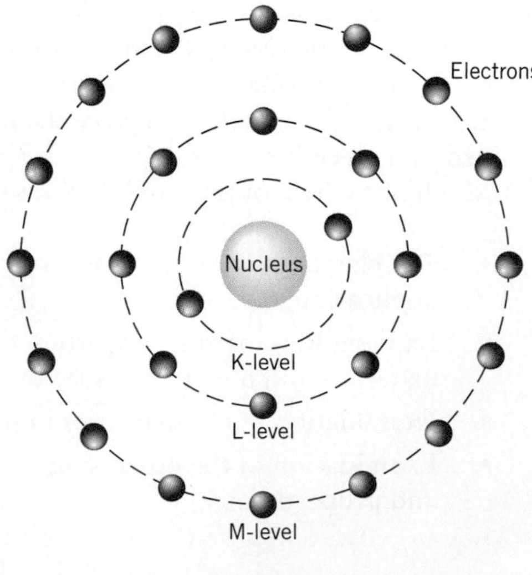

FIGURE 1-6 Bohr Model of an Atom

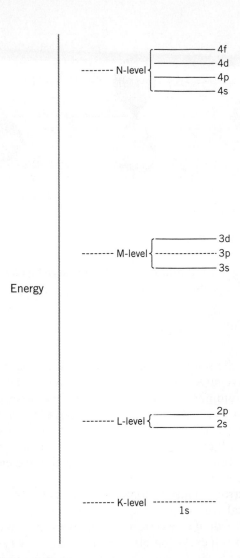

Energy

N-level
- 4f
- 4d
- 4p
- 4s

M-level
- 3d
- 3p
- 3s

L-level
- 2p
- 2s

K-level
- 1s

Suborbital Designations

n	1	2		3			4			
Primary Quantum Number Designation	K	L		M			N			
l	0	0	1	0	1	2	0	1	2	3
Suborbital Designations	1s	2s	2p	3s	3p	3d	4s	4p	4d	4f

FIGURE 1-7   Suborbital Designations

m_λ of −1, 0, or +1. The shapes of the electron clouds vary with the sublevel in which they are located. The clouds from s-sublevels are spherical, while those from p-sublevels form dumbbell shapes like the ones shown in Figure 1-8.

The *fourth quantum number* (M_s) represents the *spin* of the electron. Spin is a theoretical concept that derives from complex quantum mechanics and allows individual electrons within sublevels to be distinguished from each

| *Fourth Quantum Number* |
Number representing the spin of an electron.

| *Spin* |
A theoretical concept that enables individual electrons within sublevels to be distinguished from each other.

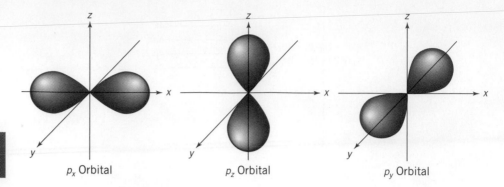

FIGURE 1-8 Shapes of Electron Clouds in p-Sublevels

p_x Orbital p_z Orbital p_y Orbital

other. The fourth quantum number has no relationship to the other quantum numbers and can have only two possible values:

$$M_s = +\frac{1}{2} \text{ or } -\frac{1}{2}. \tag{1.1}$$

Electrons with the same value of Ms have *parallel spins* while those with opposite values have *opposed spins*.

The four quantum numbers allow every electron in an atom to be characterized uniquely. In 1925, Wolfgang Pauli showed that no two electrons in an atom can have the exact same set of four quantum numbers. This observation is known as the *Pauli exclusion principle*, and its main effect is that no more than two electrons can fit in any orbital and that the two electrons in a suborbital must have opposed spins.

In general, electrons fill up the lowest available energy states with two electrons per suborbital with opposed spins, until the atom has run out of electrons. An atom with all of its electrons in the lowest possible energy levels that do not violate the Pauli exclusion principle is in its *ground state*. When excited by energy or magnetic fields, some electrons can move temporarily to higher energy levels. This is the basis for all electronic materials and is discussed in far greater detail in Chapter 8.

When atoms interact with each other, the electrons in the outermost energy levels (the valence electrons) interact with each other first and are the most important in determining the bonding between the atoms. When the outermost energy level is completely filled (e.g., the eight p-electrons found in the noble gases), there is no thermodynamic reason for the atom to bond with a neighbor. When the outermost shells are unfilled, atoms often gain, lose, or share electrons with other atoms, a process that serves as the basis for chemical bonding.

The interaction between atoms is a blend of attractive and repulsive forces. Atoms that are far apart have almost no interaction, but as they draw closer together, a blend of attractive and repulsive forces begins. Valence electrons are repelled by the negatively charged electron cloud of the adjacent atom but are attracted to the positive nucleus. The specific nature of the interaction between atoms depends on the state of the valence electrons and the type of bonding that forms.

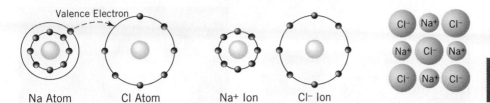

FIGURE 1-9 Ionic Bonding in Sodium Chloride (NaCl)

Ionic bonding is conceptually the simplest type of bonding between atoms. An electropositive material that has one or more extra electrons beyond its last completed sublevel approaches an electronegative material that is one or more electrons short of filling its outermost sublevel. Electrostatic forces make it energetically favorable for the electropositive material to donate its valence electron (or electrons) to the electronegative material. Metals from Groups I and II of the periodic table often form ionic bonds with the halogens from Group VII. The halogens are all one electron short of having eight valence electrons, so they easily adopt the extra electron from the metal. Compounds such as NaCl and CaF_2 are classic ionic-bonded materials. Figure 1-9 presents a schematic showing ionic bonding in sodium chloride.

For any two atoms, there is an optimal distance that represents the minimum potential energy. Consider two hydrogen atoms, each with an unpaired valence electron in its s-sublevel. When the two atoms are far apart, they do not interact in any meaningful way, but, as Figure 1-10 shows, the attractive forces between the atoms increase relative to the repulsive forces, reaching a maximum at a distance of 0.074 nm. The most energetically favorable state for the atoms occurs when potential energy is a minimum. In this case, the minimum occurs at -436 kJ. As a result, the two hydrogen atoms will form an H_2 molecule with a bond energy of -436 kJ.

Unlike the donated electrons in ionic bonds, electrons in covalent bonds may be located at any point around the two nuclei, but they are more likely to be found between them, as shown in Figure 1-11. The existence of the energy minimum provides the basis for the covalent bonding, in which the electrons are shared.

| *Ionic Bonding* |

The donation of an electron from an electropositive material to an adjacent electronegative material.

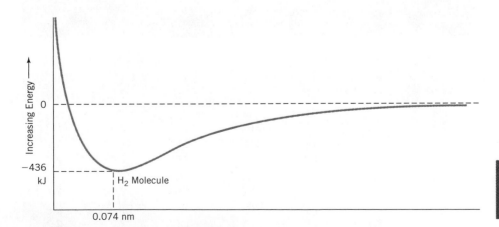

FIGURE 1-10 Potential Energy versus Atomic Distance for Two Hydrogen Atoms

FIGURE 1-11   Electron Cloud around a Nonpolar Covalent Bond

Courtesy James Newell

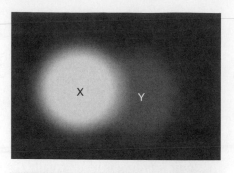

FIGURE 1-12   Electron Cloud around a Polar Covalent Bond (The more electronegative atom is on the left.)

Courtesy James Newell

| *Nonpolar* |
Interaction in which the electron density around adjacent atoms is symmetric.

| *Polar* |
Interaction in which the electron density around adjacent atoms is asymmetric.

| *Electronegativity* |
The ability of an atom in a covalent bond to attract electrons to itself.

When identical molecules (such as two hydrogens) are bonded together, the probability is exactly equal of finding a bonding electron near one molecule as the other, and the bond is called *nonpolar*. However, when different molecules interact, one is likely to have a somewhat greater affinity for electrons than the other. As a result, the electron density around the atoms will be asymmetric, and the bond is referred to as *polar*. The electron density for a polar bond is illustrated in Figure 1-12.

The ability of an atom to attract electrons to itself in a covalent bond is called its *electronegativity*. Fluorine (at the top of Group VII) is the most electronegative of all of the elements and is assigned an electronegativity value of 4.0. Cesium (in Group I) is the least electronegative and is assigned a value of 0.7. Table 1-1 summarizes the electronegativity values of many materials.

The amount of polarity in a covalent bond is directly related to the difference in electronegativities between the atoms. When the different is large (as in H—F), the bonding will be highly polar, but when the difference is small

TABLE 1-1 Electronegativity Values for Common Elements						
Electronegativity Values						
H 2.1						
Li 1.0	Be 1.5	B 2.0	C 2.5	N 3.0	O 3.5	F 4.0
Na 0.9	Mg 1.2	Al 1.5	Si 1.8	P 2.1	S 2.5	Cl 3.0
K 0.8	Ca 1.0	Sc 1.3	Ge 1.8	As 2.0	Se 2.4	Br 2.8
Rb 0.8	Sr 1.0	Y 1.2	Sn 1.8	Sb 1.9	Te 2.1	I 2.5
Cs 0.7	Ba 0.9	La 1.0	Pb 1.9	Bi 1.9	Po 2.0	At 2.2

(as in C—H), the bond will be only slightly polar. The polarity in a covalent bond is often thought of as a partial ionic character of the bond. A highly polar bond has electrons that spend significantly more time near the electronegative atom, much like an ionic bond. As a result, characterizing a bond as ionic or covalent is really an oversimplification. Most real bonds have characteristics of both ionic and covalent bonds. The difference in electronegativity corresponds directly to the percent ionic character of the bond, as shown in Figure 1-13.

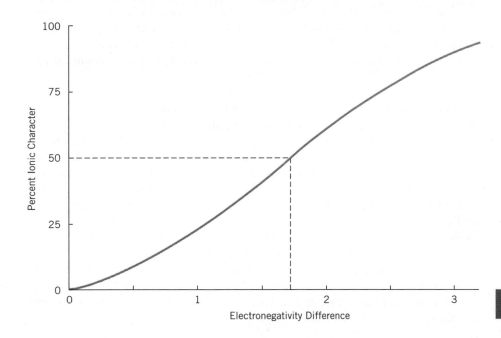

FIGURE 1-13   Percent Ionic Character

Example 1-1

Determine how ionic the bonds between the following atoms would be:

a. Sodium and chlorine
b. Carbon and nitrogen
c. Potassium and sulfur

SOLUTION

a. The electronegativity of sodium (Na) is 0.9 and chlorine (Cl) is 3.0, according to Table 1-1. The difference is $3.0 - 0.9 = 2.1$, which is about 68% ionic.
b. The electronegativity of carbon (C) is 2.5 and nitrogen is 3.0. The difference is $3 - 2.5 = 0.5$, which corresponds to about 10% ionic (90% covalent).
c. The electronegativity of potassium (K) is 0.8 and sulfur (S) is 2.5. The difference is $2.5 - 0.8 = 1.7$, which corresponds to 50% ionic.

Both ionic and covalent bonds deal with the interactions between atoms, but the molecules formed by bonded atoms also interact with each other. These interactions, referred to as secondary bonds, can have significant influence on the behavior of solids. Polar molecules in a crystalline structure align so that the positive pole of one molecule is closest to the negative pole of the adjacent molecule, as shown in Figure 1-14. As a result, an electrostatic interaction between the molecules, called a *dipole force*, forms. These interactions create added strength in the material and raise the boiling point of liquids.

The most extreme case of dipole forces involves a hydrogen atom interacting with an atom of fluorine (F), oxygen (O), or nitrogen (N) from an adjacent molecule. In such a case, the hydrogen atom of one molecule is strongly attracted to the electronegative atom in the adjacent molecule, resulting in a *hydrogen bond*. The strength of a hydrogen bond is stronger than other dipole forces because the electronegativity difference between hydrogen and F, O, or N is large and because the small size of the hydrogen molecule allows the electronegative atom to approach the hydrogen atom closely.

The final type of secondary bonding occurs in all substances and increases with molecular weight. These forces, called *dispersion forces* or *Van Der Waals forces*, are caused by temporary dipole interactions that result from momentary concentration variations in the electron clouds of adjacent molecules. For example, two adjacent molecules of H_2 have no polarity overall, but, as shown in Figure 1-15, at a given moment in time, the electrons of molecule 1A may be both on the left side while those in 2A are also on the left. As a result, the right side of molecule 1A has a momentary effective positive polarity (in the instant before the electron returns) that is drawn to the momentary negative polarity from molecule 2A. Although these interactions are extremely brief, they happen over and over again. Larger molecules have more opportunities for these interactions to occur.

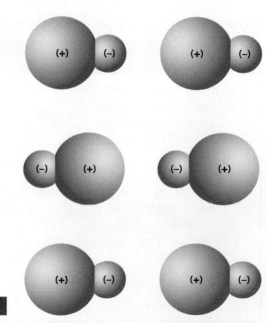

FIGURE 1-14   Illustration of Dipole Forces

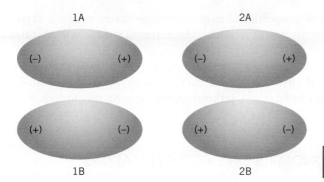

FIGURE 1-15 Dispersion Forces between Adjacent H_2 Molecules

The last significant type of bonding that greatly impacts material properties is specific to metals and is called *metallic bonding*. When two metallic atoms are bonded together, there is little to no difference in electronegativity, so the bonding is clearly not ionic. However, the valence electrons in metals behave as a delocalized electron sea in which individual electrons flow easily from atom to atom, as shown in Figure 1-16. This behavior allows for the high conductivity of metals.

| *Metallic Bonding* |
The sharing of electrons among atoms in a metal, which gives the metal excellent conducting properties because the electrons are free to move about the electron cloud around the atoms.

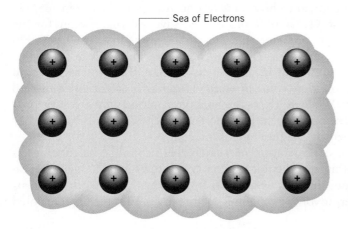

Sea of Electrons

FIGURE 1-16 Schematic Representation of the Delocalized Electron Cloud in Metals

1.4 CHANGES OF PROPERTIES OVER TIME

The previous discussion of the selection of appropriate materials proceeded as if all materials have a single set of inherent properties that remain constant over the useful life of the materials. Even the Ashby-style chart presented in Figure 1-5 compared "fresh" materials. In reality, however, the properties of materials change over time for a variety of reasons. Fatigue, corrosion, chemical scission, erosion, and a variety of other mechanisms can reduce the performance of a material. When scientists select a material for use in a specific application, they must take great care to be certain that the important properties of the material will remain acceptable throughout the expected life of the application. How properties change over time varies with the class of material and with the environment to which the material is exposed. Metals may rust or corrode, polymers may shrink or lose some of their strength, and composites may delaminate. Later chapters provide details on how the properties change with time and environment for each class of material. For now, recognize that no selection can be made without understanding the importance of these changes.

1.5 IMPACT OF ECONOMICS ON DECISION MAKING

The selection of materials is not driven exclusively by the best combination of chemical and physical properties. Cost is almost always a major factor in materials selection. Identifying the more economical alternative is complicated by the fact that not all expenses occur at once. Is it better to select a more durable material that is more expensive up front but will last longer, or would it be better to purchase a cheaper alternative even though it might need to be replaced more frequently? The issue is further complicated by the fact that money obtained now is more valuable than money in the future because money not spent today could be invested. Spending money now costs you not only the money but all of the interest that you could have earned by waiting to spend it.

For example, if a company chose to wait one year to replace some copper piping that would cost $1,000,000 and instead invested that money for the year in an account that paid 5% interest, the company would have $1,050,000 after one year. The concept that an amount of future money is less valuable than the same amount of money in the present because of the interest it could have earned is called the *time value of money*. *Interest* is essentially rent paid for the use of money. When you give the bank your money for a year, it pays you a set rate of interest as rent on that money, which bankers then invest or lend to other customers at a higher rate of interest.

| *Time Value of Money* |
The concept that future money is worth less than money in the present because of the interest it could have earned.

| *Interest* |
Rent paid to the owner of money for the temporary use of that money.

Although detailed consideration of engineering economic calculations is beyond the scope of this text, the role of economic factors in materials selection will appear throughout the various chapters in this text. Homework problems in subsequent chapters involving economic factors in decision making are noted by a $ symbol.

1.6 SUSTAINABILITY AND GREEN ENGINEERING

Historically, the analysis that determined the optimal materials selection for a given application ended after the useful life of the product. If a motor was expected to last for 20 years, the materials in that motor had to maintain workable properties for 20 years with the best economics possible. Little thought was given to what would happen to the motor when its useful life had passed, with the possible exception of factoring in any costs associated with disposing of it. Used materials were dumped into landfills or discarded into waterways and forgotten about.

By the late 1990s, most experts accepted the argument that many human endeavors were having a significantly detrimental impact on the environment. Both ethics and self-interest demand that technological developments continue in a way that is more beneficial to both society and the environment. The 1987 Bruntland Report defines *sustainability* as "meeting needs of the present generation without compromising the ability of future generations to meet their needs."[1] From the perspective of materials science, sustainable design includes several key points:

- Examine methods of conserving energy and water resources.
- Look for opportunities for reuse or recycling of existing materials.
- Select renewable resources when practical.
- Consider the end use of the material (landfill, recycle, etc.) as part of the design process.

These issues directly impact both the design and the selection of materials. Is there a way to produce the same material with less energy consumption or less waste? Can renewable resources be used instead? Can the material be reused or recycled when its purpose is completed, or must it be disposed of in a landfill?

The path taken by a material from its initial formation until its ultimate disposal is called a *life cycle*. All products begin with the harvesting of raw materials, followed by their conversion to products through a series of manufacturing steps, and ultimately their sale to a user. Historically, the goals of a materials scientist ended there, but the product itself was not finished. After the useful life of the material passes, it must be recycled, reused, or discarded. The life cycle of a material includes the entire time from the harvesting of the raw material until the ultimate disposal of the product. The American Chemistry Council (among other professional societies) has recognized the responsibility of designers to consider the entire life cycle of the material in their designs.

For some life cycle considerations, a qualitative assessment of materials selections and processes may be sufficient to reduce the environmental impact of decisions. Eliminating cyanide-containing compounds from pressure-treated lumber fits into this category. The most detailed method of analyzing the life cycle of a material involves performing a *life cycle assessment (LCA)*. The analysis begins by defining boundaries, which are the level of detail the examination will include. Within those boundaries, a detailed inventory of inputs and outputs are developed. Next, a list is generated that specifies the materials used and emitted. Finally, the list is analyzed to look for design modifications that could reduce emissions and waste. Figure 1-17 shows a schematic representation of a typical life cycle.

With the goal of reducing the harmful consequences of production, scientists and engineers are reexamining products and processes to look for more benign methods of manufacture, use, reuse, and disposal. The Environmental Protection Agency (EPA) has supported a movement toward *green engineering*, which it defines as "the design, commercialization, and use of processes and products, which are feasible and economical while minimizing generation of pollution at the source and risk to human health and the environment."[2] Nine principles of green engineering have been established:

1. Engineer processes and products holistically, use systems analysis, and integrate environmental impact assessment tools.
2. Conserve and improve natural ecosystems while protecting human health and well-being.

| *Sustainability* |
The length of time a material will remain adequate for use.

| *Life Cycle* |
The path taken by a material from its initial formation until its ultimate disposal.

| *Life Cycle Assessment (LCA)* |
The most detailed method of analyzing the life cycle of a material.

| *Green Engineering* |
A movement supporting an increase in the knowledge and prevention of environmental hazards caused during the production, use, and disposal of products.

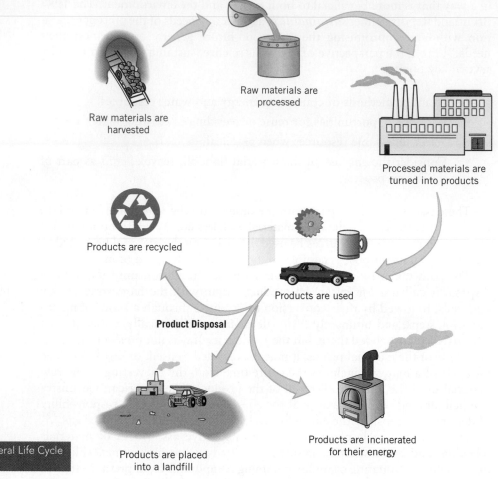

Raw materials are
processed

Raw materials are
harvested

Processed materials are
turned into products

Products are recycled

Products are used

Product Disposal

Products are incinerated
for their energy

Products are placed
into a landfill

FIGURE 1-17 General Life Cycle Schematic

3. Use life cycle thinking in all engineering activities.

4. Ensure that all material and energy balance inputs and outputs are inherently safe and as benign as possible.

5. Minimize depletion of natural resources.

6. Strive to prevent waste.

7. Develop and apply engineering solutions, while being cognizant of local geography, aspirations, and cultures.

8. Cultivate engineering solutions beyond current or dominant technologies; improve, innovate, and invent (technologies) to achieve sustainability.

9. Actively engage communities and stakeholders in development of engineering solutions.

These principles directly impact both the design and selection of appropriate materials for given applications and should be factored into the decision-making process. In this text, homework problems involving a green engineering aspect are labeled with a ◉ symbol.

1.7 CLASSES OF MATERIALS

Material properties depend on the atoms present, the bonding between atoms, and the three-dimensional arrangement of atoms within the material. The types and arrangement of atoms help classify materials as polymers, metals, composites, ceramics, or carbon as shown in Table 1-2. Materials may also be classified based on specific applications, such as electronic materials and biomaterials.

TABLE 1-2 Classes of Materials		
Material	*Definition*	*Application*
Metals	Of a category of electropositive elements that usually have a shiny surface, are generally good conductors of heat and electricity, and can be melted or fused, hammered into thin sheets, or drawn into wires.	*Courtesy James Newell*
Polymers	Naturally occurring or synthetic compounds consisting of large molecules made up of a linked series of repeated simple monomers covalently bonded together.	*Courtesy James Newell*
Ceramics	Any of various hard, brittle, heat- and corrosion-resistant materials made typically of metallic elements combined with oxygen or with carbon, nitrogen, or sulfur. Most ceramics are crystalline and are poor conductors of electricity, though some recently discovered copper-oxide ceramics are superconductors at low temperatures.	*Courtesy James Newell*
Composites	Complex materials, such as wood or fiberglass, in which two or more distinct, structurally complementary substances, especially metals, ceramics, glasses, and polymers, combine to produce structural or functional properties not present in any individual component.	*Courtesy James Newell*

| Polymers |
Covalently bonded chains
of molecules with the small
monomer units repeated
from end to end.

Polymers are covalently bonded chains of molecules with the same units (mers) repeated over and over. The overwhelming majority of polymers have carbon as the primary atom in the chain, with hydrogen, oxygen, more carbon, nitrogen, and/or fluorine attached to the sides. Figure 1-18 shows the periodic table with elements commonly found in polymers highlighted. Plastic has become a generic term for polymers, but many natural materials are also polymeric, including polysaccharides (cellulose and starches); rubber; proteins in hair, wool, and silk; and nucleic acids (RNA and DNA). Because so many types of polymers exist, their properties vary widely. Polyethylene is comparatively weak and is used for inexpensive produce bags in grocery stores, while other polymers such as Kevlar® (poly p-phenylene terephthalamide) and Zylon® (poly p-phenylene benzobisoxazole) are used as ballistic fibers in bullet-resistant vests like the one shown in Figure 1-19.

| Thermoplastic |
Polymer with a low melt-
ing point due to the lack of
covalent bonding between
adjacent chains. Such polymers
can be repeatedly melted
and re-formed.

Many polymers are flexible and lightweight, making them ideal materials for applications where great strength is not required. Polymers are classified by whether they can be remelted and reshaped. *Thermoplastic* polymers tend to have low melting points because of the lack of bonding between adjacent chains and can be repeatedly remelted and re-formed. Thermoplastic polymers are easily recycled but have less strength than many other materials. *Thermoset* materials have considerable bonding between chains, which makes them stronger than thermoplastics but also more difficult to recycle.

| Thermoset |
Polymer that cannot be
repeatedly melted and
re-formed because of
strong covalent bonding
between chains.

Metals are materials whose atoms share delocalized electrons such that any given electron is equally likely to be associated with a large number of atoms, as shown in Figure 1-20. This metallic bond gives metals exceptional electronic conductivity because the electrons are free to float through a broad electron cloud around the atoms. Metals tend to have exceptional strength but are capable of being shaped, which makes them useful for construction. Metals tend to be opaque and have a shiny surface when polished.

| Metals |
Materials possessing atoms that
share delocalized electrons.

FIGURE 1-18   Elements Commonly Found in Polymers

FIGURE 1-19   Ballistic Vest Containing Kevlar®

Courtesy James Newell

Periodic Table of the Elements

□ = metals

IA **1**																	**VIIIA** **18**
1 H	**IIA** **2**											**IIIA** **13**	**IVA** **14**	**VA** **15**	**VIA** **16**	**VIIA** **17**	2 He
3 Li	4 Be											5 B	6 C	7 N	8 O	9 F	10 Ne
11 Na	12 Mg	**IIIB** **3**	**IVB** **4**	**VB** **5**	**VIB** **6**	**VIIB** **7**	**VIIIB** **8**	**VIIIB** **9**	**VIIIB** **10**	**IB** **11**	**IIB** **12**	13 Al	14 Si	15 P	16 S	17 Cl	18 Ar
19 K	20 Ca	21 Sc	22 Ti	23 V	24 Cr	25 Mn	26 Fe	27 Co	28 Ni	29 Cu	30 Zn	31 Ga	32 Ge	33 As	34 Se	35 Br	36 Kr
37 Rb	38 Sr	39 Y	40 Zr	41 Nb	42 Mo	43 Tc	44 Ru	45 Rh	46 Pd	47 Ag	48 Cd	49 In	50 Sn	51 Sb	52 Te	53 I	54 Xe
55 Cs	56 Ba	57 La*	72 Hf	73 Ta	74 W	75 Re	76 Os	77 Ir	78 Pt	79 Au	80 Hg	81 Tl	82 Pb	83 Bi	84 Po	85 At	86 Rn
87 Fr	88 Ra	89 Ac**	104 Unq	105 Unp	106 Unh	107 Uns	108 Uno	109 Une	110 Uun	111 Uuu							

6	58 Ce*	59 Pr	60 Nd	61 Pm	62 Sm	63 Eu	64 Gd	65 Tb	66 Dy	67 Ho	68 Er	69 Tm	70 Yb	71 Lu
7	90 Th**	91 Pa	92 U	93 Np	94 Pu	95 Am	96 Cm	97 Bk	98 Cf	99 Es	100 Fm	101 Md	102 No	103 Lr

FIGURE 1-20   Elements Classified as Metals

Most metals are found in nature as metal oxides, which must be refined to produce the pure metals. Metals (or metals and nonmetals) are often blended to form *alloys*, which enable the material to achieve a wider range of properties. Common alloys include steel (iron and carbon), brass (copper and zinc), and

| *Alloys* |
Blends of two or more metals.

| *Composites* |

Material formed by blending two materials in distinct phases causing a new material with different properties than either parent.

| *Particulate Composites* |

Composites that contain large numbers of coarse particles, such as the cement and gravel found in concrete.

| *Fiber-Reinforced Composite* |

A composite in which the one material forms the outer matrix and transfers any loads applied to the stronger, more brittle fibers.

| *Laminar Composites* |

Composites that are made by alternating the layering of different materials.

| *Ceramics* |

Compounds that contain metallic atoms bonded to nonmetallic atoms such as oxygen, carbon, or nitrogen.

| *Graphite* |

An allotropic carbon material consisting of six-member aromatic carbon rings bonded together in flat planes, allowing for the easy occurrence of slip between planes.

| *Diamond* |

An allotropic, highly crystalline form of carbon that is the hardest known material.

| *Carbon Fibers* |

A form of carbon made by converting a precursor fiber into an all-aromatic fiber with exceptional mechanical properties.

| *Carbon Nanotubes* |

Synthetic tubes of carbon formed by folding one graphene plane over another.

| *Fullerenes* |

Allotropic forms of carbon made up of a network of 60 carbon atoms bonded together in the shape of a soccer ball. Also known as buckyballs after the architect Buckminster Fuller, who developed the geodesic dome.

bronze (copper and tin). The aluminum used in the sides of aluminum cans is actually an alloy of aluminum and magnesium.

Composites are mixtures of two materials in which each material continues to exist in distinct phases. The fiberglass used for insulation in most houses is a composite with glass fiber encased in a polymeric matrix. The three main classes of composites include particulate, fiber-reinforced, and laminar. *Particulate composites* contain large numbers of coarse particles, like the blend of cement and gravel used in concrete. They tend to enhance properties such as toughness or wear-resistance rather than strength. In *fiber-reinforced composites*, the outer matrix material orients the fibers and transfers any applied loads to the stronger, more brittle fibers. The applications of fiber-reinforced composites range from silicon-carbide fiber reinforced metal-matrix composites used in the engines of advanced fighter jets to more mundane and ancient applications, including the use of straw in bricks. *Laminar composites* consist of alternating layers of different materials bonded together. Plywood is a laminar composite consisting of layers of wood veneer bonded with epoxy layers between them. Regardless of type, composites offer the opportunity to blend two materials together to form a new material with properties that neither starting material could achieve alone.

Ceramics are compounds that contain metallic atoms bonded to nonmetallic items, most commonly oxygen, nitrogen, or carbon. Metal oxides fit into the category, but so do cements and glasses. The strong ionic bonding between the atoms makes ceramic materials excellent electronic insulators and resistant to chemical erosion. The properties of ceramics vary, but most tend to be strong and hard yet quite brittle. There are exceptions. Modern high-performance ceramics used in body armor are certainly not brittle, while other engineered ceramics display superconductivity. Despite a growing presence in the high-performance materials markets, the dominant ceramic materials in industrial use continue to be glasses, bricks, abrasives, and cements. Because the defining feature of ceramic materials is a blend of metals and nonmetals, atoms from the entire periodic table are found in ceramics.

Carbon materials include the naturally occurring forms of carbon—*graphite* and *diamond*—and also *carbon fibers*, *carbon nanotubes*, and *fullerenes*. Graphite consists of six-member aromatic carbon rings bonded together in flat planes. The strong covalent bonds in these aromatic rings make the layers extremely strong, but only weak Van Der Waals interactions connect the planes, making it easy for them to slide across each other. Everyone is familiar with the use of graphite in pencils, but it also serves as insulating material in nuclear reactors. Similarly, diamonds are best known for their use in jewelry, but they have commercial importance because of their exceptional hardness. The development of processes for the production of synthetic diamonds has made industrial diamond usage more affordable.

Carbon fibers, carbon nanotubes, and fullerenes are more recent carbon developments. Carbon fibers are made by converting a precursor fiber (usually pitch or polyacrylonitrile) to an essentially all-aromatic carbon fiber that approaches synthetic graphite. These highly ordered fibers are used in a variety of applications, from racecar frames, to artificial limbs, to golf clubs. Fullerenes are networks of carbon atoms bonded in the shape of a sphere, tube, or ellipsoid. For example, one fullerene consists of 60 carbon atoms bonded together in the shape of a soccer ball. Fullerenes are named after the architect Buckminster Fuller and are often called buckyballs in tribute to his use of geodesic domes, like the large Spaceship Earth in the heart of Epcot Center in Walt Disney World. The hollow inside of the buckyball intrigues scientists who foresee a variety of uses in composites. Carbon nanotubes are synthetic tubes formed by essentially folding one graphite plane over another. Nanotubes, with their unique blend of properties, have great potential for electrical applications.

In addition to the four primary classes, sometimes it is useful to categorize materials based on specific properties. Electronic materials, optical materials, and biomaterials include subsets from metals, polymers, ceramics, and composites, but often appear as separate categories because of functionalities.

The communications revolution that has connected the world is a direct result of developments in *electronic materials*. These materials are classified primarily because of their ability to conduct electrons. *Semiconductors* have conductivities between the range of insulators and conductors. *Intrinsic semiconductors* are pure materials, but most commercial semiconductors result from deliberately adding an impurity called a *dopant*. Doped silicon wafers provide the basis for most integrated circuits that control all of the computers, cell phones, and other technological wonders of the modern world.

Biomaterials are materials designed specifically for use in biological applications. Biomaterials fall into two primary categories depending on their intended use.

Structural biomaterials are designed to bear loads and provide support for living systems. Artificial limbs and replacement hip joints fit into this category.

Functional biomaterials serve a purpose for an organism. Artificial blood, membranes used for dialysis, and synthetic skin fit into this category.

Often people discuss only synthetic biomaterials, but bone, muscle, skin and a host of other natural items are also biomaterials. Only by understanding the specific properties of natural biomaterials can engineers and scientists hope to develop suitable synthetics.

Some of the classes of materials just described have subclasses containing thousands of members with somewhat different properties. It is unrealistic to hope to learn every property of every material. The next two chapters focus on underlying structures common to many classes of materials and the mechanical and chemical properties that can be used to compare and contrast materials. Beginning with Chapter 4, each class of materials is explored in detail to enable students to understand its unique benefits and limitations and how these properties are determined.

| *Electronic Materials* |
Materials that possess the ability to conduct electrons, such as semiconductors.

| *Semiconductors* |
Materials having a conductivity range between that of conductors and insulators.

| *Intrinsic Semiconductors* |
Pure materials having a conductivity ranging between that of insulators and conductors.

| *Dopant* |
An impurity deliberately added to a material to enhance the conductivity of the material.

| *Biomaterials* |
Materials designed specifically for use in biological applications, such as artificial limbs and membranes for dialysis, as well as aiding in the repair of bones and muscle.

| *Structural Biomaterials* |
Materials designed to bear loads and provide support for a living organism, such as bones.

| *Functional Biomaterials* |
Materials that interact or replace biological systems with a primary function other than providing structural support.

Summary of Chapter 1

In this chapter we examined:

- The importance of materials science
- How the needs of the intended application govern the selection of materials
- Why a thorough understanding of materials science is necessary to cope with changing and increasing demands for improved properties
- The use and limitation of heuristics and Ashby-style charts in materials selection
- The fundamental chemical principles that underscore all materials science
- The recognition that properties change over time
- The need to evaluate the economic impact of materials decisions over time
- The importance of sustainable design and selection of materials
- The green engineering principles that should factor into materials design and selection
- The various classes of materials that will be covered in depth in subsequent chapters

References

[1] G. Bruntland, ed., *Our Common Future: The World Commission on Environment and Development* (Oxford University Press, 1987).

[2] U.S. Environmental Protection Agency, Proceedings of the Green Engineering Conference: Defining the Principles, Sandestin, Florida, May 2003.

Key Terms

alloys *p. 23*
Ashby-style charts *p. 9*
biomaterials *p. 25*
Bohr model *p. 10*
carbon fibers *p. 24*
carbon nanotubes *p. 24*
ceramics *p. 24*
composites *p. 24*
diamond *p. 24*
dipole force *p. 16*
dopant *p. 25*
electronegativity *p. 14*

electronic materials *p. 25*
fiber-reinforced composites *p. 24*
fourth quantum number *p. 11*
fullerenes *p. 24*
functional biomaterials *p. 25*
graphite *p. 24*
green engineering *p. 19*
hydrogen bond *p. 16*
interest *p. 18*
ionic bonding *p. 13*
intrinsic semiconductors *p. 25*
laminar composites *p. 24*

life cycle *p. 19*
life cycle assessment (LCA) *p. 19*
metallic bonding *p. 17*
metals *p. 22*
nonpolar *p. 14*
opposed spins *p. 12*
orbitals *p. 10*
parallel spins *p. 12*
particulate composites *p. 24*
Pauli exclusion principle *p. 12*
Pittsburgh process *p. 5*
polar *p. 14*

Homework Problems

1. For each of the applications listed below, develop a list of necessary properties and decide how significant a role economics would play in the final selection of materials.

   a. Asphalt for road paving

   b. Brake pads for a car

   c. Wings on an aircraft

   d. Piping in a house

2. For each of the applications listed below, develop a list of necessary properties and decide how significant a role economics would play in the final selection of materials.

   a. Bicycle frames

   b. Tires for NASCAR cars

   c. Synthetic leather for briefcases

   d. Scissors

3. Consider the evolution of recording media from vinyl records to CDs. How did the materials' challenges change?

$ 4. An engineer must decide whether to use plain carbon steel valves or a more expensive stainless steel alternative. The plant will use 1000 valves at a time. The carbon steel valves cost $400 each and will last two years before they must be replaced. The stainless steel valves cost $1000 each but will last for six years. Assuming that any valve replacements would occur during routine annual maintenance, what factors must be considered in reaching an appropriate economic decision?

5. Classify the following bonds as primarily ionic, primarily covalent, or metallic:

   a. Carbon—Oxygen

   b. Sodium—Potassium

   c. Silicon—Carbon

   d. Potassium—Chlorine

6. Why do valence electrons play such an important role in the bonding between atoms?

7. Distinguish between primary and secondary bonds, and describe three examples of secondary bonds.

8. Identify two commercial products made from ceramics. Describe the type of ceramic used and why that ceramic was the best choice for the product.

9. Identify two commercial products made with polymers. Describe the specific polymer used and why that polymer was the best choice for the product.

10. Identify two commercial products made with metals. Describe the specific metal used and why that metal was the best choice for the product.

11. Given the choice of using a thermoplastic or thermoset polymer with similar properties for a specific application, why might the thermoplastic be a better choice?

12. Classify the following materials as a polymer, metal, ceramic, or composite:

    a. Boron nitride

    b. Bricks

    c. Plexiglas

    d. Concrete

    e. Manganese

13. Classify the following materials as a polymer, metal, ceramic, or composite:

    a. Fiberglass

    b. Silicon carbide

    c. Aluminum foil

    d. Teflon®

    e. Silk

14. During World War II, the American supply of rubber became limited. Major League baseball

responded by using a material called balata in the center of the baseball. The number of home runs declined sharply. Find out how a baseball is made and why the inclusion of a balata core would so greatly impact the number of home runs.

15. Find out what materials are collected for recycling by your college or university. What happens to these materials when they leave campus?

16. Describe the benefits and negative consequences of sending out a company newsletter electronically instead of through the U.S. Postal Service.

17. Develop a list of inputs and outputs for paper and plastic grocery bags, starting with trees growing in a forest (paper) and oil in an oil field (grocery bags). Is there a better alternative than paper or plastic?

18. Consider the life cycles for paper and plastic grocery bags. What are the environmental and economic advantages and disadvantages of each choice?

19. List the four quantum numbers for all of the electrons in the following atoms:
    a. Lithium
    b. Helium
    c. Carbon

20. How does the fourth quantum number relate to the Pauli exclusion principle?

2

Structure in Materials

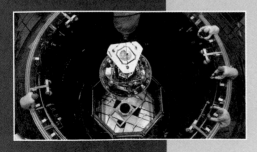

CONTENTS

Learning Objectives

By the end of this chapter, a student should be able to:

- Explain what is meant by crystallinity and a unit cell.

- Identify the 14 Bravais lattices.

- Understand the meaning of basic crystallographic terms including lattice parameter and interplanar spacing.

- Calculate the distance between atoms in a crystal.

- Determine the indices of a crystallographic direction, given a vector representing that direction and determine the vector defining a crystallographic direction given the indices.

- Determine the Miller indices of a crystalline plane if shown the plane, and draw the crystalline plane given the Miller indices.

- Use an X-ray diffraction diagram to identify the crystal structure of a material and to calculate interplanar spacing, lattice parameters, and crystallite thickness.

- Understand the uses of optical and electron microscopy and the differences between them.

- Explain the two processes that comprise the growth of crystals.

- Identify and explain the defects present in crystalline materials.

- Explain what is meant by a grain boundary and how it impacts physical properties.

- Understand how defects move through a crystal.

- Determine the critical resolved shear stress for a given slip system under stress and explain its impact on slip.

- Distinguish between the crystal mosaic structure, single crystals, and nanocrystals.

- Calculate a theoretical density for a crystalline material and explain why this value is likely to be different than an experimentally measured value.

How Are Atoms Arranged in Materials?

2.1 INTRODUCTION

In crystallography, many students are initially unnerved by the topic of how atoms are arranged in materials. Some of the issues center on needing to use geometry and visualize three-dimensional images, but often the biggest problem is seeing the relevance of the topic. Students can learn to calculate Miller indices and draw the appropriate plane, but if they have only memorized a procedure without gaining any appreciation for how it ties in to the larger issues in the course, they have learned little of lasting value. Students quite rightly want to know how this information will help them select the right material for a specific application or to redesign an existing material to make it more suitable.

The need to understand crystal structure (and all of the geometry and nomenclature that accompanies it) stems from the concept that the structure of a material governs its properties. Materials are selected because they have the appropriate properties for a given function, and those properties determine whether a material is suitable or not. Understanding material structures provides the gateway to understanding both the material properties that these structures spawn and the processing procedures that can be used to alter the structures and, as a result, the properties of the material. Figure 2-1 provides a graphical representation of the unbreakable interrelationship among structure, properties, and processing.

The development of structure of materials provides a perfect point of entry into the broader realm of materials science and engineering. The properties of any material are determined by its structure at four distinct levels:

1. *Atomic structure.* What atoms are present and what properties do they possess?
2. *Atomic arrangement.* How are the atoms positioned relative to each other, and what type of bonding, if any, exists between them?
3. *Microstructure.* What sequencing of crystals exists at a level too small to be seen with the eye?
4. *Macrostructure.* How do the microstructures fit together to make the larger material?

Table 2-1 shows how these levels of order apply to salt crystals. The properties of a material are determined by the combined effects of all four levels and can be altered using a variety of processing techniques. This chapter focuses on the development of structure in crystalline materials and includes a relatively detailed examination of X-ray diffraction. This examination serves as a tool to clarify the real meaning and relevance of the crystallographic terms in the chapter.

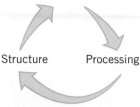

Properties

Structure   Processing

FIGURE 2-1 Structure-Properties-Processing Relationship

| *Atomic Structure* |
The first level of the structure of materials, describing the atoms present.

| *Atomic Arrangement* |
The second level of the structure of materials, describing how the atoms are positioned in relation to one another as well as the type of bonding existing between them.

| *Microstructure* |
The third level of structure in materials, describing the sequencing of crystals at a level invisible to the human eye.

| *Macrostructure* |
The fourth and final level of structure in materials, describing how the microstructures fit together to form the material as a whole.

| *Amorphous Materials* |
Materials whose order extends only to nearest neighbor atoms.

| *Crystal Structure* |
The size, shape, and arrangement of atoms in a three-dimensional lattice.

| *Bravais Lattices* |
The 14 distinct crystal structures into which atoms arrange themselves in materials.

| *Simple Cubic* |
A Bravais lattice that has one atom in each corner of the unit cell.

2.2 LEVELS OF ORDER

Most of the materials in this text possess significant order, but that is not true for all materials. The lowest level of order involves monatomic gas molecules randomly filling space, which have limited relevance in the study of materials science. Instead, most materials have at least some short-range order. Water molecules, shown in Figure 2-2, provide a classic example of such short-range order. Materials with order extending only to the nearest neighbors are referred to as *amorphous materials* (*a-* meaning without, *morph-* for form). Some solids may also be amorphous. For example, the silica glasses that are examined in Chapter 5 have no three-dimensional order.

Most solids have significant long-range three-dimensional ordering and form a regular lattice. The *crystal structure* of a solid is the size, shape, and arrangement of atoms within this three-dimensional lattice. In fact, lattices organize themselves in one of the 14 patterns, called *Bravais lattices*, shown in Figure 2-3.

Although all of the Bravais lattices appear in nature, the three cubic lattices are the easiest to visualize and are used as the basis for most of the discussion in the next section. *Simple cubic* unit cells have one atom in each of the eight corners of the cube. Although the simplest to visualize, the simple cubic structure is less common than either of the two other cubic forms.

TABLE 2-1   Levels of Order Applied to Salt Crystals

Atomic Structure	Sodium (Na) and chlorine (Cl) atoms ionically bond.	
Atomic Arrange-ment	Multiple NaCl molecules bond together to form a face-centered cubic lattice.	
Micro-structure	The edge of the lattice is called a grain boundary. These boundaries are visible under a microscope.	
Macro-structure	The plain eye sees NaCl crystals as clear solids, though they can be colored with impurities.	

| *Body-Centered Cubic (BCC)* |
One of the Bravais lattices that contains one atom in each corner of the unit cell as well as one atom in the center of the unit cell.

| *Face-Centered Cubic (FCC)* |
One of the Bravais lattices that has one atom in each corner of the unit cell and one atom on each face of the unit cell.

| *Hexagonal Close-Packed (HCP)* |
The most common of the noncubic Bravais lattices, having six atoms forming a hexagon on both the top and bottom and a single atom positioned in the center, between the two hexagonal rings.

Body-centered cubic (BCC) unit cells also have one atom in each of the eight corners, but an additional atom is present in the center of the cube. *Face-centered cubic (FCC)* unit cells have one atom in each of the eight corners, plus one atom in each of the six cube faces. Of the noncubic lattices, *hexagonal close-packed (HCP)* is the most common. The top and bottom of the lattice consist of six atoms forming a hexagon that surrounds a single atom

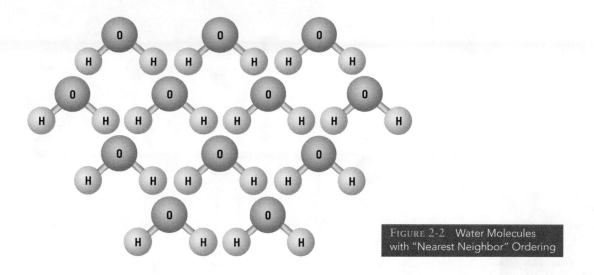

FIGURE 2-2   Water Molecules with "Nearest Neighbor" Ordering

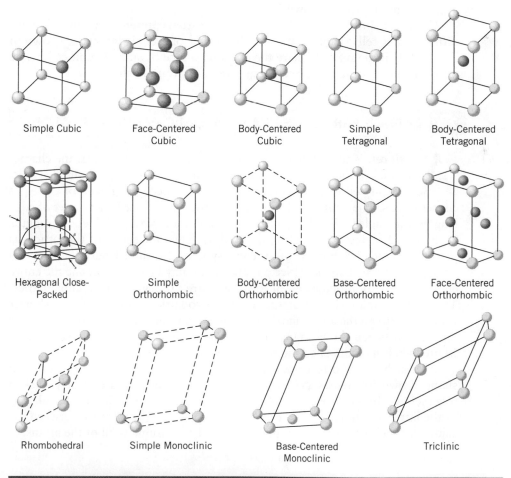

Simple Cubic

Face-Centered Cubic

Body-Centered Cubic

Simple Tetragonal

Body-Centered Tetragonal

Hexagonal Close-Packed

Simple Orthorhombic

Body-Centered Orthorhombic

Base-Centered Orthorhombic

Face-Centered Orthorhombic

Rhombohedral

Simple Monoclinic

Base-Centered Monoclinic

Triclinic

FIGURE 2-3   Fourteen Bravais Lattices

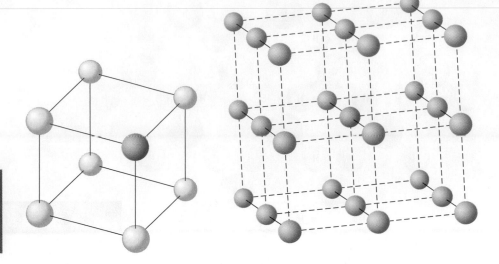

FIGURE 2-4 Schematic Representation of Multiple Lattices Sharing Atoms (The colored atom shows its position in a unit cell and in the overall lattice.)

in the center. A triangular cluster of three atoms resides in between the top and bottom planes, as shown in Figure 2-3.

Obviously, other lattices share most of the atoms present in a unit cell. Eight different unit cells share an atom in the corner of a cubic lattice, as shown in Figure 2-4. Two unit cells share the atom on a face of an FCC unit cell, while only one unit cell claims the center atom in a BCC unit cell.

2.3 LATTICE PARAMETERS AND ATOMIC PACKING FACTORS

| Unit Cell |
The smallest subdivision of a lattice that still contains the characteristics of the lattice.

| Lattice Parameters |
The edge lengths and angles of a unit cell.

A *unit cell* is the smallest subdivision of a lattice that retains the characteristics of the lattice. In the case of the most basic lattice, simple cubic, the unit cell is just a cube with one atom occupying each of the eight corner points. The sizes and shapes of the lattices are described by a set of edge lengths and angles called *lattice parameters*. For any system, a blend of lengths (a, b, and c), along with angles (α, β and γ) define the shape of the lattice, as shown in Figure 2-5. For any of the cubic systems, the three lengths (a, b, and c) are equivalent, so a single lattice parameter (a) may be used to define the entire cubic lattice. Additionally, all of the angles are 90 degrees for a cubic lattice. The more complex HCP structure requires two lattice parameters. The shorter distance between the atoms in the hexagon is represented by **a**, while the longer direction between the hexagons is represented by **c**, as shown in Figure 2-5. Many metals including beryllium and magnesium have HCP lattices.

Although we tend to draw lattices as if they were the type of large open shape shown in Figures 2-3 and 2-4, the atoms actually are closely packed together, as shown in Figure 2-6. This close packing permits the calculation of the lattice parameter. For the simple cubic system, the distance between the centers of the two atoms is simply the sum of the atomic radii of the atoms:

$$a = r_1 + r_2. \tag{2.1}$$

If a single type of atom is present in the lattice, the atomic radii are the same. The lattice parameter of a pure material (a_0) is designated with a subscript.

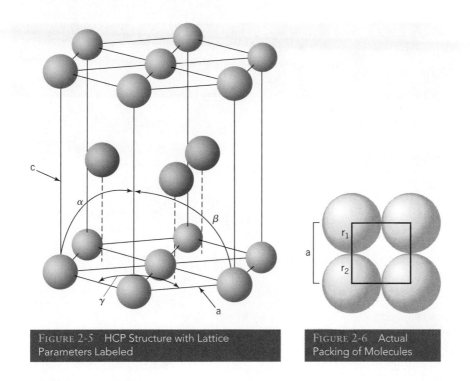

FIGURE 2-5   HCP Structure with Lattice Parameters Labeled

FIGURE 2-6   Actual Packing of Molecules

Basic geometry can be used to calculate the lattice parameters for different lattice configurations, as shown in Table 2-2.

These formulas allow for the calculation of the lattice parameter for any material that has a known crystal structure, provided that the atomic radius is also known. Table 2-3 summarizes atomic radii for several common metals.

The other relevant parameter determined by the crystal structure is the **atomic packing factor (APF)**, the amount of the unit cell occupied by atoms as opposed to void space. Table 2-4 summarizes APFs for the common lattices.

| *Atomic Packing Factor (APF)* |
The amount of the unit cell occupied by atoms as opposed to void space.

TABLE 2-2 Lattice Parameters Based on Atomic Radii for Common Lattice Systems		
Lattice Type	*Lattice Parameter* (a_0)	*Graphical Representation*
Simple cubic	$a_0 = 2r$	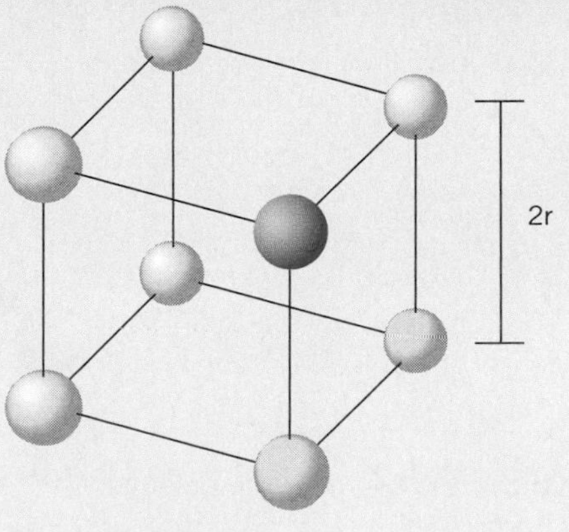 2r

(continues)

TABLE 2-2 Lattice Parameters Based on Atomic Radii for Common Lattice Systems (continued)

Lattice Type	Lattice Parameter (a_0)	Graphical Representation
Body-centered cubic (BCC)	$a_0 = \dfrac{4r}{\sqrt{3}}$	
Face-centered cubic (FCC)	$a_0 = \dfrac{4r}{\sqrt{2}}$	
Hexagonal close-packed (HCP)	$a_0 = 2rc_0$ $= 3.266r$	

Table 2-3 Atomic Radii for Common Metals

Material	Lattice Type	Atomic Radius (nm)
Aluminum	FCC	0.143
Chromium	BCC	0.125
Cobalt	HCP	0.125
Copper	FCC	0.128
Gold	FCC	0.144
α-Iron	BCC	0.124
γ-Iron	FCC	0.124
Lead	FCC	0.175
Magnesium	HCP	0.160
Nickel	FCC	0.125
Platinum	FCC	0.139
Silver	FCC	0.144
Titanium	HCP	0.144
Tungsten	BCC	0.137
Zinc	HCP	0.133

Table 2-4 Atomic Packing Factors for Different Lattice Types

Lattice Type	APF
BCC	0.68
FCC	0.74
HCP	0.74

//

Example 2-1

Calculate the lattice parameter for a lattice of lead atoms.

SOLUTION

Lead atoms form a face-centered cubic (FCC) lattice and have an atomic radius of 0.175 nm (nanometers). The lattice parameter for an FCC system is given in Table 2-2 as $a_0 = \dfrac{4r}{\sqrt{2}}$. Therefore, $a_0 = \dfrac{4*(0.175 \text{ mm})}{\sqrt{2}}$. $a_0 = 0.495$ nm.

| *Theoretical Density* |
The density a material would
have if it consisted of a single
perfect lattice.

With a complete knowledge of the lattice structure, the *theoretical density* of a material may be calculated from the equation

$$\rho = \frac{nA}{N_A V_c},$$ (2.2)

where ρ is the theoretical density of the material, n is the number of atoms per unit cell, A is the atomic weight of the material, N_A is Avogadro's number $(6.022 \times 10^{23}$ atoms/mol), and V_c is the volume of a unit cell. For a cubic system, the volume of the unit cell is the cube of the lattice parameter (a_0^3). This formula comes from the basic definition of density as mass over volume.

To determine the number of atoms per unit cell, we must first determine how many unit cells share each atom. A corner cell in a cubic lattice is shared by eight different unit cells, as seen in Figure 2-4, while a cell on a face is shared by only two. The body cell in a BCC structure belongs entirely to one cell. As a result, points are assigned based on location. Corner atoms get $\frac{1}{8}$ of a point, atoms on faces receive $\frac{1}{2}$ of a point, while body atoms receive 1 point. The number of atoms per cell is the sum of the points assigned to the cell. A simple cubic unit cell has only eight corner atoms, each worth $\frac{1}{8}$ point for a total of one atom per cell. An FCC unit cell has eight corner atoms worth $\frac{1}{8}$ point each, plus six atoms on faces, each worth $\frac{1}{2}$ point. An FCC unit cell has four atoms

Example 2-2

Determine the theoretical density of chromium at 20°C.

SOLUTION

From Table 2-3, we learn that chromium has a BCC structure with an atomic radius of 0.125 nm. We also know that the theoretical density of a material is given by Equation 2.2:

$$\rho = \frac{nA}{N_A V_c}.$$

For a BCC system, n = 2 atoms (one for the body atom, one from the corner atoms), A = 52 g/mol (from the periodic table of elements), and $N_A = 6.022 \times 10^{23}$ atoms/mol.

To calculate I, we need to know the lattice parameter (I_0). Table 2-2 tells us that the lattice parameter for a BCC system is given by

$$a_0 = \frac{4r}{\sqrt{3}}, \text{ so } a_0 = \frac{4(0.125 \text{ nm})}{\sqrt{3}} \rightarrow a_0 = \frac{4(0.125 \text{ nm})}{\sqrt{3}} = 0.289 \text{ nm}$$

$$V_c = a_0^3 = (0.289 \text{ nm})^3 = 0.0241 \text{ nm}^3$$

$$\rho = \frac{nA}{N_A V_c} = 7.17 \times 10^{-21} \text{ g/nm}^3 \text{ or } 7.17 \text{ g/cm}^3.$$

per cell. Similarly, BCC cells have eight corner atoms (each worth $\frac{1}{8}$ point) plus a body atom worth 1 point, for a total of two atoms per unit cell.

Theoretical density calculations assume a perfect lattice, but as we will learn later in the chapter, lattices contain many different types of imperfections. As a result, the real materials tend to be slightly less dense than the theoretical calculations predict.

2.5 CRYSTALLOGRAPHIC PLANES

When it is necessary to determine a direction in a crystal, a simplified system of indices is used to represent the vector defining the direction. One point in a lattice is arbitrarily selected as the origin, and all other points are labeled relative to that point. Figure 2-7 shows a simple cubic lattice with points labeled. In each case, the number of the point represents the number of lattice parameters moved in the x-, y-, and z-directions. If the point is less than one lattice parameter from the origin in a given direction, a fractional value is used. Identifying the points is an essential step in determining crystallographic dimensions and planes.

Determining the indices for a direction is a straightforward four-step procedure:

1. Using a right-handed coordinate system, determine the coordinates of two points that lie on a line in the direction of interest.

2. Subtract the coordinates of the tail point from the coordinates of the head point to determine the number of lattice parameters traveled in the direction of each axis.

3. Clear fractions and reduce the result to the nearest integer (so that 1.25 would become 1).

4. Enclose the numbers in brackets with a line above negative numbers (e.g., [1 $\bar{2}$ 0] would correspond to 1 in the x-direction, -2 in the y-direction, and 0 in the z-direction).

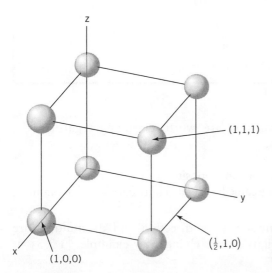

(1,1,1)

($\frac{1}{2}$,1,0)

(1,0,0)

FIGURE 2-7 Crystallographic Labeling of Atoms

Example 2-3

Determine the indices for the directions A, B, and C shown in the figure.

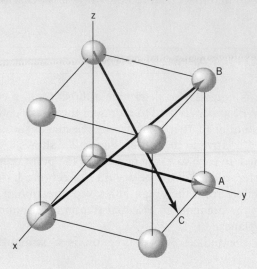

SOLUTION

A: Tail point $(0,0,0)$; head point $(0,1,0)$

$0,1,0 - 0,0,0 = 0,1,0$

No fractions to clear

$[0\ 1\ 0]$ are the indices for Direction A

B: Tail point $(1,0,0)$; head point $(0,1,1)$

$0,1,1 - 1,0,0 = -1,1,1$

No fractions to clear

$[\bar{1}\ 1\ 1]$ are the indices for Direction B

C: Tail point $(0,0,1)$; head point $(\frac{1}{2},1,0)$

$\frac{1}{2},1,0 - 0,0,1 = \frac{1}{2},1,-1$

Multiply by 2 to clear the fraction

$[1\ 2\ \bar{2}]$ are the indices for Direction C

Some points of potential confusion are worth clarifying regarding the indices for directions. Specifically,

- The directions are unit vectors so opposite signs are unequal. The are opposite directions along the same line (called anti-parallel); for example, $[0\ 0\ 1] \neq [0\ 0\ \bar{1}]$.

- Directions have no magnitude, so a direction and its multiple are equal provided that the signs do not change; for example, $[1\ 2\ 3] = [2\ 4\ 6] = [3\ 6\ 9]$.

Example 2-4

Determine the Miller indices for the planes shown in the figure.

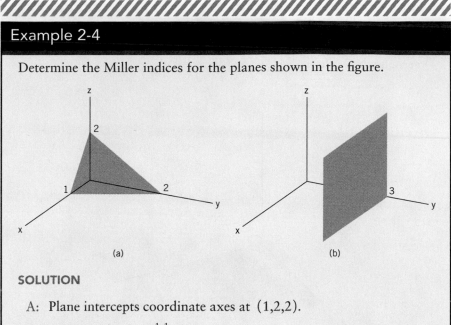

(a)   (b)

SOLUTION

A: Plane intercepts coordinate axes at $(1,2,2)$.

Reciprocals are $1, \frac{1}{2}, \frac{1}{2}$.

Multiply by 2 to clear fractions 2,1,1.

$(2\ 1\ 1)$ are the Miller indices of the plane.

B: Plane intercepts axes at $(\infty, 3, \infty)$.

Reciprocals are $0, \frac{1}{3}, 0$.

Multiply by 3 to clear fractions: $3 * 0, \frac{1}{3}, 0, \rightarrow 0, 1, 0$

$(0\ 1\ 0)$ are the Miller indices of the plane.

2.6 | MILLER INDICES

In crystals, while directions are important, we are often more concerned with their planes. Any three lattice sites in a crystal can be used to define a plane, so another system of indices was developed to clarify which plane is being discussed. The *Miller indices* for a plane passing through any three points in a lattice are determined using a four-step procedure that is similar to that used to find the indices of a direction. The procedure for determining the Miller indices of a plane is:

| *Miller Indices* |
A numerical system used to represent specific planes in a lattice.

1. Identify where the plane intercepts the x-, y-, and z-coordinate lines in terms of number of lattice parameters.
2. Take the reciprocal of these three points.
3. Clear fractions but do not reduce the results.
4. Enclose the results in parentheses.

Note: If the plane never crosses an axis, it is taken to intercept at infinity.

Example 2-5

Determine the Miller indices of the following plane:

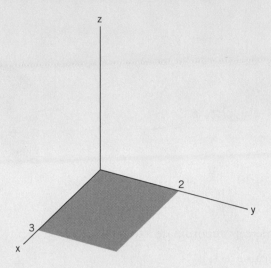

SOLUTION

The plane passes through the origin, so it intercepts the x- and y-axes everywhere and the z-axis at zero, but a zero intercept would result in an infinite z-index. So we move the origin down by one lattice parameter in the z-direction:

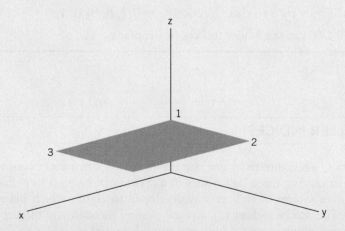

Now the z-intercept is at one, while the x- and y-intercepts become infinity.

Intercepts at $(\infty, \infty, 1)$.

The reciprocals are $0, 0, 1$.

No fractions to clear.

$(0\ 0\ 1)$ are the Miller indices of the plane.

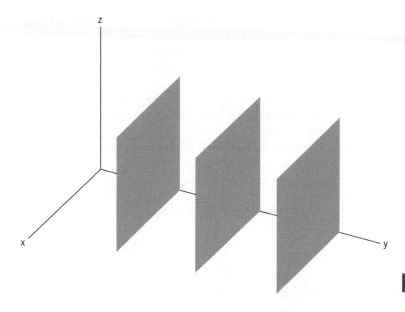

Miller indices of planes have special properties somewhat different from directions. When dealing with planes:

- Miller indices and their negatives are identical; for example, (0 1 2) = (0 $\overline{1}$ $\overline{2}$).

- Miller indices and their multiples are different; for example, (1 2 3) $\neq$ (2 4 6).

There is one other potential complication in calculating the Miller indices for a plane. If the plane passes through a coordinate axis, the intercept for that dimension would be zero. This would result in an undefined, essentially infinite Miller index, which cannot exist. Fortunately, the selection of a point as the origin was arbitrary. Any atom in the lattice can be set as the origin, so we can move the origin so that the intercept no longer passes through the coordinate axis. Example 2-5 shows this more clearly.

Within a given crystal, the same planes are repeated many times. Figure 2-8 shows a series of planes, all with the same Miller indices. Although these planes are perfect replicas, the distance between them is significant. The distance between repeated planes in a lattice is called the *interplanar spacing* (d).

| *Interplanar Spacing* |
The distance between repeated planes in a lattice.

How Are Crystals Measured?

2.7   X-RAY DIFFRACTION

X-ray diffraction is a powerful tool used to measure crystallinity and other lattice-dependent variables. X-ray diffraction also helps clarify the physical significance of planes and Miller indices. The principle of X-ray diffraction has developed from the study of optics. Electromagnetic radiation (including X-rays and visible light) travels in waves. Each type of

| *Constructive Interference* |
The increase in amplitude
resulting from two or more
waves interacting in phase.

| *Destructive Interference* |
A nullification caused by two
waves interacting out of phase.

| *Braggs' Equation* |
Formula that relates
interplanar spacing in a
lattice to constructive
interference of diffracted
X-rays. Named after the
father and son (W. H. and
W. L. Bragg) who proved
the relationship.

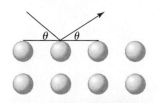

FIGURE 2-9 X-rays
Striking a Crystal Lattice

electromagnetic wave has a characteristic wavelength (λ). The wavelengths in the X-ray range are roughly the same size as most interatomic distances. When a wave strikes a solid object (e.g., an atomic nucleus), it bounces off the object with an angle of reflection equal to the angle of incidence. Figure 2-9 shows an X-ray beam striking atoms in a lattice.

Diffraction describes the interaction of waves. Figure 2-10 shows how two waves in phase add through *constructive interference* and two waves out of phase cancel through *destructive interference*.

In an X-ray diffraction machine like the one shown schematically in Figure 2-11, a source shoots X-rays into a sample and a detector collects the diffracted beams. The source and detector move together through different angles but always maintain the same "angle of incidence equals the angle of reflection" relationship to each other. Because many different atoms are present in the lattice, most waves cancel. Net constructive interference results only when *Braggs' equation* is satisfied.

$$N\lambda = 2d \sin\theta, \qquad (2.3)$$

where N is the order of reflections (taken to be 1), λ is the wavelength of the X-ray beam, d is the interplanar spacing, and θ is the angle of incidence. Orders of reflection greater than 1 are accounted for by the Miller indices.

The data generated by an X-ray diffraction experiment consist of a measurement of the intensity readings in the detector as a function of angle of incidence. The angle is generally reported as 2θ, since both the source and detector are at an angle θ. When there is no constructive interference, nothing but background scatter is detected. At 2θ values at which constructive interference occurs, an increased radiation level is detected. A typical X-ray diffraction reading is shown in Figure 2-12.

Each peak in the diffractogram corresponds to a different plane in the crystal. Many calculations can be done using X-ray diffractograms and the Bragg

(a)

(b)

FIGURE 2-10 Interference Patterns for X-rays: (a) Constructive Interference; (b) Destructive Interference

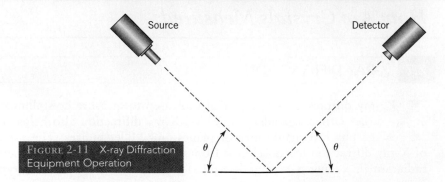

Source

Detector

FIGURE 2-11 X-ray Diffraction Equipment Operation

θ

θ

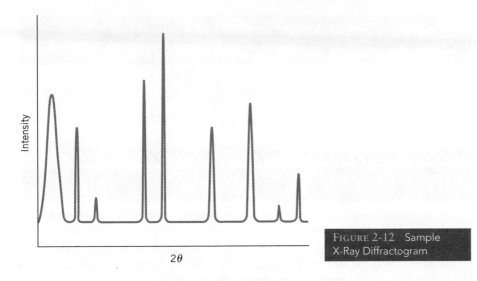

Intensity

2θ

equation. If the wavelength of the X-ray source is known, then the Bragg equation may be rearranged to determine the interplanar spacing of the plane corresponding to each peak,

$$d = \frac{n\lambda}{2\sin\theta}. \tag{2.4}$$

Because many planes are present in a given crystal, they are identified by their corresponding Miller indices ($h\ k\ l$). Thus Equation 2.3 could be written more properly as

$$d_{hkl} = \frac{n\lambda}{2\sin\theta}. \tag{2.5}$$

The interplanar spacing (d_{hkl}) of any given plane in a pure cubic system can be related to the lattice parameter by the following equation:

$$d_{hkl} = \frac{a_0}{\sqrt{h^2 + k^2 + l^2}}. \tag{2.6}$$

Regardless of the specific atoms, the existence of any specific plane is a function of the lattice types. A specific set of $h^2 + k^2 + l^2$ combinations (called *extinction conditions*) exists that is the same for every simple cubic lattice; a different set is present for every face-centered cubic lattice; and a third set exists for every body-centered cubic lattice. Table 2-5 summarizes the reflections present for each cubic lattice type.

Table 2-5 shows that the sum of the Miller indices ($h^2 + k^2 + l^2$) for the plane that produces the first diffraction peak in a BCC system must equal 2. Therefore, Equation 2.6 can be used to calculate the lattice parameter for each plane, if the lattice type is known. For any cubic lattice, the lattice parameter should be the same in each direction, which provides a check of other calculations.

The diffractogram also can be used to determine the type of lattice present in the material. If Equations 2.5 and 2.6 are combined, this relationship results:

$$\sin^2\theta = \frac{\lambda^2}{4a_0^2}(h^2 + k^2 + l^2). \tag{2.7}$$

| *Extinction Conditions* |
The systematic reduction in intensity of diffraction peaks from specific lattice planes.

TABLE 2-5   Reflections Present for Each Cubic Lattice Type

Lattice Type	$h^2 + k^2 + l^2$
BCC	2, 4, 6, 8, 10, 12, 14, 16
FCC	3, 4, 8, 11, 12, 16
Simple	1, 2, 3, 4, 5, 6, 8

//

Example 2-6

An X-ray diffraction source with a wavelength of 0.7107 angstroms is beamed through a sample to generate the following peaks. If the material has a BCC lattice structure, determine the interplanar spacing, the lattice parameter, and the sum of the squares of the Miller indices for each plane.

Peak	2θ
1	20.20
2	28.72
3	35.36

SOLUTION

For any given peak, the interplanar spacing is given by Equation 2.5, $d_{hkl} = n\lambda/2\sin\theta$, so $d_{hkl} = (1)(0.7107 \text{ angstroms})/2\sin(10.10°) = 2.026$ angstroms.

For a BCC system, Table 2-5 shows that the sum of the squares of the Miller indices for the first plane should be 2. Thus, Equation 2.6 may be used to calculate the lattice parameter (a_0):

$$d_{hkl} = \frac{a_0}{\sqrt{(h^2 + k^2 + l^2)}}$$

$$(2.026 \text{ angstroms}) = \frac{a_0}{\sqrt{2}}$$

$$a_0 = 2.868 \text{ angstroms}$$

to give the results summarized next:

Peak	2θ (°)	d_{hkl} (angstroms)	$h^2 + k^2 + l^2$	a_0 (angstroms)
1	20.20	2.026	2	2.867
2	28.72	1.432	4	2.865
3	35.36	1.170	6	2.867

Equation 2.7 can be used to analyze the relationship between peaks. Because the X-ray wavelength does not change and the lattice parameter is the same for all planes in a cubic lattice, Equation 2.7 can be applied to two peaks to provide

$$\frac{\sin^2\theta_2}{\sin^2\theta_1} = \frac{(h^2 + k^2 + l^2)_2}{(h^2 + k^2 + l^2)_1}. \tag{2.8}$$

The ratio of the $\sin^2\theta$ terms on the left side of Equation 2.8 gives the relative ratio of the sums of the squares of the Miller indices of the two peaks. Along with the information in Table 2-5, this information can identify the peak. Example 2-7 illustrates this more clearly.

Example 2-7

Determine the type of lattice in the material responsible for the following diffractogram information:

Peak	2θ
1	20.20
2	28.72
3	35.36
4	41.07
5	46.19
6	50.90
7	55.28
8	59.42

SOLUTION

We begin by applying Equation 2.8 to the first two peaks. Note that the value listed in the table is 2θ, while θ is needed for the equation.

$$\frac{\sin^2(14.36)}{\sin^2(10.10)} = \frac{(h^2 + k^2 + l^2)_2}{(h^2 + k^2 + l^2)_1},$$

which provides

$$\frac{0.0615}{0.0308} = \frac{(h^2 + k^2 + l^2)_1}{(h^2 + k^2 + l^2)_2} = 2.$$

This tells us that the ratio of the sum of the squares of the Miller indices of the first two peaks is 2. There is no way to know the sum of the reflections for the first peak, but we have just determined that the $(h^2 + k^2 + l^2)$ for the second peak is twice that of the first. According to Table 2-5, the $(h^2 + k^2 + l^2)$ values for the first two peaks of a BCC system are 2 and 4. The ratio of $\frac{4}{2}$ is 2, so the system could be a BCC system. Similarly, simple cubic has $(h^2 + k^2 + l^2)$ values of 1 and 2 for its first two peaks, so simple cubic remains a possibility. However, the first two FCC peaks have 3 and 4 as their values. This provides a ratio of 1.33 rather than 2, telling us that the diffractogram could not have been generated by an FCC lattice.

Now compare each peak in the diffractogram to the first peak.

Peak	2θ	$\mathrm{Sin}^2\theta$	$\mathrm{Sin}^2\theta/\mathrm{Sin}^2\theta_1$
1	20.20	0.0308	1
2	28.72	0.0615	2
3	35.36	0.0922	3
4	41.07	0.1230	4
5	46.19	0.1539	5
6	50.90	0.1847	6
7	55.28	0.2152	7
8	59.42	0.2456	8

If the lattice were simple cubic, peak 7 would have an $(h^2 + k^2 + l^2)$ value that was eight times that of peak 1, according to Table 2-5. Instead, peak 7 has an $(h^2 + k^2 + l^2)$ value that is seven times that of peak 1. The only pattern that would yield this exact ratio would be body-centered cubic, for which the $(h^2 + k^2 + l^2)$ is 2 for peak 1 and 14 for peak 7.

| Crystallites |
Regions of a material in which the atoms are arranged in a regular pattern.

| Grain Boundaries |
The areas of a material that separate different crystallite regions.

| Crystal Mosaic |
A hypothetical structure accounting for irregularities in the boundaries between crystallites.

| Scherrer Equation |
A means of relating the amount of spreading in a X-ray diffractogram to the thickness of the crystallites in the sample.

| Full-Width Half-Maximum (FWHM) |
A standard used to measure the spread in the peak of a diffractogram, measured at the intensity value corresponding to the half highest value in the peak.

Until now, much of the discussion has treated materials as if they were comprised of a single, perfectly aligned crystal lattice. Instead, real materials consist of crystalline regions, or *crystallites*, separated from each other by *grain boundaries*. Therefore, a real material exhibits a structure much more like the *crystal mosaic* shown in Figure 2-13.

The same X-ray diffractogram provides information about the average crystallite size. If the material was a pure crystal, each peak would be extremely thin and have virtually no spreading. In reality, each peak spreads across a range of 2θ values. For relatively small grains, the amount of spreading is related to the thickness of crystallites in a plane by the *Scherrer equation*:

$$t = \frac{0.9\lambda}{B \cos \theta_B},$$ (2.9)

where t represents the crystallite thickness, λ is the wavelength of the X-ray source, B is the spread in the peak, and θ_B is the θ value at the top of the peak. Because a diffraction peak narrows as it approaches the top, the spread in the peak depends on where it is measured. As a standard, *full-width half-maximum (FWHM)* is used, meaning that the spread in the peak is measured at the intensity value corresponding to half the highest value in the peak. Figure 2-14 shows FWHM for a sample peak.

By reading the values of $2\theta_1$ and $2\theta_2$ at FWHM, B can be determined from the equation

$$B = 0.5(2\theta_2 - 2\theta_1) = \theta_2 - \theta_1$$ (2.10)

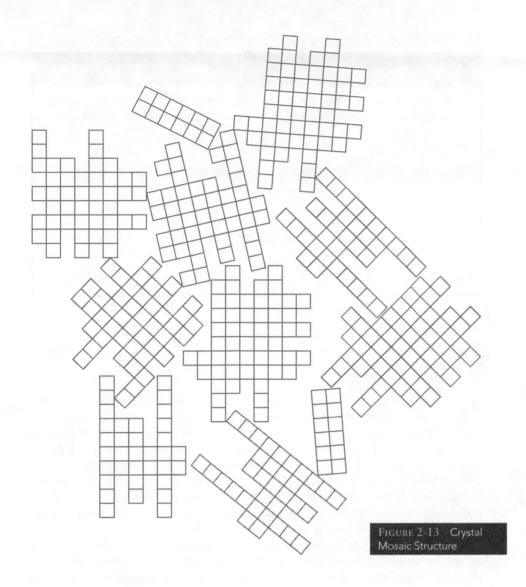

FIGURE 2-13 Crystal
Mosaic Structure

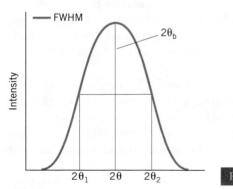

FIGURE 2-14 Full-Width Half-Maximum Measurement

Example 2-8

Estimate the thickness of the crystallites from the planes corresponding to peak 2 in Example 2-6 given that $2\theta_1 = 28.46$ and $2\theta_2 = 28.98$.

SOLUTION

Crystallite thickness is estimated using the Scherrer equation (2.9), so

$$t = \frac{0.9\lambda}{B \cos \theta_B}$$

$$t = \frac{0.9(.7107 \text{ nm})}{0.5(28.98 - 28.46) \cos (28.72/2)}$$

$$t = 2.54 \text{ nm}$$

2.8 MICROSCOPY

Seeing the features directly often would be the best way to understand the structure of a material. Some features, including grain sizes, large flaws in the material, cracking, and structures present in alloys, at times are visible to the naked eye. Often, however, the features of interest are too small to be seen directly. In such cases, the use of microscopes becomes valuable.

Several different types of microscopes exist and are classified by their source of light (or other radiation). The most common microscopes (present in essentially every science lab) are optical microscopes. If the material is opaque (e.g., metals, ceramics, and most polymers and composites), only a surface can be examined microscopically, and reflected light passing through the lens must reveal the image. For most materials, the surface must be polished before any meaningful features will be revealed. Many materials require surface treatment by an etching agent to reveal information. The reactivity between etching agents and some materials varies depending on the orientation of their grains. Specific etching agents are chosen so that adjoining grains will be affected differently and the contrast between the grains will become visible under the optical microscope.

| *Optical Microscopy* |
The use of light to magnify objects up to 2000 times.

Optical microscopy offers several advantages. The equipment is inexpensive and easy to operate. Large features, such as grains and cracks, are often apparent. Commercial software can calculate the size of each visible grain. However, optical microscopes are limited to about 2000× magnification, and many of the features that govern behavior exist at a much smaller scale.

When optical microscopy is insufficient, materials scientists turn to electron microscopy. Here, instead of visible light, a focused high-energy beam of electrons serves as the source for the image. The effective wavelength of an electron beam is 0.003 nm, allowing for resolution of finer details. Two distinct types of electron microscopes are available to provide different information.

Scanning electron microscopes (SEMs) collect the back-scattered beam of electrons and use it to project a blown-up image on a monitor, much like a television or computer screen. Resolution of details to the submicron level is possible with an SEM, and most systems are capable of capturing the digital image for printing or analysis. Surface features are directly visible, which makes the SEM ideal for rough surfaces, even at lower magnification. When using the SEM, some skill in positioning and focusing the beam is required. Moreover, researchers must be careful to ensure that the portion of the material they are examining is sufficiently representative of the entire material. Unlike an optical microscope, a high-end SEM costs several hundred thousand dollars.

Transmission electron microscopy (TEM) involves passing the electron beam through the sample and using the differences in beam scattering and diffraction to resolve an image. TEM is especially effective for examining microstructural defects. TEMs can magnify an image 1,000,000 times, but, like SEMs, they are extremely expensive and require some skill to operate. Additionally, sample preparation presents challenges since most materials absorb electron beams. To compensate, an extremely thin film of material must be prepared for examination to allow enough of the electron beam to pass through. Because the TEM uses diffraction of an electron beam to gather its imaging information, examination of the diffraction and scattering patterns can provide additional structural information. The principles of electron diffraction are highly similar to those of X-ray diffraction, discussed earlier in this chapter.

| Scanning Electron Microscopes (SEMs) |
Microscopes that focus a high-energy beam of electrons at the source and collect the back-scattered beam of these electrons.

| Transmission Electron Microscopy (TEM) |
An electron microscope that passes the electron beam through the sample and uses the differences in the beam scattering and diffraction to view the desired object.

How Do Crystals Form and Grow?

2.9 NUCLEATION AND GRAIN GROWTH

Crystallite growth is a two-stage phenomenon. First, tiny regions of order must form. These nano-crystallites are called *nuclei*, and the process by which they form is *nucleation*. Nuclei form randomly. Most are too small to sustain themselves and quickly disappear, but when a nucleus happens to be larger than some critical value (typically about 100 atoms), the thermodynamics of the system change and additional growth is favored.

Homogeneous nucleation occurs when a pure material cools sufficiently to support the formation of stable nuclei. *Heterogeneous nucleation* results when impurities provide a surface for the nucleus to form. The presence of this surface makes the nucleation phenomenon much easier, and only a few atoms are needed to achieve the critical radius. Stable nuclei in frozen water would not form until $-40°C$ via homogeneous nucleation, but impurities allow the nucleation to occur at much higher temperatures. With either homogeneous or heterogenous nucleation, the rate of nucleation is a function of temperature.

Once stable nuclei have formed, they begin the second step in the process: *grain growth*. Like many things in material science, the growth of grains follows a temperature-dependent relationship called the Arrhenius equation. For growth of grains,

$$\frac{dG}{dt} = A_0 \exp\left(\frac{-E_A}{RT}\right),$$

(2.11)

| Nuclei |
Tiny clusters of arranged atoms that serve as the framework for subsequent crystal growth.

| Nucleation |
The process of forming small aligned clusters of atoms that serve as the framework for crystal growth.

| Homogeneous Nucleation |
Clustering that occurs when a pure material cools sufficiently to self-support the formation of stable nuclei.

| Heterogeneous Nucleation |
The clustering of atoms around an impurity that provide a template for crystal growth.

| Grain Growth |
The second step in the formation of crystallites, which is dependent on temperature and can be described using the Arrhenius equation.

where G is the size of a growing crystal, t is time, A_0 is a pre-exponential constant that varies with material, R is the gas constant, E_A is the activation energy for diffusion, and T is absolute temperature.

For crystallite growth to occur, both the nucleation and grain growth processes must take place. The overall transformation rate is the product of the nucleation rate and the growth rate.

What Kinds of Flaws Are Present in Crystals and What Do They Affect?

2.10 POINT DEFECTS

| *Point Defect* |
A flaw in the structure of a material that occurs at a single site in the lattice, such as vacancies, substitutions, and interstitial defects.

Most discussions of lattice structures have focused on perfectly built lattices with no flaws of any kind, but all crystal lattices have flaws. When the flaw occurs at a single specific site in the lattice, it is called a *point defect*. Three different types of point defects can occur: vacancies, substitutions, and interstitial defects.

| *Vacancies* |
Point defects that result from the absence of an atom at a particular lattice site.

Vacancies result from the absence of an atom at a lattice site. Vacant lattice sites reduce the strength and stability of the overall lattice. Fortunately, few vacancies exist at room temperature, but the number of vacancies increases with increasing temperature. The elevated temperature provides more energy for atoms in the lattice to break their bonds and diffuse away. The temperature dependence of vacancies is governed by a form of the *Arrhenius equation*,

| *Arrhenius Equation* |
Generalized equation used to predict the temperature dependence of various physical properties.

$$N_v = N_0 \exp\left(\frac{-Q_v}{RT}\right),$$ (2.12)

| *Substitutional Defects* |
Point defects that result when an atom in the lattice is replaced with an atom of a different element.

where N_v is the number of vacancies, N_0 is the material specific pre-exponential constant, R is the gas constant, T is absolute temperature, and Q_v is the energy required to initiate a vacancy. As materials approach their melting temperature, vacancy rates often approach one lattice site in 10,000.

Substitutional defects result when an atom in a lattice site is replaced with an atom of a different element. Substitutions can be either beneficial or harmful, depending on the type of substitution and the desired properties. Sometimes substitutions are deliberately induced in a material through a process called *doping*. As we will discover in Chapter 9, doping is particularly important in the manufacture of electronic materials.

| *Interstitial Defects* |
Point defects that occur when an atom occupies a space that is normally vacant.

Interstitial defects result when an atom occupies a space in a lattice that is normally vacant. Generally the invading atom must be small enough to fit in a gap in the lattice. Ionic and ceramic materials are subject to a special form of interstitial defects related to charged particles. A *Frenkel defect* results from the diffusion of a cation into an interstitial site on the lattice, as shown in Figure 2-15(a). The result of the diffusion is a cation interstitial defect and a cation vacancy. A *Schottky defect* occurs when a cation and anion vacancy form in the lattice, as shown in Figure 2-15(b). Because the material must remain electrically neutral, one vacancy cannot form without the other.

| *Frenkel Defect* |
A point defect found in ceramic materials that occurs when a cation diffuses onto an interstitial site on the lattice.

| *Schottky Defect* |
A point defect that occurs in ceramics when both a cation and an anion are missing from a lattice.

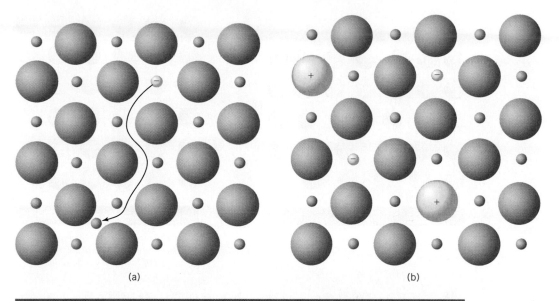

(a)   (b)

FIGURE 2-15   (a) Frenkel Defect; (b) Schottky Defect

2.11   DISLOCATIONS

In addition to point defects, entire sections of the lattice itself can be deformed. These larger-scale lattice defects are called *dislocations* because they result from bends or waves in the lattice itself. There are three primary types of dislocations: edge, screw, and mixed. *Edge dislocations* result from the addition of an extra partial lattice plate, as shown in Figure 2-16. Atoms that directly contact the edge dislocation are squeezed too close together while those immediately beyond it are pushed too far apart. The line that extends along the extra partial plane of atoms is referred to as the *dislocation line*.

A *screw dislocation* results from the lattice being cut and shifted by one atomic spacing, as shown in Figure 2-17.

Mixed dislocations result when a lattice contains both edge and screw dislocations with a discernible transition region in between, as shown in Figure 2-18.

The magnitude and direction of lattice distortion caused by dislocations is represented by a *Burgers vector* (b) defined by

$$\|b\| = \frac{a}{2}\sqrt{h^2 + k^2 + l^2} \tag{2.13}$$

For most metals and other close-packed systems, the magnitude of the Burgers vector is the same as the interplanar spacing of the material since the lattice is usually offset by one unit. The Burgers vector and the dislocation line form a right angle in edge dislocations and are parallel to each other in screw dislocations.

| *Dislocations* |
Large-scale lattice defects that occur from alterations to the structure of the lattice itself.

| *Edge Dislocations* |
Lattice defects caused by the addition of a partial plane into an existing lattice structure.

| *Dislocation Line* |
The line extending along the extra partial plane of atoms in an edge dislocation.

| *Screw Dislocation* |
Lattice defect that occurs when the lattice is cut and shifted by a row of atomic spacing.

| *Mixed Dislocations* |
The presence of both screw and edge dislocations separated by a distance in the same lattice.

| *Burgers Vector* |
A mathematical representation of the magnitude and direction of distortions in a lattice caused by dislocations.

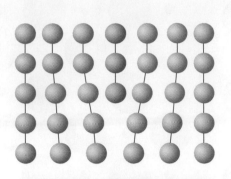

 FIGURE 2-16 Edge Dislocation

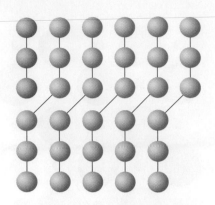

FIGURE 2-17 Screw Dislocation

From William D. Callister, Materials Science and Engineering, *6th edition. Reprinted with permission of John Wiley & Sons, Inc.*

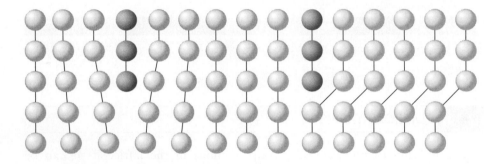

FIGURE 2-18 Mixed Dislocations

Dislocations form from three primary sources:

1. *Homogeneous nucleation.* Bonds in the lattice backbone rupture, and the lattice is sheared, creating two dislocation planes facing in opposite directions.

2. *Grain boundaries.* Steps and ledges present at the boundary between adjacent grains propagate during the early stages of deformation.

3. *Lattice/surface interactions.* Localized steps on the surface of the crystal concentrate stress in small regions making propagation of the dislocation far more likely.

Homogeneous nucleation requires concentrated stress to fracture the lattice bonds and seldom occurs spontaneously; grain boundary initiation and surface initiation are easier and more common.

2.12 SLIP

When placed under shear stress, dislocations can move through a material. The shear forces cause bonds between atoms in a plane to break. The cut plane shifts slightly and bonds to adjoining atoms, causing the dislocation to slide by one atomic length. The process repeats and results in a deformed crystal, as shown in Figure 2-19. The movement of

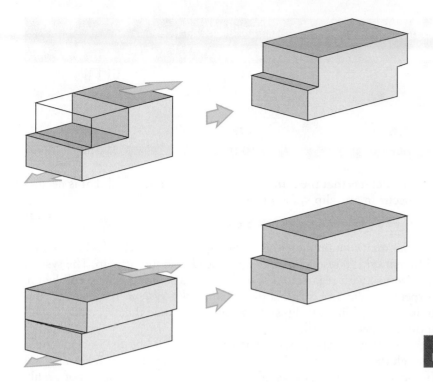

dislocations through a crystal is called *slip*. Several factors impact the likelihood of slip:

- Slip occurs more readily when the atoms are close together. The stress required to induce slip increases exponentially with increasing interplanar spacing.
- Slip requires the breaking of bonds, so materials with strong covalent bonds (such as polymers) are resistant.
- Materials with ionic bonds (such as metal oxides) are resistant because of larger interplanar distances and the repulsions caused when particles with like charge are forced to pass close to each other.

The direction that the dislocation moves is called the *slip direction,* and the planes impacted by the slip are called the *slip planes*. Together the slip planes and slip direction make up the *slip system*. Sufficient energy to break the bonds and move the atoms is required for slip to occur, and the dislocation will move in the direction that requires the least energy. Different lattice configurations have specific slip systems that are most likely to form. Table 2-6 summarizes the slip systems for the FCC, BCC, and HCP lattices.

For slip to occur, a sufficient amount of stress must be applied to permanently deform the material. This level of stress, called the *yield stress* (σ_y) is discussed more thoroughly in Chapter 3. When such a stress is placed on a material, it is unlikely that the stress is acting in the direction of the slip plane. Until a critical stress threshold is reached for a given set of slip planes, the dislocations cannot slip. Figure 2-20 helps illustrate the concept. If ϕ represents the angle between the applied stress and the normal to the slip plane and λ represents the angle between the force and the slip direction,

| *Slip* |
The movement of dislocations through a crystal, caused when the material is placed under shear stress.

| *Slip Planes* |
New planes formed after the material has undergone slip.

| *Slip System* |
Composed of both the slip plane and the slip direction.

TABLE 2-6 Slip Systems Present in Different Lattices		
Lattice Type	Slip Planes	Slip Direction
BCC	(1 1 0)	[1 1 1]
	(1 1 2)	
	(1 2 3)	
FCC	(1 1 1)	[1 1 0]
HCP	(0 0 0 1)	[1 0 0]

| Schmid's Law |
The equation used to
determine the critical resolved
shear stress in a material.

| Critical Resolved
Shear Stress |
The lowest stress level
at which slip will begin
in a material.

| Primary Slip System |
The first set of planes in a
material to experience slip
under an applied stress.

| Hall-Petch Equation |
Correlation used to estimate
the yield strength of a given
material based on grain size.

Schmid's law dictates that the *critical resolved shear stress* (τ_c) that is needed for slip to occur in any slip system is defined as

$$\tau_c = \sigma_y \cos \phi \cos \lambda \tag{2.14}$$

A material may contain many different slip systems, each of which would have a different critical resolved shear stress needed to initiate slip. The system with the lowest τ_c with regard to a specific applied stress direction will be the first to experience slip and is called the *primary slip system*. If the stress continues to increase, additional slip systems may overcome their τ_c, and slip will begin in those systems as well.

If an entire material was composed of a single crystal, the dislocations could move through the entire material. However, most real materials have a crystal mosaic structure with smaller crystallites, or grains. The atoms are not evenly spaced at the boundary between adjacent grains, and slip cannot continue. When a defect is propagating through a material, it stops (or at least is significantly slowed) when it reaches a grain boundary. Materials with larger grains are more affected by slip, so smaller grains are generally desirable. Smaller grains lead to greater strength. The *Hall-Petch equation* can be used to estimate the yield strength (σ_y) of a given material based on grain size:

$$\sigma_y = \sigma_0 + \frac{K_y}{\sqrt{d}}, \tag{2.15}$$

where σ_0 and K_y are material specific constants and d is the average diameter of a grain. The American Society for Testing and Materials (ASTM) provides

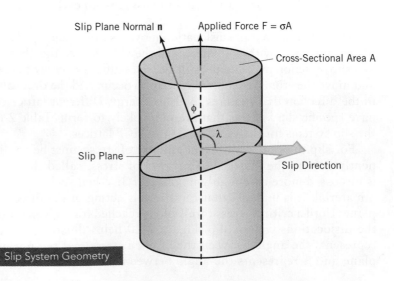

FIGURE 2-20   Slip System Geometry

a standard method for characterizing grain sizes. They define the *grain size number* (G) as

$$N - 2^{G-1}, \tag{2.16}$$

where N is the number of grains observed in a one square inch area at $100\times$ magnification. Equation 2.14 can be written to solve directly for G:

$$G = 1.443 \ln(N) + 1. \tag{2.17}$$

2.13 DISLOCATION CLIMB

*D*islocation climb is another mechanism by which a dislocation can propagate through a lattice. Unlike slip, dislocation climb allows the dislocation move in directions that are perpendicular to the slip plane. Vacancies in the lattice are the key to dislocation climb. As discussed in Section 2.10, vacancies can move throughout the lattice. When a vacancy moves into a site adjacent to the partial plane of atoms in an edge dislocation, the nearest atom in the partial plane can fill the vacant site, and a new vacancy forms in the partial plane as shown in Figure 2-21. The atom is said to have made a *positive climb*. The crystal shrinks in the direction perpendicular to the extra partial plane because of the removal of the extra atom. The vacancy now present in the partial plane can be replaced by an atom on the other side. In this case, the crystal grows in the direction perpendicular to the partial plane because a new atom has been added to the partial plane. This is called a *negative climb*. Both positive and negative climb are strongly affected by temperature because increasing the temperature increases the rate at which vacancies move within the lattice. Compressive stresses favor positive climb while tensile stresses favor negative climb.

| Grain Size Number |
A numerical quantity developed by the American Society for Testing and Materials (ASTM) to characterize grain sizes in materials.

| Dislocation Climb |
Mechanism by which dislocations move in directions that are perpendicular to the slip plane.

| Positive Climb |
The filling of a vacancy in the partial plane of an edge dislocation by an adjacent atom resulting in a shrinking of the crystal in the direction perpendicular to the partial plane.

| Negative Climb |
The filling of a vacancy in the partial plane of an edge dislocation by an adjacent atom resulting in a growth of the crystal in the direction perpendicular to the partial plane.

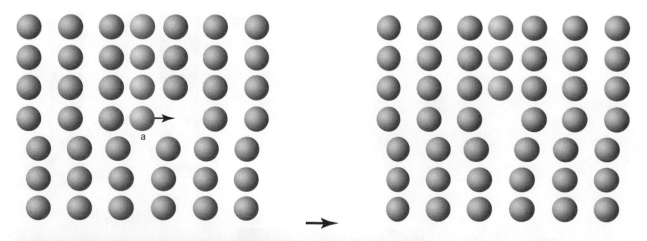

FIGURE 2-21   Positive Dislocation Climb

2.14 MONOCRYSTALS AND NANOCRYSTALS

The materials talked about in this chapter so far have had many regions of crystallinity separated by grain boundaries. The regions between crystallites block slip and increase the strength of materials but also can have adverse effects. Misalignments of crystallites impact the local electrical properties of semiconductor materials and would make many of the advances in microelectronics impossible. To get around this difficulty, scientists and engineers have developed single crystals, or *monocrystals*, in which the entire material consists of a single unbroken grain.

| *Monocrystals* |
Materials in which the entire structure is a single unbroken grain.

Single crystals are formed by building the crystal one layer of atoms at a time. Typically, a very small ordered solid called a *seed crystal* is immersed in a molten solution of the relevant material. The seed crystal provides a framework for the new atoms to follow. By carefully controlling temperature gradients and other processing variables, relatively large single crystals, or *boules*, may be produced. Although high-end applications, such as the formation of single-crystal semiconductors, require clean-room conditions, high-purity materials, and exceptional control of processing variables, the same basic concept is used in more mundane applications, such as the development of rock candy. Perfect single crystals are extremely uncommon in nature, but synthetic boules can be grown to more than one meter in length.

| *Boules* |
Large, artificially produced monocrystals.

In addition to their improved electrical properties, single crystals have significant optical advantages as well. Single-crystal sapphire fibers with optical losses as low as 0.3 dB/m are in commercial production. The single crystals do have significant drawbacks. They are expensive to make and are extremely susceptible to defects. Because there are no grain boundaries, defects are able to propagate throughout the entire material, significantly enhancing the likelihood of failure. Slip in these systems occurs in parallel planes, as shown in Figure 2-22.

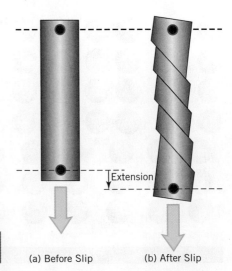

FIGURE 2-22 Illustration of Slip in Single-Crystal Systems

(a) Before Slip   (b) After Slip

Extension

Nanocrystals—crystalline materials with sizes of nanometers in length—represent a potential technological revolution. These materials typically range from a few hundred to several thousand atoms in size. As a result, they are larger than most molecules but far smaller than typical crystalline solids. Nanocrystals tend to have exceptional thermodynamic and electrical properties that fall between those of individual molecules and larger solids. Their potential applications range from optic to electronics, catalysis, and imaging.

Because the nanocrystals are so small, they are much less susceptible to defects than larger materials, and most of their properties can be controlled by controlling their size. Single atoms may be trapped inside of nanocrystals, significantly enhancing the luminescent properties of the atom. This opens the door for the creation of a host of optical and magnetic applications, many of which are just being developed. Nanocrystals are already used in photovoltaic solar cells that appear to be more efficient than traditional cells. Nanocrystals also offer extremely high surface-to-volume ratios, which make them ideal for many catalysis applications. Some oil companies now use nanocrystals in their diesel fuel production processes.

| *Nanocrystals* |
Crystalline materials with sizes of nanometers in length.

Summary of Chapter 2

In this chapter we examined:

- The structures present in crystalline materials
- How to characterize the structures
- How to represent planes and directions using Miller indices
- How to use these concepts to calculate densities and other material properties
- How X-ray diffraction reveals the crystal structure of materials
- How microscopy can help identify additional features
- How crystals form and grow
- How real lattices differ from ideal ones
- What defects exist in lattices
- How defects move through lattices through slip and dislocation climb
- The role and importance of grain boundaries
- The differences among crystals, single crystals, and nanocrystals

Key Terms

amorphous materials *p. 33*
Arrhenius equation *p. 54*
atomic arrangement *p. 33*
atomic packing factor (APF) *p. 37*
atomic structure *p. 33*
body-centered cubic (BCC) *p. 34*
boules *p. 60*
Bravais lattices *p. 33*
Braggs' equation *p. 46*
Burgers vector *p. 55*
constructive interference *p. 46*
critical resolved shear stress *p. 58*
crystal mosaic *p. 50*
crystal structure *p. 33*
crystallites *p. 50*
destructive interference *p. 46*
diffraction *p. 46*
dislocations *p. 55*
dislocation climb *p. 59*
dislocation line *p. 55*
edge dislocations *p. 55*
extinction conditions *p. 47*

face-centered cubic (FCC) *p. 34*
Frenkel defect *p. 54*
full-width half-maximum (FWHM) *p. 50*
grain boundaries *p. 50*
grain growth *p. 53*
grain size number *p. 59*
Hall-Petch equation *p. 58*
heterogeneous nucleation *p. 53*
hexagonal close-packed (HCP) *p. 34*
homogeneous nucleation *p. 53*
interplanar spacing *p. 45*
interstitial defects *p. 54*
lattice parameters *p. 36*
macrostructure *p. 33*
microstructure *p. 33*
Miller indices *p. 43*
mixed dislocations *p. 55*
monocrystals *p. 60*
nanocrystals *p. 61*
negative climb *p. 59*
nucleation *p. 53*

nuclei *p. 53*
optical microscopy *p. 52*
point defect *p. 54*
positive climb *p. 59*
primary slip system *p. 58*
scanning electron microscopes (SEMs) *p. 53*
Scherrer equation *p. 50*
Schmid's law *p. 58*
Schottky defect *p. 54*
screw dislocation *p. 55*
simple cubic *p. 33*
slip *p. 57*
slip planes *p. 57*
slip system *p. 57*
substitutional defects *p. 54*
theoretical density *p. 40*
transmission electron microscopy (TEM) *p. 53*
unit cell *p. 36*
vacancies *p. 54*

Homework Problems

1. The theoretical density of iridium is 22.65 g/cm³ while osmium's theoretical density is 22.61 g/cm³. However, experimentally osmium is denser than iridium. Explain briefly how this phenomenon might occur.

2. Calculate the theoretical density of silver at 20°C.

3. The lattice parameter for molybdenum is 0.314 nm and its density at 20°C is 10.22 g/cm³. Determine the crystal structure of molybdenum.

4. Determine the indices for the following directions:

a.

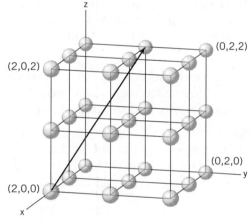

b.

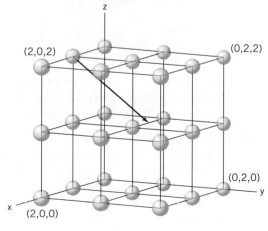

c.

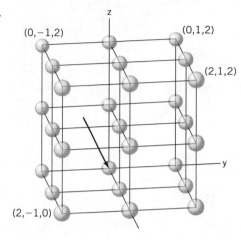

5. Determine the indices for the following directions:

a.

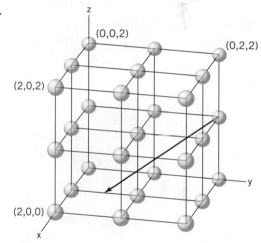

b.

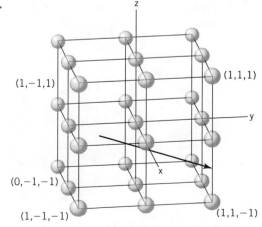

c.

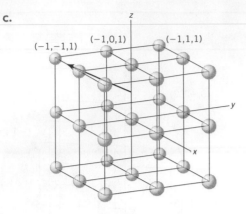

6. Draw the crystallographic directions defined by the following indices:

a. $[1\,0\,2]$ b. $[0\,\bar{1}\,1]$ c. $[\bar{2}\,3\,\bar{1}]$

7. Draw the crystallographic directions defined by the following indices:

a. $[0\,1\,2]$ b. $[\bar{1}\,\bar{1}\,1]$ c. $[2\,0\,\bar{2}]$

8. Determine the Miller indices for the following planes:

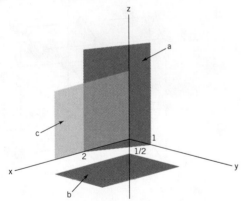

9. Determine the Miller indices for the following planes:

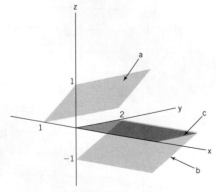

10. On a set of coordinate axes, draw the planes that correspond to the following sets of Miller indices:

a. $(1\,0\,0)$ b. $(\bar{2}\,0\,1)$ c. $(1\,2\,1)$

11. On a set of coordinate axes, draw the planes that correspond to the following sets of Miller indices:

a. $(0\,2\,0)$ b. $(\bar{2}\,0\,1)$ c. $(\bar{1}\,\bar{2}\,1)$

12. The rate of grain growth for a given material is 10 nm/min at 293 K and 150 nm/min at 323 K.

a. Determine the activation energy and pre-exponential constant.

b. Estimate the rate of grain growth at 310 K.

13. The activation energy for grain growth for a specific material is 15 kj/mol and the pre-exponential constant is 1000 nm/min. What temperature would be needed to achieve a grain growth rate of 2.44 nm/min?

14. A graduate student tells you that he has observed a Schottky defect in a sample of pure aluminum. Explain why it may take a long time for this graduate student to get his Ph.D.

15. The following square represents one square inch at a magnification of 100×. Determine the grain size number for the material.

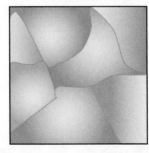

16. How many grains would be present in a one square inch sample at 100× magnification for a material with a grain size number of 4?

17. During an X-ray diffraction experiment (λ = 0.7107 A), a diffraction peak appears at 37.3° and another at 46.2°. Lacking any other information, which of the two planes would you expect to be more susceptible to slip? Explain your answer.

18. The yield strength of a material is 400 psi when the average grain diameter is 0.22 inches and increases to 450 psi when the grain diamter is 0.15 inches. What grain diameter would give a yield strength of 500 psi?

19. Would a material with a BCC, FCC, or HCP lattice structure be more prone to interstitial defects? Explain your answer.

20. An X-ray source ($\lambda = 0.7107$ A) is used to examine a powdered sample. If the interplanar spacing for a given plane is 1.35 angstroms, the distance between atoms in the plane is 1.91 angstroms and the average crystallite thickness is 12 angstroms.
    a. Predict the angle of incidence that will correspond to the diffraction peak.
    b. Estimate the spreading of the peak at full-width half-maximum.
    c. Provide a potential set of Miller indices for this plane.
    d. Draw the plane (on a set of coordinate axes).

21. An X-ray source ($\lambda = 0.7307$ A) is beamed into a powdered sample of a material with an FCC structure. The interplanar spacing that corresponds to the first diffraction peak is 3.40 angstroms.
    a. Predict the 2θ values for the first five diffraction peaks for this sample.
    b. If the spreading in the second peak at FWHM is $2\theta = 0.20°$, determine the average crystallite thickness.

22. An X-ray source ($\lambda = 0.7307$ A) is used to examine an unknown metal alloy. The resulting diffraction peaks are summarized in the next table. Testing of the alloy using Archimedes principle shows that the density of the metal is 17.3 g/cm³. Determine the average molecular weight of the alloy.

Peak Number	2θ
1	21.04
2	24.34
3	34.70
4	40.93
5	42.84
6	49.88

23. The Miller indices for a plane (one of many) in a sample are (2 1 0).
    a. Draw the plane that corresponds to the **first** X-ray diffraction peak that would appear for that sample. *Note:* The (2 1 0) plane may not be the first.
    b. If the lattice parameter of the material is 1.54 angstroms and a 0.7307 angstrom source is used, determine the interplanar spacing of the (2 1 0) plane.
    c. If the density of the material is 7.88 g/cm³, what is the molecular weight of the sample?

24. If a tensile stress is applied to a crystalline sample, which of the slip planes described in the next table would be the primary slip plane?

Plane	Angle between the Applied Stress and the Normal to the Slip Plane	Angle between the Stress and the Slip Direction
A	20.7°	77.0°
B	4.2°	33.7°
C	63.5°	20.4°

25. Explain why different planes within the same crystal lattice would have different values of τ_c.

26. Why does the angle of the applied stress change the primary slip plane of a lattice?

3 Measurement of Mechanical Properties

CONTENTS

Learning Objectives

By the end of this chapter, a student should be able to:

- Know how to find and read an ASTM standard for a specific test.

- Calculate tensile strength, tensile modulus, breaking strength, modulus of resilience, Poisson's ratio, yield strength, and true stress and strain from tensile test data.

- Identify and define the regions of elastic stretching and plastic deformation.

- Calculate the offset yield strength for a material with no clear transition between elastic stretching and plastic deformation.

- Explain the operation of a hardness test.

- Convert hardness test results between different scales.

- Explain in his or her own words the difference between brittle and ductile fracture.

- Calculate stress intensity factors and determine whether a crack will propagate or not.

- Discuss the factors that impact the intensification of stress on a crack tip.

- Explain the physical basis for creep and the procedure for creep testing.

- Explain the procedure for compression testing and why recoil testing must sometimes be used.

- Calculate flexural strength and flexural modulus from a bend test.

- Calculate appropriate error bars for experimental data.

- Determine whether two means are statistically different.

- Calculate the endurance limit and fatigue life from fatigue testing.

- Explain the process for conducting an accelerated aging study and the limitations of its results.

How Do I Know How to Measure Properties?

3.1 ASTM STANDARDS

The measurement of mechanical properties is an essential factor in determining the suitability of a specific material for a specific function. However, when properties are measured by different investigators in different laboratories, there is the potential for inconsistencies in technique and results. To reduce this problem, standards for conducting tests, measuring data, and reporting results have been established.

ASTM International, formerly known as the American Society for Testing and Materials (ASTM), has published over 12,000 standards for the testing of materials. Although compliance with these standards is voluntary, they provide a detailed description of testing procedures that ensure that results from different laboratories are directly comparable. *ASTM standards* may be located and purchased online (www.astm.org), through a 77-volume annual book of standards, or through CD-ROM compilations. A representative list of standards for the testing techniques discussed in this chapter is presented in Table 3-1.

ASTM standards open with a discussion of their scope, followed by a list of referenced documents. They define terminology and summarize the test method, including significance, use, and interferences. Most include a detailed description of testing apparatus with illustrations. Directions for preparing test specimens, calibrating the equipment, and conditioning the environment are also provided. Detailed experimental procedures and instructions for performing calculations also are given.

| *ASTM Standards* |
Guidelines published by the American Society for Testing and Materials that provide detailed testing procedures to ensure that tests performed in different laboratories are directly comparable.

TABLE 3-1 Representative ASTM Standards for Test Methods Described in Chapter 3

Test Type	Relevant ASTM Standard
Tensile—Concrete surfaces	C1355
Tensile—Metallic materials	E8M
Tensile—Metal matrix composites	D3552
Tensile—Polymer matrix composites	D4762
Tensile—Single textile fibers	D3822
Compression—Metals	E209
Compression—Fiber-reinforced ceramics	WK3484
Compression—Concrete	C116
Compression—Composites	D3410
Bend Test—Ceramics	C1421
Brinell Hardness	E10
Rockwell Hardness	E18
Creep—Ceramics	C1291
Creep Rupture—Metals	E139
Creep Crack Growth—Metals	E1457
Izod Impact—Notched plastics	D256
Charpy Impact—Notched plastics	D6110
Fatigue Testing of Homogenous Materials	E606

What Properties Can Be Measured and What Do They Tell Me?

There are far too many tests to list (much less describe) in an introductory text. Instead, this section focuses on the eight most common and important tests performed on a wide variety of materials: tensile testing, compressive testing, bend testing, hardness testing, creep testing, impact testing, fatigue testing, and accelerated aging. Even within these eight fundamental tests, there are countless variations in operation depending on the equipment available, the material to be tested, and many other factors. For each test method, the basic operating principle is described along with comments on what the data tell about the material. A brief synopsis of each test is provided in Table 3-2.

3.2 TENSILE TESTING

The tensile test provides a wealth of information about a material. Although a variety of specific ASTM standards govern the exact testing procedures for different types of materials, as shown in Table 3-1, all use the same basic operating principle. The material sample is secured between a pair of clamps. The upper clamp is attached to a fixed bar and a load cell.

TABLE 3-2	Summary of Test Methods	
Tensile test	The material sample is secured between a pair of clamps. The upper clamp is attached to a fixed bar and a load cell. The lower grip is attached to a movable bar that slowly pulls the material downward. The load cell records force, and an extensiometer records the elongation of the sample.	
	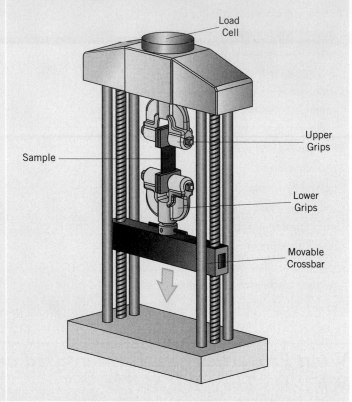	
Compressive test	Uses the same apparatus as tensile testing, but instead of pulling the sample apart, the sample is subjected to a crushing load. Many materials display similar tensile and compressive moduli and strengths, so compressive tests are often not performed except in cases where the material is expected to endure large compressive forces. However, the compressive strengths of many polymers and composites are significantly different from their tensile strengths.	
Bend test	Used to test brittle materials. As the sample begins to deflect under an applied force, the bottom experiences a tensile stress while the top experiences compressive stress.	
	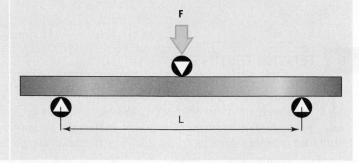	

Hardness test	Although there are dozens of techniques to measure hardness, the most common is the Brinell test, in which a tungsten-carbide sphere of 10 mm diameter is pushed into the surface of the test material using a controlled force. The size of the indentation is used to determine the hardness of the material.

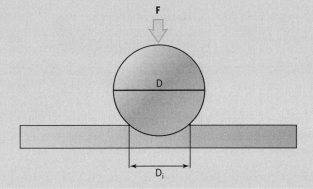

| *Hardness Test* |
A method used to measure the resistance of the surface of a material to penetration by a hard object under a static force.

Creep test	Creep refers to plastic deformation of a material over time (usually at elevated temperatures). When a continuous stress is applied to a material at elevated temperature, it may stretch and ultimately fail below the yield strength. Ultimately, creep occurs because of dislocations in the material. Many materials, including some polymers and solder, experience creep at relatively low temperatures.

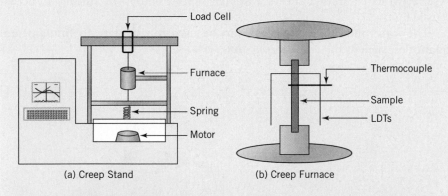

(a) Creep Stand   (b) Creep Furnace

Impact test	Toughness defines a material's resistance to a blow. In an impact test, a hammer is secured to a pendulum at some initial height and released. The orientation of the sample varies depending on specific testing techniques.

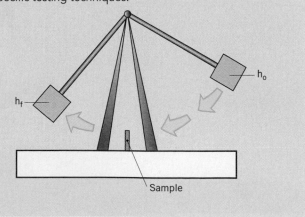

(continues)

TABLE 3-2 Summary of Test Methods *(continued)*	
Fatigue test	A material is passed through many cycles of tension and compression below the yield strength until failure ultimately results. 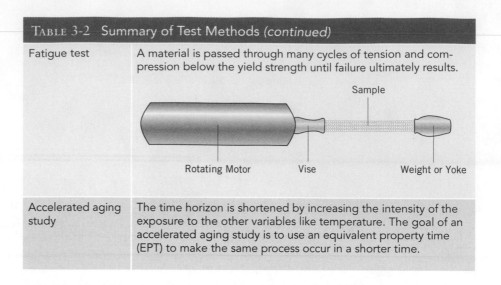
Accelerated aging study	The time horizon is shortened by increasing the intensity of the exposure to the other variables like temperature. The goal of an accelerated aging study is to use an equivalent property time (EPT) to make the same process occur in a shorter time.

The lower grip is attached to a movable bar that slowly pulls the material downward. The load cell records force, and an extensiometer records the elongation of the sample. Figure 3-1 shows a schematic drawing of a tensile testing system, while Figure 3-2 shows a Kevlar®-epoxy composite during an actual tensile test.

The force and elongation data can be used to calculate to fundamental quantities, such as the ***engineering stress*** (σ),

| *Engineering Stress* |
The ratio of applied load to cross-sectional area.

$$\sigma = \frac{F}{A_0},\tag{3.1}$$

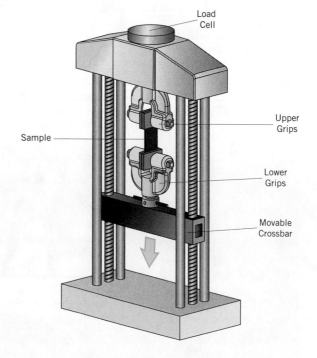

FIGURE 3-1 Schematic Tensile Test Apparatus

FIGURE 3-2 Photograph of a Tensile Test on a Kevlar®-Epoxy Composite

Courtesy James Newell

where F is the measured force and A_0 is the initial cross-sectional area of the sample, and the *engineering strain* (ϵ),

$$\epsilon = \frac{1 - l_0}{l_0}, \qquad (3.2)$$

where l_0 is the initial sample length and l is the elongated sample length. The test is run until the sample breaks. The data are reported in the form of a stress-versus-strain plot as shown in Figure 3-3 and Table 3-3.

This relatively simple plot provides insight to many key properties. During the early stages of the tensile test, the material would return to its original state if the strain was released. This region where no permanent changes occur to material is called the *elastic stretching* region. The material will return completely to its previous state once the strain has been released provided the material remains in the elastic stretching region, but as soon as the first change occurs from which that the material cannot completely recover, *plastic deformation* begins. For most materials, the stress-strain curve provides a straight line in the elastic stretching region, but the slope changes noticeably when plastic deformation begins. The stress at the point of transition between elastic stretching and plastic deformation is called the *yield strength* (σ_y). Once the stress on a material has exceeded the yield strength, it will not return completely to its original form.

Even when plastic deformation has begun, many materials are capable of handling additional stress. The stress at the highest applied force (the maximum on the stress-strain curve) is called the *tensile strength* (σ_s) of the material. The stress at which the material finally breaks completely is called its *breaking strength* (σ_B).

| *Engineering Strain* |
A property determined by measuring the change in the length of a sample to the initial length of that sample.

| *Elastic Stretching* |
The region on a stress-strain curve in which no permanent changes to the material occur.

| *Plastic Deformation* |
The region on a stress-strain curve in which the material has experienced a change from which it will not completely recover.

| *Yield Strength* |
The stress at the point of transition between elastic stretching and plastic deformation.

| *Tensile Strength* |
The stress at the highest applied force on a stress-strain curve.

| *Breaking Strength* |
The stress at which the material breaks completely during tensile testing.

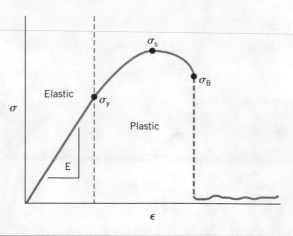

FIGURE 3-3   Representative Stress-Strain Curve

TABLE 3-3   Regions of Stress-Strain Curve

Elastic stretching region—material returns to its original form when stress is released, no lasting damage, region is usually a straight line	
Plastic deformation region—material cannot completely recover when stress is released because of permanent microstructure changes	
Breaking region—the material has failed here after going through plastic deformation; brittle materials fail immediately after elastic deformation	

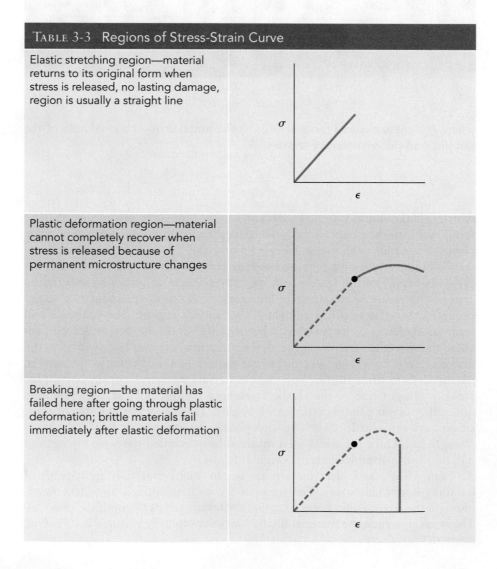

Not all materials can experience plastic deformation without breaking. *Ductile* materials can deform without breaking, while materials that fail completely at the onset of plastic deformation are called *brittle*. Section 3.9 on fracture mechanics provides a more thorough discussion of the distinction between brittle and ductile materials. For a brittle material, the yield strength, tensile strength, and breaking strength are the same, as shown in Figure 3-4.

The slope of the stress-strain curve in the elastic region provides a quantity that is shown in Equation 3.3. This quantity goes by three different names: *elastic modulus* (E), *tensile modulus*, or *Young's modulus*. Materials with high bond energies also have high elastic moduli because more force is required to stretch them. While grain size has a significant impact on tensile strength, Young's modulus is not impacted by the microstructure of the material and remains the same regardless of grain size. Graphite has the highest theoretical modulus (1080 GPa) of all materials. The all-aromatic nature of the chemical bonds of graphite keeps the plane essentially perfectly aligned.

$$E = \frac{\Delta\sigma}{\Delta\epsilon} \quad\quad (3.3)$$

The area contained under the elastic portion of a stress-strain curve is the *elastic energy* (e_E) of the material, which represents how much energy the material can absorb before permanently deforming. The elastic energy always can be determined by integrating the curve

$$e_E = \int_0^{\epsilon_y} \sigma \, d\epsilon. \quad\quad (3.4)$$

For materials displaying linear elastic behavior, the relationship simplifies to

$$e_E = \frac{\sigma_y}{2}. \quad\quad (3.5)$$

The elastic energy is used to calculate the *modulus of resilience* (E_r), which is the ratio of the elastic energy to the strain at yielding:

$$E_r = \frac{e_E}{\epsilon_y}. \quad\quad (3.6)$$

The modulus of resilience determines how much energy will be used for deformation and how much will be translated to motion. Golf ball manufacturers strive to enhance the modulus of resilience to improve the performance of their product.

As a material deforms longitudinally (stretches), it also undergoes a simultaneous lateral deformation (shrinking). *Poisson's ratio* (μ), shown in Equation 3.7, relates the magnitude of these concurrent deformations.

$$\mu = \frac{-\epsilon_{\text{lateral}}}{\epsilon_{\text{longitudinal}}}. \quad\quad (3.7)$$

For most materials, Poisson's ratio hovers around 0.3.

| **Ductile** |
Materials that can plastically deform without breaking.

| **Brittle** |
Materials that fail completely at the onset of plastic deformation. These materials have linear stress-versus-strain graphs.

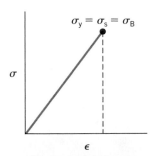

FIGURE 3-4 Stress-Strain Curve for a Brittle Material

| **Elastic Modulus** |
The slope of the stress-strain curve in the elastic region. Also called *Young's modulus* or *tensile modulus*.

| **Elastic Energy** |
The area contained under the elastic portion of a stress-strain curve, which represents how much energy the material can absorb before permanently deforming.

| **Modulus of Resilience** |
The ratio of the elastic energy to the strain at yielding, which determines how much energy will be used for deformation and how much will be translated to motion.

| **Poisson's Ratio** |
Relates the longitudinal deformation and the lateral deformation of a material under stress.

Example 3-1

For the stress-strain curve provided, determine:

a. Yield strength
b. Tensile strength
c. Breaking stress
d. Young's modulus
e. Strain at failure
f. Whether the material is brittle or ductile

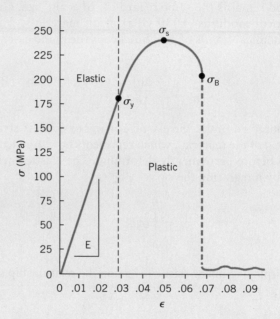

SOLUTION

a. The yield stress occurs at the point where the stress-strain curve stops being linear, about 180 MPa.
b. The tensile strength is the highest point on the stress-strain curve, about 240 MPa.
c. The breaking stress is the stress at complete failure, about 205 MPa.
d. The Young's modulus is the slope of the stress-strain curve in the elastic region, about 4500 MPa.
e. The strain at failure is about 0.068.
f. The material is ductile since plastic deformation occurs before failure.

The amount of deformation a material can withstand without breaking is called its *ductility*. The more ductile a material is, the easier it is to shape and machine but the less likely it is to retain its shape under stress. Two measures are used to evaluate ductility, the percent elongation and the percent reduction in area:

$$\% \text{ elongation} = 100\% * \frac{(l_f - l_0)}{l_0} \tag{3.8}$$

and

$$\% \text{ reduction in area} = 100\% * \frac{(A_f - A_0)}{A_0} \tag{3.9}$$

where f represents values at failure and 0 represents the initial values.

We have already discussed the idea that as the material is stretched, its cross-sectional area decreases. Because the strain on the material is a function of area, the reduced cross-section leads to greater strain. In the example shown in Figure 3-3, the stress on the material is dropping between the tensile strength and the breaking strength, even though the force is increasing. In fact, the true cross-sectional area is decreasing while the formula given for calculating engineering stress in Equation 3.1 uses the initial cross-sectional area. The *true stress* (σ_t) and *true strain* (ϵ_t) on a material, accounting for the change in cross-sectional area, are given by

$$\sigma_t = \frac{F}{A_i} \tag{3.10}$$

and

$$\epsilon_t = \ln\frac{l_i}{l_0}. \tag{3.11}$$

Because of the relative complexity of accounting for the change in area during a tensile test, and because of the limited effect the change usually has on the parameters of interest, the engineering stress and engineering strain are generally used instead of true stress and strain.

In some ductile materials, including some polymers and soft metals such as lead, the narrowing occurs in a highly localized area called *necking*. At this point, the cross-sectional area diminishes far more rapidly than in the rest of the sample. Most children have observed this phenomenon when they stretch chewing gum from their mouths. The gum narrows rapidly at one point and ultimately fails. Necking usually results from flaws that were present at formation or induced under the tensile load.

For some materials, the transition between elastic stretching and plastic deformation cannot be defined clearly from a stress-strain curve. In such cases, an *offset yield strength* (σ_y) is calculated instead. The offset yield strength is determined by calculating the initial slope of stress-strain curve, then drawing a line of equal slope that begins at a strain value of 0.002. The stress level at which this new line intersects the stress-strain curve is the offset yield strength. Figure 3-5 illustrates this process.

For the data shown in Figure 3-5, the offset yield strength would be approximately 200 MPa, the point at which the offset line crosses the stress-strain curve.

| **Ductility** |
The ease with which a material deforms without breaking.

| **True Stress** |
A ratio of the force applied to a sample and the instantaneous cross-sectional area of the sample.

| **True Strain** |
Represents the ratio of the instantaneous length of the chain to the initial length of the chain.

| **Necking** |
The sudden decrease in cross-sectional area of a region of a sample under a tensile load.

| **Offset Yield Strength** |
An estimate of the transition between elastic stretching and plastic deformation for a material without a linear region on a stress-strain curve.

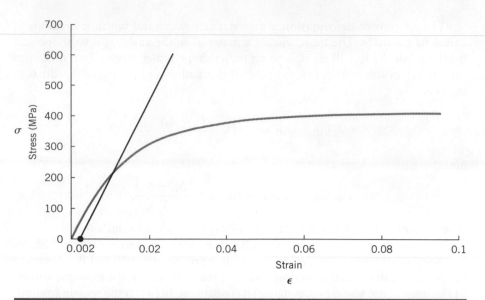

σ
Stress (MPa)

Strain
ε

FIGURE 3-5   Offset Yield Strength

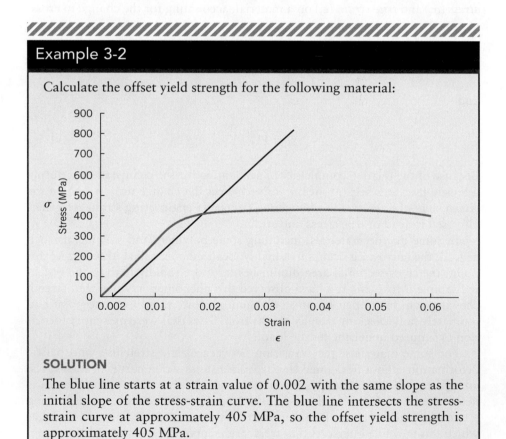

Example 3-2

Calculate the offset yield strength for the following material:

σ
Stress (MPa)

Strain
ε

SOLUTION

The blue line starts at a strain value of 0.002 with the same slope as the initial slope of the stress-strain curve. The blue line intersects the stress-strain curve at approximately 405 MPa, so the offset yield strength is approximately 405 MPa.

3.3 COMPRESSIVE TESTING

In many ways, compressive testing is a direct analog to tensile testing. The same apparatus is often used, but instead of pulling the sample apart, the sample is subjected to a crushing load. Equations 3.10 and 3.11 apply to the determination of both compressive strength and compressive modulus, although the compressive modulus is negative since l_i is less than l_0. Many materials display similar tensile and compressive moduli and strengths, so compressive tests are often not performed except in cases where the material is expected to endure large compressive forces. However, the compressive strengths of many polymers and composites are significantly different from their tensile strengths.

Compression testing also enables a direct examination of the mode of deformation, as shown in Table 3-4. The deformation is classified as buckling when $L/D > 5$, shearing when $2.5 < L/D < 5$, homogeneous compression when $L/D < 2$, barreling when friction exists at the contact surface and $L/D < 2$, and double barreling when friction exists and $L/D > 2$. Compressive instabilities can also be observed.

TABLE 3-4 Summary of Modes of Deformation

Deformation Mode	Resultant Shape	Condition
Buckling		$L/D > 5$
Shearing		$2.5 < L/D < 5$
Double barreling		Friction exists at contact surface $L/D > 2$
Barreling		Friction exists at contact surface $L/D < 2$
Homogenous compression		$L/D < 2$
Compressive instability		

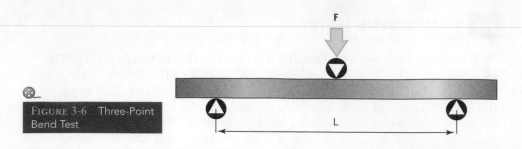

FIGURE 3-6   Three-Point
Bend Test

3.4   BEND TESTING

Very brittle materials do not withstand tensile testing and tend to fracture when being secured into the grips. Many brittle ceramics also fail at very low strain levels, making any slight misalignment in the grips very significant. In such cases, a bend test is used to examine the deformation behavior of the material. Most bend tests involve a three-point loading, as shown in Figure 3-6, although four-point test systems are used in some cases. In the three-point system, a force (F) is applied to the top surface of the sample, placing the top in compression. A pair of circular rollers, separated by a distance (L), supports the bottom of the sample. As the sample begins to deflect, the bottom experiences a tensile stress, whose maximum is concentrated just above the rollers. For most ceramics, the compressive strength is around an order of magnitude greater than the tensile strength, so failure begins on the bottom surface. The *flexural strength* (σ_F) of the sample is defined as

| *Flexural Strength* |
The amount of flexural stress a material can withstand before breaking. Measured through the bend test.

$$\sigma_F = \frac{3F_f L}{2wh^2},$$ (3.12)

where F_f is the load at failure, L is the distance between the rollers, w is the width of the sample, and h is the thickness of the sample. The flexural modulus (E_F) in the elastic region is defined as

$$E_F = \frac{F_f L^3}{4wh^3\delta},$$ (3.13)

where δ is the amount of deflection experienced by the material during bending.

3.5   HARDNESS TESTING

| *Hardness* |
The resistance of the surface of a material to penetration by a hard object under a static force.

Hardness is the resistance of the surface of a material to penetration by a hard object and closely relates to the wear resistance of materials. Hard materials abrade weaker materials and outlast them. Although there are dozens of techniques to measure hardness, the most common is the Brinell test, in which a tungsten-carbide sphere of 10 mm diameter is pushed into the surface of the test material using a controlled force. The *Brinell hardness (HB)* is determined by the equation

$$HB = \frac{F}{\left(\dfrac{\pi}{2}\right) D(D - \sqrt{D^2 - D_i^2})}, \qquad (3.14)$$

in which F is the applied load in kilograms, D is the diameter of the sphere (10 mm), and D_i is the diameter of the impression left by the sphere in the test material in millimeters. Harder materials have higher HB values. The Brinell hardness for metals ranges from around 50 to about 750. The value is reported as the hardness number followed by three letters (HBW), such that a Brinell hardness of 300 would by reported as 300 HBW. The W indicates that tungsten-carbide spheres were used. Originally steel balls were used, but they are no longer considered acceptable because they tend to flatten when the Brinell hardness of the sample approaches 400. A schematic of the equipment for the Brinell test is shown in Figure 3-7.

The Brinell test is quick, easy, fairly accurate, and by far the most commonly used. It is also nondestructive since the material is not broken during the test. The resultant dimple resembles the mark left on a car door when it is struck by a grocery cart. The *Rockwell hardness test* is a variant of the Brinell test in which a diamond cone or steel ball is used instead of a tungsten-carbide sphere. The measurement centers on how the impression depth changes with different forces, but the operating principle remains the same.

Materials scientists often convert the Brinell values to a spot on the *Moh hardness* scale, which was developed in 1822 by Friedrich Moh. The softest material, talc, was assigned a value of 1, while diamond receives a value of 10. Eight other common minerals were assigned values between 2 and 9 in order of increasing hardness. Table 3-5 summarizes the Moh hardness scale.

The Moh scale provides a qualitative feel to the hardness values but is nonlinear. As a result, topaz is not twice as hard as fluorite. Figure 3-8 provides a comparison of Brinell, Rockwell, and Moh hardness values along with a comparison to common metals. In general, ceramics are usually harder than metals, which are usually harder than polymers.

| *Brinell Hardness (HB)* |
One of the many scales used to evaluate the resistance of a material's surface to penetration by a hard object under a static force.

| *Rockwell Hardness Test* |
A specific method of measuring the resistance of a material's surface to penetration by a hard object under a static force.

| *Moh Hardness* |
A nonlinear, qualitative scale used to evaluate the resistance of a material's surface to penetration by a hard object.

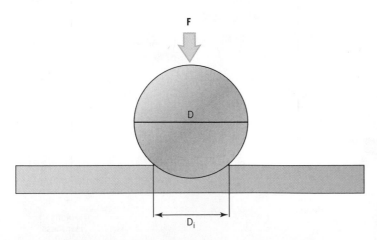

FIGURE 3-7 Schematic of a Brinell Test

TABLE 3-5 Moh Hardness Scale	
Moh Hardness Value	*Mineral*
1	Talc
2	Gypsum
3	Calcite
4	Fluorite
5	Aptite
6	Feldspar
7	Quartz
8	Topaz
9	Corundum
10	Diamond

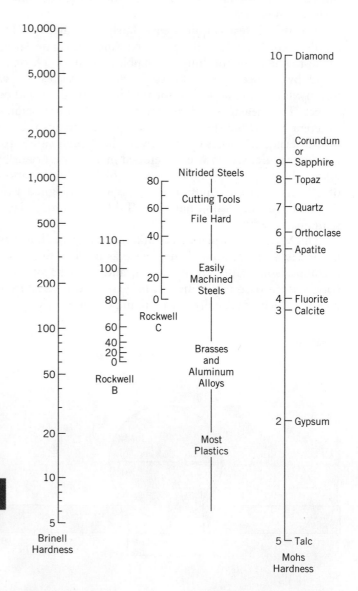

FIGURE 3-8 Comparison of Hardness Scales

From William D. Callister, Materials Science and Engineering, *6th edition.* *Reprinted with permission of John Wiley & Sons, Inc.*

Because hardness and tensile strength both relate to the ability of a material to withstand plastic deformation, there is a rough correlation between the two properties. A rough estimate of tensile strength can be determined from Brinell hardness by the correlations

$$\sigma_s \text{ (psi)} = 500 * HB \tag{3.15}$$

and

$$\sigma_s \text{ (MPa)} = 3.45 * HB. \tag{3.16}$$

Some materials, including many mineralogical samples and glasses, cannot be tested in the manner just described because the large localized forces imposed by the penetrating sphere shatter the samples. In such cases, a microhardness test is required. Light loads (often only a few grams) are applied to a diamond indenter, and the corresponding impression is converted to a hardness value. The shape of the indenter is either a pyramid (in Vickers scale testing) or a narrow rhombus (for Knoop scale testing). Details for microhardness testing can be found in ASTM E-384.

3.6 CREEP TESTING

Creep refers to plastic deformation of a material exposed to a continuous stress over time. Most materials experience creep only at elevated temperatures. When a continuous stress is applied to a material at elevated temperature, it may stretch and ultimately fail below the yield strength. Ultimately, creep occurs because of dislocations in the material. At higher temperatures, the dislocations can diffuse and propagate more readily, making failure more likely.

The measurement of creep is fairly straightforward. In each case, the equipment consists of a frame, a furnace to maintain the elevated temperature around the sample, and a motor with a lever system to apply load to the sample. In older systems, users may have to add weights to apply the load. A thermocouple is placed in direct contact with the sample to record temperature, while linear displacement transducers (LDTs) recorded elongation. Figure 3-9 shows a schematic of a creep test system.

| *Creep* |
Plastic deformation of a material under stress at elevated temperatures.

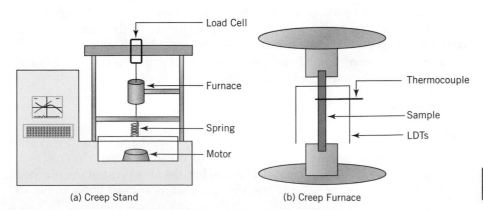

(a) Creep Stand

(b) Creep Furnace

FIGURE 3-9 Schematic of a Creep Test System

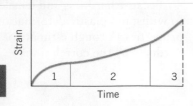

FIGURE 3-10   Representative Strain-Time Plot from Creep Testing

The data obtained from a creep test are plotted as strain (ϵ) versus time. The graph, like the sample shown in Figure 3-10, reveals that creep occurs in three distinct stages. During *primary creep* (stage 1), dislocations slip and move around obstacles. In this stage, the creep rate ($\dot{C}$) is rapid initially but slows. *Creep rate* is defined as the change in the slope of the strain-time curve at any point.

$$\dot{C} = \frac{\Delta\epsilon}{\Delta t}. \tag{3.17}$$

During *secondary creep* (stage 2), the rate that dislocations propagate equals the rate that they are blocked, resulting in an almost linear region on the graph. As *tertiary creep* (stage 3) begins, the rate of deformation accelerates rapidly and continues until rupture. The *Larson-Miller parameter (LM)* is used to characterize creep behavior and may be calculated from the equation

$$LM = \frac{T}{1000}(A + B \ln t), \tag{3.18}$$

where T is the temperature in Kelvin, A and B are material-specific empirical constants, and t is the time to rupture in minutes.

3.7   IMPACT TESTING

*T*oughness refers to a material's resistance to a blow and is measured through an impact test. In a *Charpy impact test*, a hammer is secured to a pendulum at some initial height (h_0). A notched test sample, like the one shown in Figure 3-11, is secured in place in the path of the hammer at the bottom of the arc of the pendulum. At the onset of the test, the hammer is released and its stored potential energy transforms into kinetic energy, as shown in Figure 3-12. At the base of the arc, the hammer breaks through the sample, using some of the kinetic energy in the process. The hammer then proceeds up the other side of the arc until it stops at a final height (h_f). Neglecting friction, if the hammer met no resistance on its path, the initial and final heights would be the same.

The *impact energy* (e_I) of the sample is the same as the loss in potential energy between the initial and final states:

$$e_I = mg(h_0 - h_f), \tag{3.19}$$

where e_I is the impact energy, m is the mass of the hammer, and g is the acceleration due to gravity. Another version of the impact test, the *Izod test*, operates in fundamentally the same fashion, but the alignment of the notched sample is altered. In the Charpy test the specimen is aligned horizontally so

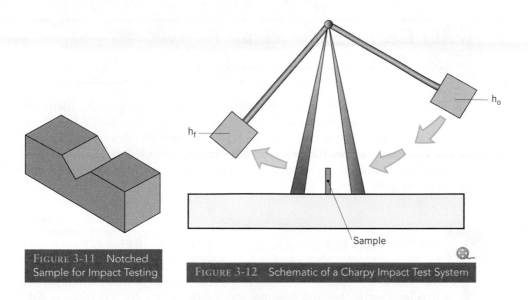

FIGURE 3-11 Notched Sample for Impact Testing

FIGURE 3-12 Schematic of a Charpy Impact Test System

that the hammer strikes the notch directly in its path, while the sample in the Izod test is aligned vertically with the notch facing away from the blow, as shown in Figure 3-13. Sometimes notched and unaltered samples are tested and compared to determine the notch sensitivity of the material. Impact tests are performed frequently at a variety of temperatures and are readily applicable to metals, ceramics, and sometimes polymers, but tend not to work well for composites. The more complex nature of failure in composites limits the value of such a simple test.

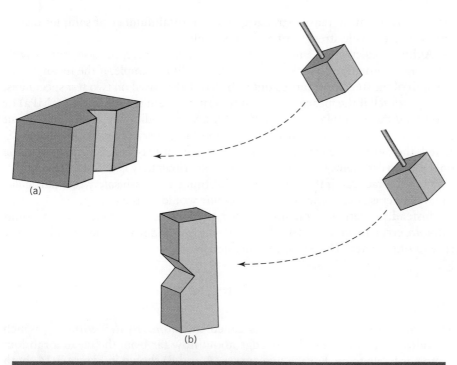

(a)

(b)

FIGURE 3-13 Alignment of Samples in (a) Izod and (b) Charpy Impact Tests

Will I Get the Same Result Every Time I Run a Specific Test?

The short answer is no. Every time a different but supposedly identical sample is tested by any of the techniques described, a result that is at least slightly different from the one before will be obtained. Statistics provides an effective tool for handling this random error in measurement and determining whether we are looking at real differences between samples or just scatter in the data.

3.8 ERROR AND REPRODUCIBILITY IN MEASUREMENT

If seven different people are handed a piece of steel rebar and asked to measure its tensile strength, they will generate seven different results. Microscopic differences in the lattices, crystalline misalignments, and a host of factors ensure that no two material samples are truly identical. Even if they were identical, all tests are subject to inherent random errors that cannot be eliminated. The most intuitive approach to bypassing this problem involves taking multiple measurements and reporting the mean,

$$\overline{\sigma} = \frac{\sum\limits_{i=1}^{N} \sigma_i}{N},$$ (3.20)

where $\overline{\sigma}$ is the mean tensile strength, N is the total number of samples tested, and σ_i is the tensile strength of a given sample.

Although calculating a mean provides a useful first step, it does not provide sufficient information to make many decisions. For example, if the mean tensile strength of the steel rebar turned out to be 812 MPa based on two test specimens, you cannot tell if the mean came from two similar results (808 and 816 MPa) or two very different results (626 and 998 MPa). Some additional information about the spread in the data is needed. More important, every time another sample is tested, the mean value will change. In some cases, it may change drastically. The only way to determine the true mean value experimentally is to test every sample, which is impractical. If the only way to determine the crash safety of a particular car was to crash every one ever made, no one would do the testing.

Instead, a statistical quantity called the *variance* (s^2) takes into account random error from a variety of sources and provides information about the spread of the data. The variance is defined as

$$s^2 = \frac{1}{N-1}\sum\limits_{i=1}^{N} (\sigma_i - \overline{\sigma})^2.$$ (3.21)

The square root of the variance is called the *standard deviation* (s), which provides even more direct knowledge about how far from the mean a random sample is likely to be. For the two curves (A and B) shown in Figure 3-14, both have a mean of $\overline{x}$, but curve B has a larger standard deviation than curve A.

$

| Variance |
A statistical quantity that takes into account the random error from a variety of sources and provides information about the spread of the data.

| Standard Deviation |
The square root of the variance. This value provides more knowledge about the distance from the mean a random sample is likely to be.

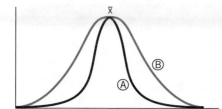

FIGURE 3-14 Two Curves with the Same Mean but Different Standard Deviations

$$s = \sqrt{\frac{\sum_{i=1}^{N}(\sigma_i - \overline{\sigma})^2}{N-1}} \qquad (3.22)$$

Standard deviations are based on probabilities and can be used to make probability statements about reported means. For any given measurement,

$$\overline{\mu} = \overline{x} \pm \delta, \qquad (3.23)$$

where $\overline{\mu}$ is the true mean that would be determined if infinite samples were measured, $\overline{x}$ is the estimate mean determined by testing n samples, and δ is a statistical quantity called the *error bar* or *confidence limit*. The size of the error bars depends on the number of samples, the standard deviation, and the desired level of confidence. For most applications, 95% confidence is used as a standard. At 95% confidence, the true mean ($\overline{\mu}$) will be between ($\overline{x} - \delta$) and ($\overline{x} \pm \delta$) 95% of the time. If 99% confidence was used, the true mean would be bracketed by the error bars 99% of the time, but the error bars would be much larger. The error bars (δ) are determined by the equation

$$\delta = \frac{t * s}{\sqrt{N}}, \qquad (3.24)$$

where t is the value from the statistical *t-table* provided in Table 3-6, s is the standard deviation, and N is the number of samples.

The axes on the t-table shown in Table 3-6 are confusing. The table consists of values of t as a function of degrees of freedom (n) and F, which is a complex function related to the level of uncertainty (α) by the equation

$$F = 1 - \frac{\alpha}{2}. \qquad (3.25)$$

For 95% confidence, the level of uncertainty is 5% (0.05), so Equation 3.25 indicates that F would be 0.975. The degrees of freedom (n) are defined as

$$n = N - 1, \qquad (3.26)$$

so there is one less degree of freedom than the number of samples tested.

Testing additional samples will reduce the size of the error bars without lowering the confidence limits because the square root of N appears in the denominator of the δ equation. Increasing N also reduces the value of t, which appears in the numerator. Ultimately, the decision on how many samples to test balances the costs of running additional tests with the increased accuracy of the results.

| **Error Bar** |
A limit placed on the accuracy of a reported mean, based on the number of samples tested, the standard deviation, and the desired level of confidence.

| **Confidence Limit** |
The degree of certainty in an estimate of a mean.

| **t-Table** |
A statistical table based on the degrees of freedom and the level of uncertainty in a set of reported sample values.

TABLE 3-6 Percentage Points, Student's t-Distribution

The table gives values such that:

$$F(t) = \int_{-\infty}^{t} \frac{\Gamma(n + 1/2)}{\sqrt{n\pi}\,\Gamma(n/2)} \left(1 + \frac{x^2}{n}\right)^{-(n+1)/2} dx$$

where n equals number of degrees of freedom.

n	\multicolumn{8}{c}{$F(t)$}
	0.6	0.75	0.9	0.95	0.975	0.99	0.995	0.9995
1	0.325	1.000	3.078	6.314	12.706	31.821	63.657	636.619
2	0.289	0.816	1.886	2.920	4.303	6.965	9.925	31.598
3	0.277	0.765	1.638	2.353	3.182	4.541	5.841	12.924
4	0.271	0.741	1.533	2.132	2.776	3.747	4.604	8.610
5	0.267	0.727	1.476	2.015	2.571	3.365	4.032	6.869
6	0.265	0.718	1.440	1.943	2.447	3.143	3.707	5.959
7	0.263	0.711	1.415	1.895	2.365	2.998	3.499	5.408
8	0.262	0.706	1.397	1.860	2.306	2.896	3.355	5.041
9	0.261	0.703	1.383	1.833	2.262	2.821	3.250	4.781
10	0.260	0.700	1.372	1.812	2.228	2.764	3.169	4.587
11	0.260	0.697	1.363	1.796	2.201	2.718	3.106	4.437
12	0.259	0.695	1.356	1.782	2.179	2.681	3.055	4.318
13	0.259	0.694	1.350	1.771	2.160	2.650	3.012	4.221
14	0.258	0.692	1.345	1.761	2.145	2.624	2.977	4.140
15	0.258	0.691	1.341	1.753	2.131	2.602	2.947	4.073
16	0.258	0.690	1.337	1.746	2.120	2.583	2.921	4.015
17	0.257	0.689	1.333	1.740	2.110	2.567	2.898	3.965
18	0.257	0.688	1.330	1.734	2.101	2.552	2.878	3.922
19	0.257	0.688	1.328	1.729	2.093	2.539	2.861	3.883
20	0.257	0.687	1.325	1.725	2.086	2.528	2.845	3.850
21	0.257	0.686	1.323	1.721	2.080	2.518	2.831	3.819
22	0.256	0.686	1.321	1.717	2.074	2.508	2.819	3.792
23	0.256	0.685	1.319	1.714	2.069	2.500	2.807	3.767
24	0.256	0.685	1.318	1.711	2.064	2.492	2.797	3.745
25	0.256	0.684	1.316	1.708	2.060	2.485	2.787	3.725
26	0.256	0.684	1.315	1.706	2.056	2.479	2.779	3.707
27	0.256	0.684	1.314	1.703	2.052	2.473	2.771	3.690
28	0.256	0.683	1.313	1.701	2.048	2.467	2.763	3.674
29	0.256	0.683	1.311	1.699	2.045	2.462	2.756	3.659
30	0.256	0.683	1.310	1.697	2.042	2.457	2.750	3.646
40	0.255	0.681	1.303	1.684	2.021	2.423	2.704	3.551
60	0.254	0.679	1.296	1.671	2.000	2.390	2.660	3.460
120	0.254	0.677	1.289	1.658	1.980	2.358	2.617	3.373
∞	0.253	0.674	1.282	1.645	1.960	2.326	2.576	3.291

Example 3-3

A series of six composite samples are tested in tension. The measured tensile strengths (in MPa) for the six replicate samples were 742, 763, 699, 707, 714, and 751. Determine the mean tensile strength with appropriate error bars based on 95% confidence.

SOLUTION

The mean of the samples is given by

$$\overline{\sigma}_s = \frac{742 + 763 + 699 + 707 + 714 + 751}{6} = 729 \text{ MPa}.$$

The error bars are calculated from Equation 3.23:

$$\delta = \frac{t * s}{\sqrt{N}}$$

N is the number of samples, in this case 6; t must be found from Table 3-6, t (F = .975, n = 5) = 2.571.

   The standard deviation comes from Equation 3.21:

$$s = \sqrt{\frac{\sum_{i=1}^{N}(\sigma_i - \overline{\sigma})^2}{N - 1}} = 26.1,$$

so $\delta = (2.571 * 26.1)/\sqrt{6} = 27.4$. Therefore, $\overline{\sigma}_s = 729 \pm 27.4$ MPa. It is 95% likely that the true mean lies between 721.6 and 756.4 MPa.

The same principle may be used to determine whether two sets of samples are statistically different. When two distinct samples are examined, the **pooled variance** $(S_{12})^2$ must be calculated, but as before, the square root of the pooled variance is more useful:

$$S_{12} = \sqrt{\frac{(N_1 - 1) * S_1^2 + (N_2 - 1) * S_2^2}{N_1 + N_2 - 2}}, \tag{3.27}$$

where N_1 is the number of samples of the first material, S_1 is the standard deviation of the first set of samples, N_2 is the number of samples from the second material, and S_2 is the standard deviation of the second set of samples.

   The standard difference between the means (S_D) is given by

$$S_D = S_{12}\sqrt{\frac{N_1 + N_2}{N_1 * N_2}}. \tag{3.28}$$

The means are statistically significantly different if and only if Equation 3.29 is satisfied,

$$|\overline{X}_1 - \overline{X}_2| > t * S_D, \tag{3.29}$$

where $\overline{X}_1$ and $\overline{X}_2$ are the calculated means of the two sample types.

| *Pooled Variance* |
A value used to determine if two distinct sets of samples are statistically different.

Example 3-4

Suppose 10 samples of a metal alloy were subjected to Brinell hardness testing and provided a mean Brinell hardness of 436 with a standard deviation of 12.5. Eight samples of a competing alloy resulted in a mean Brinell hardness of 487 with a standard deviation of 10.5. Are the samples significantly different?

SOLUTION

At first glance, the second alloy appears to be harder than the first, but the statistics must be examined to be sure. First, we must calculate the square root of the pooled variance

$$S_{12} = \sqrt{\frac{(N_1 - 1) * S_1^2 + (N_2 - 1) * S_2^2}{N_1 + N_2 - 2}}$$

$$= \sqrt{\frac{(10 - 1) * (12.5)^2 + (8 - 1) * (10.5)^2}{10 + 8 - 2}} = 11.7.$$

Next we use that value to find the standard difference between the means

$$S_D = S_{12}\sqrt{\frac{N_1 + N_2}{N_1 * N_2}} = 11.7\sqrt{\frac{10 + 8}{10 * 8}} = 5.50.$$

Now we need the appropriate value of t. Because we still wish to have 95% confidence, the value of F remains at 0.975. However, the degrees of freedom now becomes

$$n = N_1 + N_2 - 2 = 10 + 8 - 2 = 16.$$

From Table 3-6, t (F = .975, n = 16) = 2.120.
    We are now ready to apply Equation 3.28.

$$|\overline{X}_1 - \overline{X}_2| > t * S_D \text{ becomes } |436 - 487| > 2.210 * 5.50 \text{ or } 51 > 11.7.$$

Therefore, the Brinell hardness of the second alloy is significantly greater than that of the first to 95% confidence.
    The statistical analysis above brings up an incredibly important ethical dilemma faced by all companies: How much testing is enough? Testing more samples provides greater accuracy in the results, but costs money both because of man-hours spent performing the testing and the samples destroyed. If a company is crash testing Rolls-Royce automobiles, each data point could cost more than $100,000.
    Most companies try to strike a balance between cost and certainty. If a company makes a product like soda bottles and 1 in 100,000 cannot withstand the pressurization process, the loss may be acceptable. If 1 component in 100,000 of an aircraft engine is defective, the result may be catastrophic.

3.9 FRACTURE MECHANICS

Ａll material failures result from the formation and propagation of a crack, but different types of materials respond to the formation of cracks quite differently. Ductile materials experience substantial plastic deformation in the area of the crack. The material essentially adapts to the presence of the crack. The growth of the crack tends to be slow, and, in some cases, the crack becomes stable and will not grow unless increased stress is applied. Cracks in sidewalks or in the drywall of houses tend to be stable.

Ductile materials tend to fail either in the *cup-and-cone* mode, shown in Figure 3-15, in which one piece has a flat center with an extended rim like a cup while the other has a roughly conical tip, or in a shear fracture caused by a lateral shearing force.

Brittle materials behave differently. They cannot undergo the plastic deformation needed to stabilize the crack without failing. As a result, small cracks propagate spontaneously, much like a small crack in a car windshield can grow and result in the entire windshield failing. Brittle materials tend to form a simpler cleavage fracture surface, as shown in Figure 3-16. The study of the crack growth leading to failure of a material is called *fracture mechanics*.

The key to all fracture mechanics is the presence of a crack or other non-atomic scale defect, such as a pore. The development of the stress equation $\sigma = F/A_0$ is predicated on the assumption that the stress is evenly distributed across the cross-sectional area of the sample, but the presence of cracks makes this assumption invalid. Cracks, voids, and other imperfections serve as

| *Fracture Mechanics* |
The study of crack growth leading to material failure.

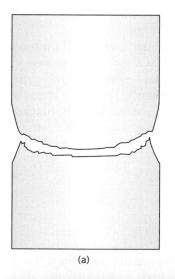

(a)

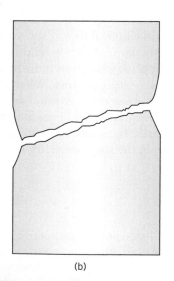

(b)

FIGURE 3-15   Ductile Failure Modes: (a) Cup-and-Cone; (b) Lateral Shear

FIGURE 3-16   Brittle Fracture

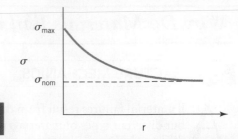

FIGURE 3-17   Relationship between σ_{nom} and σ_{max}

stress *raisers* that cause highly localized increases in stress. Thus, the traditional stress that has been discussed previously is more appropriately defined as the *nominal stress* (σ_{nom}), and the maximum stress that is concentrated at the tip of the crack is called the maximum stress (δ_{max}).

A *stress concentration factor* (k) can be defined as the ratio of the maximum stress to the applied stress,

$$k = \frac{\sigma_{max}}{\sigma_{nom}}. \qquad (3.30)$$

If a plot is made of stress versus the distance from the imperfection, like the one shown in Figure 3-17, the stress felt at the imperfection is the nominal stress times the stress concentration factor.

The magnitude of the stress concentration factor depends on the geometry of imperfection. As early as 1913, fundamental geometric arguments showed that for an elliptical flaw, the stress concentration factor related to the ratio of the length (a) of the ellipse to the width (b) by the equation,

$$k = \left(1 + 2\frac{a}{b}\right). \qquad (3.31)$$

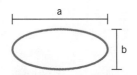

FIGURE 3-18   Representative Elliptical Flaw

A hypothetical elliptical flaw is shown in Figure 3-18.

A perfectly circular flaw would have a = b and the stress concentration factor would be 3, indicating that the flaw resulted in a threefold increase in applied stress in the local area. When the flaw becomes less circular and more elongated, the stress concentration factor increases greatly. For a thin crack in which a $\gg$ b, Equation 3.31 indicates that the stress at the crack tip approaches infinity. As such, the stress concentration factor ceases to have meaning and a new parameter, the *stress intensity factor* (K), must be employed. For a simple crack,

$$K = f\sigma\sqrt{\pi a}, \qquad (3.32)$$

where f is a dimensionless geometric factor.

The critical question that governs whether the material will fracture as the result of an applied stress is whether the stress intensity at the crack tip will surpass a critical threshold. When K exceeds this critical stress intensity factor, called the *fracture toughness* (K_c), the crack will propagate. The actual stress needed for crack propagation (σ_c) for a brittle material is defined as

$$\sigma_c = 2\sigma_{nom}\sqrt{\frac{E\gamma_s}{\pi a}}, \qquad (3.33)$$

where E is the elastic modulus of the material, γ_s is the specific surface energy at the crack surface, and a is the crack length. The fracture toughness relates to the actual stress needed to propagate a crack by Equation 3.34:

$$K_c = f\sigma_c\sqrt{\pi a}. \qquad (3.34)$$

Because the fracture toughness is a function of the width of the material, it cannot be tabulated directly. At some width, the influence of width on fracture toughness becomes less pronounced, and above a critical thickness, width no longer impacts fracture toughness at all. The fracture toughness of the material above this critical thickness is called the *plane strain fracture toughness* (K_{Ic}), which is provided for several different materials in Table 3-7.

The most important parameter in determining the stress needed for crack propagation is the flaw size. As a result, in the manufacturing process, great care is taken to reduce flaw size. Impurities are commonly filtered from liquid metals, molten polymers are passed through sieves filters prior to forming, and a complex powder pressing technique (discussed in Chapter 6) is used to reduce flaw size and improve the fracture toughness of many ceramic materials.

| *Fracture Toughness* |
The value that the stress concentration factor must exceed to allow a crack to propagate.

| *Plane Strain Fracture Toughness* |
The fracture toughness above the critical thickness in which the width of the material no longer impacts the fracture toughness.

TABLE 3-7	Plane Strain Fracture Toughness for a Variety of Materials
Material	K_{Ic} (MPa $\sqrt{m}$)
ABS terpolymer	4
Alumina (Al_2O_3)	1.5
Aluminum-copper alloys	20–30
Cast iron	6–20
Cement/concrete	0.2
Magnesia (MgO)	3
Polycarbonate	2.7
Polyethylene (high density)	2
Polyethylene (low density)	1
PMMA (polymethylmathacrylate)	0.8
Porcelain	1
Silicon carbide	3
Steel—medium carbon	51
Steel—nickel chromium	42–73
Zirconia (toughened)	9
Other ductile metals	100–350

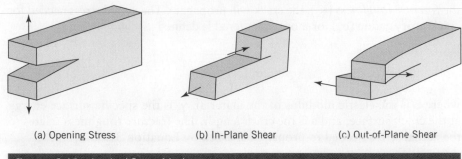

(a) Opening Stress   (b) In-Plane Shear   (c) Out-of-Plane Shear

FIGURE 3-19   Applied Stress Modes

| Ductile-to-Brittle Transition |
The transition of some metals in which a change in temperature causes them to transform between ductile and brittle behavior.

| Opening Stresses |
Stresses that act perpendicularly to the direction of a crack, causing the crack ends to pull apart and opening the crack further.

| In-Plane Shear |
The application of stresses parallel to a crack causing the top portion to be pushed forward and the bottom portion to be pulled in the opposite direction.

| Out-of-Plane Shear |
The application of stress perpendicular to a crack, which pulls the top and bottom portions in opposite directions.

Even though flaw size is the most important parameter, many other factors impact the stress concentration, including ductility, temperature, and grain size. The area around the crack tip in ductile materials can undergo plastic deformation and relieve some of the stress intensification. As a result, brittle polymers and ceramics tend to have much lower fracture toughnesses than ductile metals. Temperature impacts the fracture toughness of materials that have a transition between ductile and brittle behavior. Most polymers undergo a distinct transformation between brittle, glassy materials with little fracture toughness and softer, more rubbery materials with much higher fracture toughness. Metals with an FCC (face-centered cubic) structure typically do not undergo a *ductile-to-brittle transition,* but BCC (body-centered cubic) metals usually do. Small grain sizes also tend to enhance fracture toughness.

The behavior of the crack in the presence of stress also depends on the direction of the stress relative to the crack. *Opening stresses* act perpendicularly to the direction of the crack, as shown in Figure 3-19(a). An opening stress pulls the ends of the crack apart and causes it to open farther. An opening stress applied in the opposite direction would push the crack together and not result in propagation. *In-plane shear* involves the application of stress parallel to the crack. The top portion of the crack is pushed forward while the bottom in pulled in the opposite direction. The shear stress propagates the crack by sliding the top and bottom half along each other, but neither half leaves its original plane, as shown in Figure 3-19(b). *Out-of-plane shear* results when a stress perpendicular to the crack pulls the top and bottom halves of the crack in opposite directions, as shown in Figure 3-19(c).

How Do Mechanical Properties Change over Time?

Both environmental conditions and the type of material influence how properties change over time. Metal left in salt water for one year will behave quite differently from metal kept in a clean, dry environment. Two primary tests provide some insight to how properties are likely to change over time: fatigue testing and accelerated aging studies.

Materials ultimately will fail when exposed to repeated stresses, even if the level of stress is below the yield strength. Failure because of repeated stresses below the yield strength is called *fatigue*. Although all classes of materials can experience fatigue, it is especially important in metals. The goals of fatigue testing are to determine the number of stress cycles at a given stress level a material can experience before failing (the *fatigue life*) and the stress level below which there is a 50% probability that failure will never occur (the *endurance limit*).

The most common fatigue test uses a *cantilever beam test*, similar to that shown in Figure 3-20. A cylindrical-shaped specimen is mounted in a vise at one end and a weight or yoke is applied to the other. A motor rotates the sample producing alternating compressive and tensile forces as the sample rotates, as shown in Figure 3-21. A counter records the number of cycles until the specimen fails. For each sample, the stress amplitude (S) is measured by the equation

$$S = \frac{10.18 \, lm}{d^3}, \tag{3.35}$$

where l is the length of the test specimen, m is the mass of the applied weight or yoke, and d is the specimen diameter.

Multiple samples are tested at different stress levels and the results are compiled to form an *S–N curve*, in which the stress amplitude is plotted against the natural log of the number of cycles to failure, as shown in Figure 3-22. This curve may be used to determine the fatigue life for the material at a given stress level. These curves are also called *Wohler curves*, after the nineteenth-century engineer August Wohler who studied the causes of fracture in railroad axles. The stress level at the inflection point on the S–N curve provides the endurance limit. For many materials, the endurance limit is approximately 50% of

| *Fatigue* |
Failure because of repeated stresses below the yield strength.

| *Fatigue Life* |
The number of cycles at a given stress level that a material can experience before failing.

| *Endurance Limit* |
The stress level below which there is a 50% probability that failure will never occur.

| *Cantilever Beam Test* |
Method used to determine fatigue by alternating compressive and tensile forces on the sample.

| *S–N Curve* |
A curve plotting the results of testing multiple samples at different stress levels that is used to determine the fatigue life of a material at a given stress level.

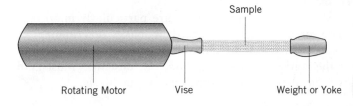

Sample

Rotating Motor   Vise   Weight or Yoke

FIGURE 3-20 Cantilever Beam Apparatus

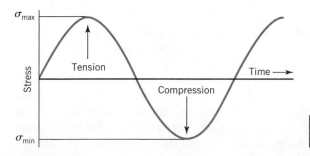

σ_{max}

Stress

Tension

Time →

Compression

σ_{min}

FIGURE 3-21 Stress Cycles during Fatigue Testing

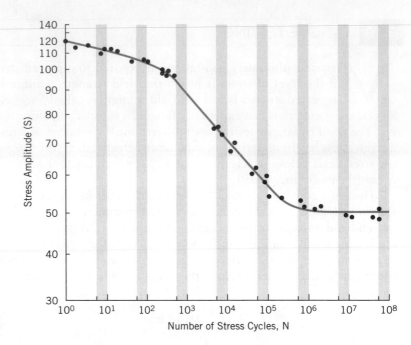

the tensile strength. Some materials, including aluminum and most polymers, have no endurance limit. If exposed to enough cycles at any stress level, they eventually will fail. This is why aircraft companies carefully record the number of stress cycles (experienced during takeoff and landing) each plane undergoes and routinely replace wings and other parts that experience stresses well before the fatigue life is approached. This lesson was learned through tragedy. In 1954 three de Havilland Comet aircraft came apart in midair and crashed. Subsequent investigations revealed that sharp corners around the windows provided initiation sites for cracks. All aircraft windows were immediately redesigned, and fatigue testing became routine in the industry.

The problems with variability in measurement are particularly important with fatigue testing. The inherent scatter in the data makes determining an exact fatigue life for a specific material difficult. Large numbers of S–N curves are pulled to improve the statistical fit of the data, but designers must take great care not to allow their material to approach its fatigue life.

3.11 ACCELERATED AGING STUDIES

The properties of most materials change with prolonged exposure to heat, light, and oxygen. The most certain way to determine how the properties of a specific material will change with environmental conditions over time is to simulate the environmental conditions in the laboratory and perform the types of tests described in this chapter at different times throughout the useful life of the material. Unfortunately, the long time horizon of such testing makes it impractical. If a polymer company developed a new paint formulation for cars, it could paint several sample cars and put them in different parts of the country for 15 years to see how they held up, but the company would generate no revenue on the paint for 15 years and another company could develop something even better in that length of time.

Instead, companies perform *accelerated aging studies* in which the time horizon is shortened by increasing the intensity of the exposure to the other variables. Studies of reaction kinetics tell us that the same fundamental processes occur at different temperatures, but temperature significantly impacts the rate of the process. The goal of an accelerated aging study is to use an *equivalent property time (EPT)* to make the same process occur in a shorter time. For example, if a metal component in an engine would be expected to withstand six months in air (21% oxygen) at 200°C, an equivalent reaction could be obtained by elevating the temperature to 300°C for a shorter time.

The Arrhenius equation can be used to relate time and temperature. As discussed in Chapter 2, most material changes follow the form:

$$\frac{dG}{dt} = A_0 \exp\left(\frac{-E_A}{RT}\right), \tag{3.36}$$

where G is the property of interest, t is time, E_A is the activation energy, R is the gas constant, A_0 is the pre-exponential constant, and T is temperature. When data for use in the Arrhenius equation are unavailable, the approximation that the rate of change doubles for every 8°C to 10°C increase in temperature is often used, but its accuracy depends on the true activation energy of the system.

Accelerated aging studies provide useful and important information in much shorter time than lifetime aging studies, but it is important to recognize that the EPT is an idealized estimate. Problems can be completely missed or appear to be far more serious than they really are. Ultimately, accelerated aging studies are most useful as a tool to identify concerns for more methodical study.

| *Accelerated Aging Studies* |
Tests that approximate the impact of an environmental variable on a material over time by exposing the material to a higher level of that variable for shorter times.

| *Equivalent Property Time (EPT)* |
A period used to force the same aging processes to occur on a sample in a shorter amount of time.

Summary of Chapter 3

In this chapter we examined:

- ASTM standards for performing tests on materials
- The operation of a tensile test
- The definition and calculation of yield strength, tensile strength, breaking strength, tensile (Young's) modulus, Poisson's ratio, offset yield strength, engineering stress, engineering strain, true stress, true strain, and modulus of resilience
- The difference between elastic stretching and plastic deformation
- Compression testing and modes of deformation
- The bend test and measurement of flexural strength
- Hardness and its measurement and importance
- The relationship among Brinnel, Rockwell, and Moh hardnesses
- Creep and the principles of creep testing
- The differences among primary, secondary, and tertiary creep
- The definition of toughness and its measurement through impact testing
- Error and reproducibility in testing
- The calculation and use of 95% confidence limits
- The use of pooled variances to determine whether two means are statistically different
- The difference between brittle fracture and ductile fracture
- How to calculate stress intensification factors and how to determine whether a crack will propagate or not
- The importance of fatigue and the principles of fatigue testing
- The determination of fatigue life and the endurance limit from an S–N plot
- The uses and limitations of accelerated aging studies

Key Terms

Homework Problems

1. Your company has been hired by the Defense Department to test a new advanced ceramic to determine its suitability for tank armor. Other than cost, describe the three physical properties (names and definitions) that you believe matter the most for this application, explain why they matter the most, and describe in detail (using equations, charts, and simple schematics when appropriate) how you would measure these properties.

2. The army is looking to cut costs, so it decides to do less testing on tank treads. One scientist suggests eliminating hardness testing while another suggests eliminating testing for toughness. Explain the difference between toughness and hardness, and discuss why each might be important to a tank tread.

3. A bend test is used to measure the flexural strength and modulus of a sample. If the relevant sample length is 10 inches, the sample is 1 inch high, and its flexural strength and modulus are 10 psi and 1000 psi, respectively, determine the amount of deflection experienced by the sample during the test.

4. Draw two representative tensile test plots, one for a brittle material and the other for a ductile material. On each plot, label the axes, tensile strength, yield strength, breaking strength, elastic modulus, and regions in which elastic stretching and plastic deformation occur.

5. An inventor claims that she can increase the tensile strength of a polymeric fiber by adding a small quantity of the rare element toughenitupneum during spinning. To prove her claim, she provides data obtained from testing samples with and without her addition. The six samples tested without the addition had tensile strengths of 3100, 2577, 2715, 2925, 3250, and 2888 GPa, respectively. The six samples tested with the additive had strengths of 3725, 3090, 3334, 3616, 3102, and 3441 GPa. Has the inventor proven her claim? If not, suggest improvements that could help her.

6. The following tensile data were collected from a standard 0.505 inch diameter test specimen of copper alloy.

Load (lb-force)	Sample Length (in)
0	2.00000
3000	2.00167
6000	2.00383
7500	2.00617
9000	2.00900
10500	2.04000
12000	2.26000
12400	2.50000
11400	3.02000 (Fracture)

After fracture, the sample length is 3.014 inches and the diameter is 0.374 inches.

Plot the data and calculate:

a. Offset yield strength

b. Tensile strength

c. Modulus of elasticity

d. Engineering stress at fracture

e. True stress at fracture

f. Modulus of resilience

Then label the regions of plastic deformation and elastic stretching on your plot.

7. Interpretation of tensile test data for most polymers is more complicated than for other materials. Unlike most materials, there is no true plastic deformation in most polymers. Because chains can flow across each other, thermal treatments often can restore a polymer to its original strength, even when it had previously been stressed to almost its tensile strength.

a. What key parameter cannot be measured for most polymers by tensile testing?

b. High-performance polymers (such as Kevlar) do not experience this complication and fail like a brittle material. Draw a tensile test plot for Kevlar and label the tensile strength, yield strength, tensile modulus, breaking strength, elastic stretching region, and plastic deformation region.

8. A Brinell test is performed on a metal sample. A 3000 kg load applied for 30 seconds on a 10 mm tungsten carbide ball leaves an indentation of 9.75 mm in the sample. Calculate the Brinell hardness and determine where it fits on the Moh hardness scale.

9. What impression diameter would result from applying a 3000 kg load to a tungsten carbide sphere in a Brinell test of a sample with 420 HBW?

10. For the following data,

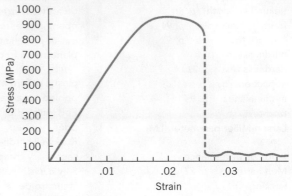

determine:

a. Yield strength

b. Tensile strength

c. Young's modulus

d. Breaking strength

e. Whether the material was brittle or ductile

f. Modulus of resilience

g. Percent elongation

11. For the following data,

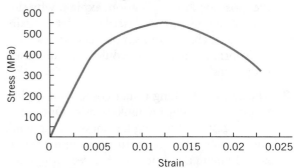

determine:

a. Yield strength

b. Tensile strength

c. Young's modulus

d. Breaking strength

e. Whether the material was brittle or ductile

f. Modulus of resilience

12. For the following data,

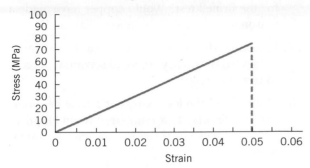

determine:
a. Yield strength
b. Tensile strength
c. Young's modulus
d. Breaking strength
e. Whether the material was brittle or ductile
f. Modulus of resilience

13. For the following data,

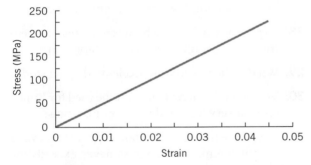

determine:
a. Yield strength
b. Tensile strength
c. Young's modulus
d. Breaking strength
e. Whether the material was brittle or ductile
f. Modulus of resilience

14. For the following data,

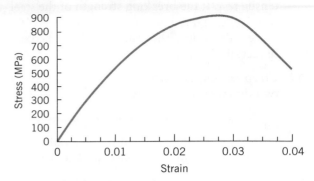

determine:
a. Yield strength
b. Tensile strength
c. Young's modulus
d. Breaking strength
e. Whether the material was brittle or ductile
f. Modulus of resilience

15. A 1 foot long, 1 inch diameter sample is placed in a cantilever beam apparatus. Different weights are attached to the sample, and the number of cycles to failure at each weight was recorded to produce the table:

Applied Weight (lb)	Cycles to Failure
1	2,000,000
5	385,000
10	75,000
15	17,000
20	5,000
25	1,100

Generate an S–N plot and estimate the fatigue life of the material.

16. A .505 inch diameter steel rod is subjected to a tensile test. If the breaking strength of the steel was 50 ksi and the final rod diameter was .460 inches, determine the true stress and percent reduction in area.

17. Creep testing is performed on a material at two different temperatures. At 473 Kelvin (K), the time to failure was 200 minutes, while the time to failure was 145 minutes at 573K. If the Larson-Miller parameter for the material is 100, determine the empirical constants A and B.

18. A series of 10 creep tests provided Larson-Miller parameters of 46, 49, 48, 48, 45, 44, 49, 50, 46, and 44 for one material. After sitting in a drawer for one year, samples of the same material were tested again for creep behavior. This time the results of the 10 trials were 42, 46, 48, 44, 42, 41, 45, 46, 44, and 42. Can the investigator state with 95% certainty that the creep behavior of the material changed during the year in storage?

19. A .505 inch diameter sample failed in compression under a load of 50 ksi. Determine the compressive strength of the material and predict the nature of the failure if L/D = 4.

20. A three-point bend test is performed on a brittle material sample that is 10 inches long, 1 inch wide, and 0.5 inches thick. If a force of 2000 psi results in a deflection of 0.05 inches before failure, determine the flexural strength and flexural modulus of the sample.

21. A sample with a flexural modulus of 400 GPa is placed in a three-point bend test. If the sample is 20 cm long, 2 cm thick, and 4 cm wide and a deflection of .08 cm results before breaking, determine the flexural strength of the material.

22. Explain why the bend test is used instead of a tensile test for brittle materials.

23. Why would ASTM specify the type of ball to use for the Brinell test? Would copper have made a reasonable choice? What about diamond?

24. How does the creep rate change as the material passes from primary creep, to secondary creep, to tertiary creep?

25. How would the local stress at a flaw be different at a circular flaw compared to an elliptical flaw that was six times longer than its maximum width?

26. For the following applications, determine which properties are most important to measure and explain why:
    a. Automobile bumpers
    b. Climbing ropes
    c. Bookcase shelves
    d. Airplane wings
    e. Brake pads

27. If running multiple trials reduces error and increases confidence in results, why not perform lots of each kind of test for every material?

28. Explain the difference between a stress concentration factor and a stress intensity factor.

29. What is the purpose an accelerated aging study?

30. Why would different-size balls be used for Brinell tests on very hard and very soft materials?

31. Explain why materials fail after many cycles even if the maximum stress never exceeds the yield strength of the material.

32. Why is the fracture toughness of metal so much higher than the fracture toughness of glass?

33. If the tensile strength of a sample dropped from 850 MPa to 775 MPa when exposed to 400°C for 100 hours and from 850 MPa to 775 MPa when exposed to 500°C for 60 hours, estimate the tensile strength of the material if it is exposed to 150°C for 60 hours and for 2000 hours.

34. Suggest an application for which the creep behavior of a polymer would be an important parameter.

35. Why is it important to filter a molten metal or polymer before forming it into a final product?

36. A relatively thick plate of either aluminum or steel must be capable of withstanding 100 MPa of applied stress. Based on the information in the following table, estimate the maximum flaw size allowable in the aluminum plate and the steel plate, then explain the cause for the difference in the numbers.

Property	Steel	Aluminum
Elastic Modulus (GPa)	200	70
Specific Surface Energy (J/m^2)	0.32	0.29

4 Metals

Digital Vision

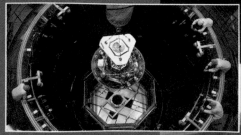

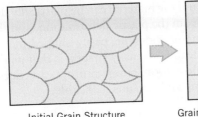

Initial Grain Structure   Grain

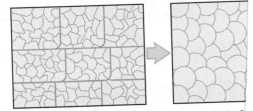

Recovery and Recrystallization   Grain G

of dislocation does not change during
remain virtually unchanged, but condu

When the material is heated to its *rec*
tion of small grains occurs at the subgra
reduction in the number of dislocations
tion temperature marks the transition
the second (recrystallization). The impa
metal reverts back to its original prop
more ductile.

Identifying the recrystallization to
Recrystallization is a kinetic phenome
and time. Some degree of recrystalliza
continues into the third stage (grain
varies with the amount of cold worki
a metal that had undergone substanti
the nucleation of new grains more fav
mum level of cold working (usually 2
occur. Below this level of deformation
to support the nucleation of new grai
varies with the initial grain size and t

The final stage of annealing, grai
ing and ultimately consuming neigh
The process is directly analogous to
Chapter 3 and is generally undesir
large grain size on mechanical prop
the annealed material closely resen
the plastic deformation. However, t
original (pre–cold-worked) size of t
too high of a temperature.

When the forming operations dis
the recrystallization temperature of t
and the impact on microstructure is

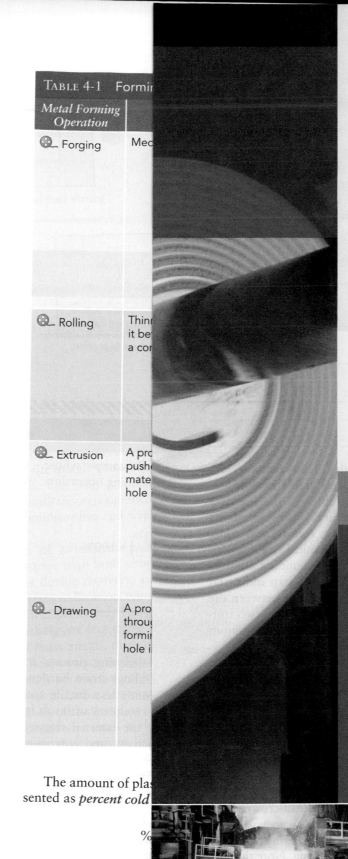

TABLE 4-1   Formin

Metal Forming Operation	
Forging	Mec
Rolling	Thin it be a co
Extrusion	A pro push mate hole
Drawing	A pro throug formin hole

The amount of plas
sented as *percent cold*

%

where A_0 is the initial
after deformation.

CONTENTS

Learning Objectives

By the end of this chapter, a student should be able to:

- Explain the four primary forging operations.

- Explain why strain hardening happens, describe its influence on properties, and calculate percent cold work.

- Explain the changes to a strain-hardened metal that occur during the three stages of annealing: recovery, recrystallization, and grain growth.

- Explain the differences between hot working and cold working.

- Read a phase diagram, including the labeling of the type of system, the liquidus line, the solidus line, any eutectic points, and any eutectoid points.

- Calculate the phase composition in any two phase region.

- Explain all symbols on the carbon-iron phase diagram, including the identity and properties of each phase.

- Explain the general steelmaking process.

- Identify the microstructures and properties associated with the nonequilibrium products of steel.

recrystallization occurs continuously, and the material can be plastically deformed indefinitely. No strengthening results from the plastic deformation, and the material remains ductile. When asked to name a ductile, easily shaped metal, many people think of lead. The reason that lead is so easily shaped is that its recrystallization temperature is significantly below room temperature. Lead continuously recrystallizes during forming operations and remains malleable.

What Advantages Do Alloys Offer?

4.2 ALLOYS AND PHASE DIAGRAMS

| *Alloys* | Blends of two or more metals.

| *Solubility* | The amount of a substance that can be dissolved in a given amount of solvent.

| *Phase* | Any part of a system that is physically and chemically homogeneous and possesses a defined interface with any surrounding phases.

Alloys are homogeneous mixtures of a metal with one or more metals or nonmetals, often forming a solid solution. By carefully selecting the composition of the alloy and the processing conditions, materials scientists and engineers can develop materials with a wider range of properties than could be achieved from pure metals alone. Table 4-2 lists common metallic alloys.

A solid solution is fundamentally the same as other solutions that are more familiar. When hot water is poured over coffee grounds, some of the components of the coffee grounds are drawn into solution with the water. The resultant solution, the coffee, is an equilibrium mixture of coffee components and water. The ***solubility*** of the coffee components in the water varies with temperature. More will leach into the water at high temperature than at low, so coffee is made from very hot water. Metals behave in much the same fashion. If molten copper and molten tin are mixed together, the tin is soluble in the copper, and a brass alloy results.

To understand solid solutions, it is first necessary to review some basic chemistry. A ***phase*** is defined as any part of a system that is physically and chemically homogeneous and possesses a defined interface with any surrounding phases.

TABLE 4-2 Common Metallic Alloys

Name	*Composition (%)*
Aluminum alloy 3S	98 Al, 1.25 Mn
Aluminum bronze	90 Cu, 10 Al
Gold 8-carat	47 Cu, 33 Au, 20 Ag
Palladium gold	40 Cu, 31 Au, 19 Ag, 10 Pd
Manganese bronze	95 Cu, 5 Mn
Bronze, speculum metal	67 Cu, 33 Sn
Red brass	85 Cu, 15 Zn
Gold 14-carat	58 Au, 14–28 Cu, 4–28 Ag
Platinum gold, white	60 Au, 40 Pt
Steel	99 Fe, 1 C
316 stainless steel	63–71 Fe, 16–18 Cr, 10–14 Ni, 2–3 Mo, 0.4 Mn, 0.03 C
Tinfoil	88 Sn, 8 Pb, 4 Cu, 0.5 Sb

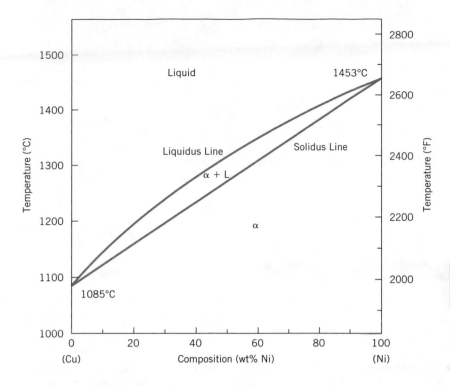

FIGURE 4-3 Binary Isomorphic Phase Diagram for Copper–Nickel Alloys

From P. Nash, ed., Phase Diagrams of Binary Nickel Alloys, *ASM International, Materials Park, OH. Reprinted with permission of ASM International.*

A phase may consist of a single component or multiple components. The phases present in most metals at equilibrium vary with composition and temperature. A *phase diagram* provides a wealth of information about these phases. Phase diagrams are classified based on the number of materials and the nature of the equilibrium between phases. The simplest is a *binary isomorphic* system, like the one shown in Figure 4-3.

The diagram is classified as binary because only two species (copper and nickel) are present in the alloy and as isomorphic because there is only one solid phase and one liquid phase present. Other metallic alloys have more complicated diagrams, but all provide fundamentally the same type of information.

The phase diagram indicates which phases are present at any temperature and composition, and can be used to determine how the phases present will change as the metal alloy is heated or cooled. Consider a 50wt% mixture of copper and nickel. The phase diagram in Figure 4-4 shows that a single metallic phase (α) will exist at all temperatures below about 2300°F. As the temperature increases, the lower line of the graph is crossed and some materials will melt, resulting in a liquid phase in equilibrium with solid α. Just above 2400°F, the last of the solid melts as the upper line is crossed and a single liquid phase (L) remains. The lower line is called the *solidus line*, below which only solids exist at equilibrium. The upper line is called the *liquidus line*, above which only liquid exists.

In regions where more than one phase exists at the same time, the phase diagram also provides information about the composition of the two phases. Consider the same 50% mixture of copper and nickel. At 1280°C, the phase diagram in Figure 4-4 indicates that both a liquid phase (L) and a solid phase (α) will exist. Although the 50wt% of the total material is nickel, the two phases will have different amounts. The compositions are revealed by

| *Phase Diagram* |
A graphical representation of the phases present at equilibrium as a function of temperature and composition.

| *Binary Isomorphic* |
A two-component alloy with one solid phase and one liquid phase.

| *Solidus Line* |
Line on a phase diagram below which only solids exist at equilibrium.

| *Liquidus Line* |
Line on a phase diagram above which only liquid exists at equilibrium.

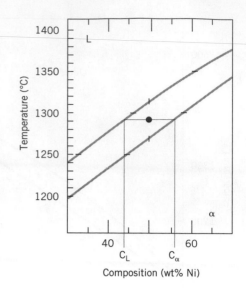

| *Tie Line* |

A horizontal line of constant
temperature that passes
through the point of interest.

constructing a *tie line*, which is simply a horizontal line of constant temperature that passes through the point of interest. The intersection of the tie line
and the equilibrium curve between phases provides the phase compositions.

As the blow-up in Figure 4-4 shows, the tie line intersects the liquidus line
at 43wt% nickel and the solidus line at 56wt% nickel. The intersection points
indicate that the liquid phase will contain 43wt% nickel and 56wt% copper,
while the solid α-phase will consist of 56wt% nickel and 44wt% copper.

A total mass balance and a species mass balance on the nickel provide the
total amount of material in each phase. The mass of the solid phase (M_α) plus
the mass of the liquid phase (M_L) yield the total mass (M_T),

$$M_\alpha + M_L = M_T. \tag{4.2}$$

Similarly, we know that all of the nickel must be in either the solid phase or the
liquid phase. Therefore, the weight fraction of the nickel in the solid phase (w_α)
multiplied by the mass of the solid phase (M_α), plus the weight fraction of the
nickel in the liquid phase (w_L) multiplied by the mass of the liquid phase (M_L)
must be equal to the total mass (M_T) multiplied by the total weight fraction of
nickel (w_T), as shown in Equation 4.3:

$$w_\alpha M_\alpha + w_L M_L = w_T M_T. \tag{4.3}$$

| *Lever Rule* |

A method for determining the
compositions of materials in
each phase using segmented
tie lines representing overall
weight percentages of the
different materials.

Although every binary phase composition problem can be solved by a total
mass balance and species mass balance, a procedure known as the *lever rule*
often is used. Figure 4-5 shows the same tie line used in the previous example,
but with the tie line broken into segments. The first segment extending from
the liquidus line to the overall weight percent of nickel is labeled A, while the
segment from the overall weight percent to the solidus line is labeled B. With
a basis of 1 gram of material, the mass of each phase can be calculated from
Equations 4.4 and 4.5:

$$M_\alpha = \frac{B}{A + B} \tag{4.4}$$

Example 4-2

Determine the amount of material in the liquid and solid phases for an alloy of 50wt% copper, 50wt% nickel at 1280°C.

SOLUTION

Select a basis of one gram of total material (M_T = 1 gram). The problem statement provides that w_T = 50%. The tie line in Figure 4-4 shows that w_α = 56% and w_L = 43%. Equations 4.2 and 4.3 become

$$M_\alpha + M_L = 1 \text{ gram}$$

and

$$(0.56)M_\alpha + (0.43)M_L = (0.5)(1 \text{ gram}).$$

This leaves two equations and two unknowns, so the equations may be solved simultaneously:

$$M_L = 1 \text{ g} - M_\alpha$$
$$(0.56)\,M_\alpha + (0.43)(1 \text{ g} - M_\alpha) = 0.5 \text{ g}$$
$$M_\alpha = 0.538 \text{ g},$$

so

$$M_L = 1 \text{ g} - M_\alpha = 0.462 \text{ g}.$$

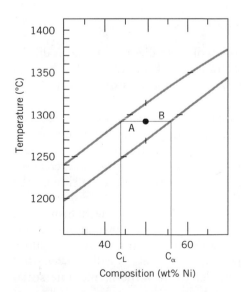

FIGURE 4-5 Copper–Nickel System with a Tie-Line Label for Use with the Lever Rule

Example 4-3

Determine the mass of liquid and solid (α) in 1 gram of a 50wt% alloy of nickel and copper at 1280°F using the lever rule.

SOLUTION

$$M_\alpha = \frac{B}{A + B} = \frac{0.56 - 0.50}{(0.50 - 0.43) + (0.56 - 0.50)} = 0.462 \text{ g}$$

$$M_L = \frac{A}{A + B} = \frac{0.50 - 0.43}{(0.50 - 0.43) + (0.56 - 0.50)} = 0.538 \text{ g}$$

Obviously, the result is the same whether calculated by the lever rule or by direct mass balances.

and

$$M_L = \frac{A}{A + B}. \tag{4.5}$$

| **Diffusion** |
The net movement of atoms in response to a concentration gradient.

The phase diagrams represent an equilibrium state, but it takes time for the lattice atoms to realign. Significant movement of molecules must take place through a process called *diffusion*, in which a concentration gradient drives a net flow of atoms from the region of greater concentration to that of lesser concentration. Everyone is familiar with diffusion in gases and liquids. The smell of food cooking diffuses from the pot to your nose, or perfume molecules diffuse from an open bottle throughout a room. In liquids, a drop of red food coloring diffuses throughout a cup of water until the entire mixture turns red. Diffusion in solids works much like diffusions in liquids or gases, but at much slower rates.

| **Fick's First Law** |
Equation determining steady-state diffusion.

For steady-state systems, *Fick's First Law* provides that

$$J_A = -D_{AB}\frac{dC_A}{dx}, \tag{4.6}$$

where J_A is the net flux (atoms per area-time) of molecules of material A diffusing in the x-direction because of a concentration gradient (dC_A/dx). The term D_{AB} is the *diffusivity* and represents how easily molecules of material A can diffuse through material B.

| **Diffusivity** |
Temperature-dependent coefficient relating net flux to a concentration gradient.

Diffusivity is strongly impacted by temperature. Higher temperatures mean more energy for molecular movement and a corresponding increase in diffusion rates. Diffusivity is also impacted by the crystal structure of the solvent lattice. For example, carbon will diffuse more readily in a body-centered cubic (BCC) structure than a face-centered cubic (FCC) structure, even though the FCC structure can hold more interstitial carbon.

| **Interstitial Diffusion** |
The movement of an atom from one interstitial site to another without altering the lattice.

Unlike liquids or gases, solids have two distinct diffusion mechanisms, and the type of mechanism also greatly impacts the diffusivity. When atoms (in this case carbon atoms) diffuse from one interstitial site to another without altering the lattice itself, as shown in Figure 4-6, the process is called *interstitial diffusion*. Sufficient activation energy is needed to allow the smaller interstitial atom to squeeze between the larger matrix atoms.

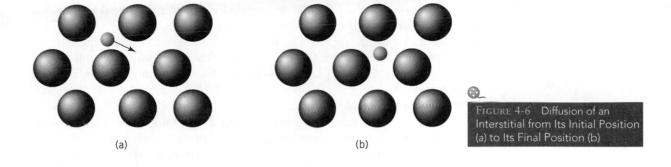
(a)   (b)

FIGURE 4-6   Diffusion of an Interstitial from Its Initial Position (a) to Its Final Position (b)

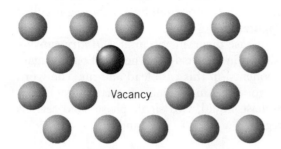

Vacancy

becomes:

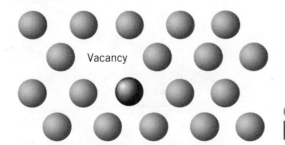

Vacancy

FIGURE 4-7   Vacancy Diffusion in Metals

When the atoms in the lattice itself move into new positions, the process is called *vacancy diffusion,* or *substitutional diffusion.* In this case, a lattice iron atom "jumps" into a vacant lattice space leaving a new vacancy behind, as shown in Figure 4-7. As temperature increases, the diffusivity increases and more vacancies become available, so the diffusion flux increases substantially.

Regardless of the type of diffusion mechanisms, all diffusion relates to the movement of atoms or molecules from regions of higher concentration to those of lower concentrations. The driving force for diffusion is not the application of a physical force causing atoms to move but, rather, a response to a concentration gradient.

The form of Fick's First Law given in Equation 4.6 represents steady state, but many thermal transitions (including those that occur during austenizing) change with time. In such cases, the unsteady-state form known as *Fick's Second Law* is required,

$$\frac{dC_A}{dt} = \frac{d[D_{AB}(dC_A/dx)]}{dx}.$$

(4.7)

| *Vacancy Diffusion* |
The movement of an atom within the lattice itself into an unoccupied site.

| *Fick's Second Law* |
Equation representing the time-dependent change in diffusion.

The solution of the resulting second-order partial differential equation is beyond the scope of this introductory text, but the idea that diffusion changes with time has a significant effect on final microstructures. Consider the copper–nickel system that we discussed earlier. If a 50wt% mixture is cooled from 1350°C to 1280°C as in Example 4-3, the phase diagram indicates that the equilibrium phase compositions of the solid phase should be 56% nickel. However, the first solid begins to form at 1310°C, as shown in Figure 4-8. At this temperature, the solid phase would contain about 64% nickel. As the temperature continues to drop, another solid layer with a slightly lower nickel concentration forms on the growing α-phase. Because the total amount of nickel in the solid phase is capped at 56%, the outermost layers of solid are depleted in nickel while the inner layers are nickel-rich. The nonuniform distribution of elements in the solid phase because of the nonequilibrium cooling is called *segregation*, and is illustrated in Figure 4-9.

| *Binary Eutectic* |
A phase diagram containing six distinct regions: a single-phase liquid, two single-phase solid regions (α and β), and three multiphase regions (α and β, α and L, and β and L.

The next level of complexity in phase diagrams occurs for *binary eutectic* systems, like the lead–tin system shown in Figure 4-10. The word *binary* again signifies the presence of two components, in this case lead and tin. The word *eutectic* comes from the Greek for "good melting." Binary eutectic systems contain six distinct regions: a single-phase liquid, two single-phase solid regions (α and β), and three multiphase regions (α + β, α + L, and β + L).

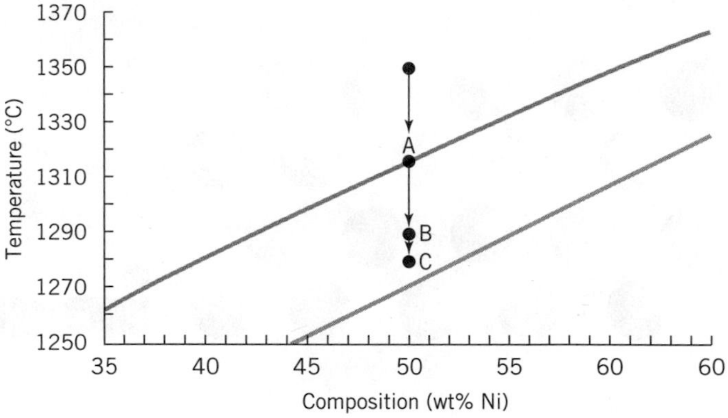

FIGURE 4-8 Nonequilibrium Cooling of a 50wt% Copper–Nickel Alloy: (A) Formation of the First Solid, (B) Intermediate Cooling, and (C) Final Equilibrium Concentrations

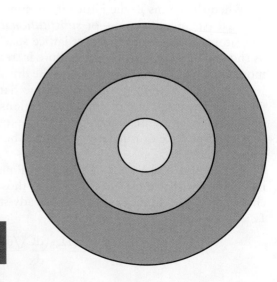

FIGURE 4-9 Segregation in a Copper–Nickel Alloy (Decreasing color indicates decreasing nickel concentration.)

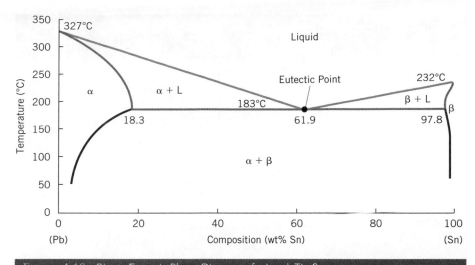

FIGURE 4-10   Binary Eutectic Phase Diagram of a Lead–Tin System

From T.B. Massalski, ed., Binary Alloy Phase Diagrams, 2nd edition. *Vol. 3. ASM International, Materials Park, OH. Reprinted with permission of ASM International.*

The defining feature of a binary eutectic system is a horizontal *eutectic isotherm* that represents the portion of the solidus line above the α + β solid solution. In the case of the lead–tin system, the eutectic isotherm occurs at 183°C. At a concentration of 61.9wt% tin, there exists a point at which the α + β phases melt directly to form a single-phase liquid without passing through any regions of α + L or β + L. The point on the phase diagram where this direct melting occurs is called a *eutectic point* and the corresponding temperature and composition are known as the *eutectic temperature* (T_E) and *eutectic composition* (C_E).

When the material passes through the eutectic point, a eutectic reaction occurs. The general form of the eutectic reaction is given by the equation

$$L(C_E) \leftrightarrow \alpha (C_{\alpha,E}) + \beta (C_{\beta,E}). \tag{4.8}$$

In the case of the lead–tin alloy, this reaction would be written as

$$L(61.9\% \text{ Sn}) \leftrightarrow \alpha (18.3\% \text{ Sn}) + \beta (97.8\% \text{ Sn}). \tag{4.9}$$

The more complicated diagram results in additional descriptors for the various equilibrium lines on the graph. The solidus line (shown in blue in Figure 4-10), below which only solids exist, extends from 327°C along the border between the α-phase and the α + L region, then continues as the eutectic isotherm at 183°C, and forming the border between the β and β + L regions. The red line defining the border between the pure α-phase and the α + β phase and the red line defining the border between the β and α + β phase are called *solvus lines.*

The eutectic point is one particular type of *invariant point*, which is any spot on a phase diagram where three phases are in equilibrium. Two other types of invariant points are *eutectoids*, in which one solid phase is in equilibrium with a mixture of two different solid phases, and *peritectics*, in which a solid and liquid are in equilibrium with a different solid phase. Table 4-3 summarizes the primary invariant points.

| **Eutectic Isotherm** |
A constant temperature line on a phase diagram that passes through the eutectic point.

| **Eutectic Point** |
The point on the phase diagram at which the two solid phases melt completely to form a single-phase liquid.

| **Solvus Lines** |
Lines defining the border between the pure solid phase and a blend of two solid phases on a phase diagram.

| **Eutectoids** |
Points in which one solid phase is in equilibrium with a mixture of two different solid phases.

| **Peritectics** |
Points at which a solid and liquid are in equilibrium with a different solid phase.

TABLE 4-3   Different Classes of Invariant Points

Invariant Type	Reaction	Example
Eutectic	$L \rightarrow \alpha + \beta$	
Eutectoid	$\gamma \rightarrow \alpha + Fe_3C$	
Peritectic	$\alpha + L \rightarrow \beta$	

All of the principles for reading the phase diagrams and calculating phase compositions apply to binary eutectics. In any two-phase region, a horizontal tie line can be drawn to determine the equilibrium concentrations of the two regions, and the lever rule (or mass balances) can be applied to determine the total amount of material in each phase.

4.3   CARBON STEEL

| Carbon Steel |
Common alloy consisting of interstitial carbon atoms in an iron matrix.

Perhaps no metal is as ubiquitous in advanced societies as *carbon steel*, which consists of interstitial carbon atoms in an iron matrix. Figure 4-11 shows a schematic of the steelmaking process. Steel is produced from three

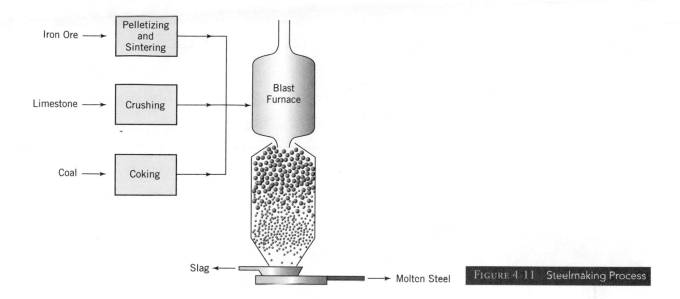

FIGURE 4-11   Steelmaking Process

basic raw materials: iron ore, coal, and limestone. The iron ore provides all of the iron in the system. It is pelletized, *sintered*, and added to a large blast furnace. The volatile components of coal are driven off through a process called *coking*, and the resulting carbon-rich material is added to the blast furnace. Upon heating, the iron ore is reduced to metallic iron, and carbon dioxide and carbon monoxide gases are given off. Crushed limestone is added to the melt and forms a layer of slag on the top of the metal. The slag helps remove impurities from the system. The resultant metal, called *pig iron*, is then treated with oxygen to remove excess carbon and become steel. In general, making 1 ton of pig iron requires about 2 tons of iron ore, 1 ton of coke, and 500 pounds of limestone.

Like any other phase diagram, the iron–carbon system could extend from 0% carbon to 100% carbon. However, the maximum carbon percentage in steel is 6.7wt%. Above this limit, iron carbide (Fe_3C) precipitates out in a hard and brittle phase called *cementite*. As a result, the phase diagram used for analyzing steel is really an iron–iron carbide diagram and ends with a carbon percentage of 6.7, as shown in Figure 4-12.

The regions of this phase diagram have distinct characteristics. On the extreme lower left portion of the phase diagram, a BCC α-ferrite phase occurs. Large iron atoms arranged in a BCC lattice leave little room for interstitial carbon atoms, so, not surprisingly, the solubility of carbon in the interstitial regions of α-ferrite is small (reaching a maximum of 0.022%). At higher carbon concentrations in the lower right portion of the diagram, a two-phase mixture of α-ferrite and cementite (Fe_3C) exists. This mixture is called *pearlite* because its coloring and glossy surface resembles the mother-of-pearl from oyster shells. The microstructure of pearlite consists of alternating layers of cementite and α-ferrite, as shown in Figure 4-13.

The δ-ferrite phase in the upper left-hand corner of the phase diagram represents a BCC structure quite similar to α-ferrite but with a much greater carbon solubility (maximum 0.09%). In between the α- and δ-ferrite phases lies an FCC phase called *austenite* (g) with a much higher carbon solubility (maximum 2.08%).

| *Sintered* |
Formed into a solid from particles by heating until individual particles stick together.

| *Pig Iron* |
Metal remaining in the steelmaking process after the impurities have diffused into the slag. When treated with oxygen to remove excess carbon, pig iron becomes steel.

| *Cementite* |
A hard, brittle phase of iron carbide (Fe_3C) that precipitates out of the steel past the solubility limit for carbon.

| *Pearlite* |
Mixture of cementite (Fe_3C) and α-ferrite named for its resemblance to mother-of-pearl.

| *Austenite* |
Phase present in steel in which the iron is present in an FCC lattice with higher carbon solubility.

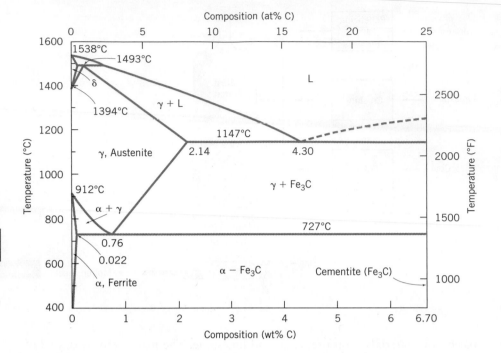

FIGURE 4-12 Phase Diagram for Steel

From T.B. Massalski, ed.,
Binary Alloy Phase Diagrams,
2nd edition. Vol. 3. ASM
International, Materials Park,
OH. Reprinted with permission
of ASM International.

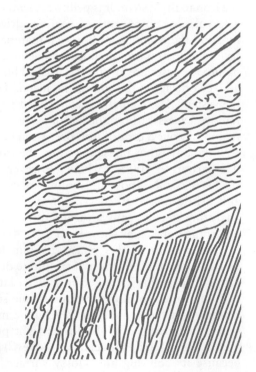

FIGURE 4-13 Microstructure of Pearlite

The steel phase diagram contains three distinct invariant points:

1. A eutectoid at 723°C and 0.76% carbon
2. A eutectic at 1147°C and 4.30% carbon
3. A peritectic at 1493°C and 0.53% carbon

The eutectoid reaction is of particular interest and can be written as

$$\gamma(0.76\% \text{ C}) \rightarrow \alpha(0.022\% \text{ C}) + \text{Fe}_3\text{C}(6.67\% \text{ C}).$$

Carbon steel is classified based on which side of the eutectoid concentration (0.76%) its carbon concentration lies. Steel with exactly 0.76wt% carbon is called *eutectoid steel*. That with less than 0.76wt%C is called *hypoeutectoid steel*, while that with more than 0.76wt% is called *hypereutectoid steel*.

When eutectoid steel is heated to temperatures greater than 727°C for sufficient time, the material transforms to pure austenite (γ), shown in Figure 4-14, through a process called *austenizing*. The transformation of pearlite to austenite requires a significant reorganization of the iron lattice from a BCC structure to an FCC structure.

When austenite cools, the phase diagram in Figure 4-12 indicates that pearlite should form. This transformation requires diffusion out of the austenite FCC lattice back into the ferrite BCC lattice. If the cooling between 727°C and 550°C happens slowly enough for the diffusion to take place, pearlite does form. However, when the cooling is done more rapidly, nonequilibrium products are produced. These materials have significant commercial importance but do not appear on an equilibrium phase diagram. Table 4-4 summarizes the most important nonequilibrium products.

If the austenite is rapidly quenched to near-ambient temperatures by plunging the metal into cold water, a diffusionless transformation occurs to a nonequilibrium phase called *martensite*. In this *martensitic transformation*, elongated BCC cells

| *Hypoeuctectoid Steel* |
Iron-carbon solid solution with less than 0.76wt% carbon.

| *Hypereuctectoid Steel* |
Iron-carbon solid solution with more than 0.76wt% carbon.

| *Austenizing* |
Process through which the iron lattice in steel reorganizes from a BCC to an FCC structure.

| *Martensite* |
Nonequilibrium product of steel formed by the diffusionless transformation of austenite.

| *Martensitic Transformation* |
Diffusionless conversion of a lattice from one form to another brought about by rapid cooling.

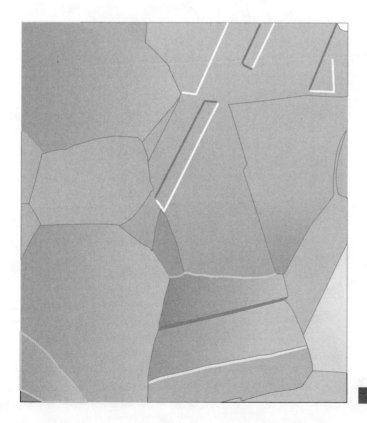

FIGURE 4-14   Microstructure of Austenite

form, and all of the carbon atoms remain as interstitial impurities in the iron lattice. Figure 4-15 shows the microstructure of martensite.

Martensite is significantly harder and stronger than any other carbon steel product, but it is not ductile, even at low carbon concentrations. Martensite is not a stable structure and will convert to pearlite if heated; at ambient temperatures, however, it can last indefinitely.

In the movie *Conan the Barbarian*, Arnold Schwarzenegger heats his sword in a fire (presumably to temperatures greater than 727°C for long enough to austenize the steel), then immediately plunges the sword into the snow. Although he likely did not know it, he was performing a martensitic transformation to improve the strength and hardness of his sword.

The martensitic transformation is not unique to steel and refers to any diffusionless transformation in metals. The *Jomiziny quench test* is used to

| *Jomiziny Quench Test* |
Procedure used to determine the hardenability of a material.

TABLE 4-4	Nonequilibrium Products of Carbon Steel		
Phase	Microstructure	Formed by	Mechanical Properties
Martensite	Body-centered tetragonal cells with all carbon as interstitial impurities	Rapid quenching of austenite to room temperature	Hardest and strongest but difficult to machine; very low ductility
Bainite	Elongated needlelike particles of cementite in an α-ferrite matrix	Quenching of austenite to between 550°C and 250°C and holding at that temperature	Second in hardness and strength to martensite but more ductile and easier to machine
Spherodite	Spheres of cementite in an α-ferrite matrix	Heating bainite or pearlite for 18 to 24 hours near 700°C	Least hard and strong of the nonequilibrium products but most ductile and easiest to machine
Coarse pearlite	Thick alternating layers of cementite and α-ferrite	Isothermal treatment just below the eutectoid temperature	The least strong and hard except for spheroidite; second to spheroidite in ductility
Fine pearlite	Thinner layers of cementite and α-ferrite	Heat treatment at lower temperatures	Between bainite and coarse pearlite in both strength and ductility

FIGURE 4-15 Microstructure of Martensite

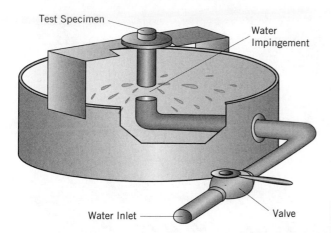

Test Specimen

Water Impingement

Water Inlet

Valve

determine how alloy composition impacts the ability of the metal to undergo a martensitic transformation at a specific treatment temperature. In the test pictured in Figure 4-16, a 1 inch diameter cylindrical bar that is 4 inches long is austenized then rapidly mounted in a fixture. One end of the cylindrical sample is quenched by a spray of cold water. The rapidly quenched area undergoes the martensitic transformation and becomes harder. A series of hardness tests are performed to determine how far from the quench point the improved hardness extends. The depth of this increased hardness from the quenching of one end of the sample is called the *hardenability* of the metal.

If austenite is quenched rapidly to temperatures between 550°C and 250°C and allowed to remain at that temperature, another nonequilibrium structure, called *bainite*, forms with properties and microstructures between those of martensite and pearlite. The conversion of austenite to either bainite or pearlite is competitive. Once one has formed, it cannot be transformed to the other without austenizing the material again.

Bainite is a ferrite matrix with elongated cementite particles. The microstructure, shown in Figure 4-17, shows a distinctive needlelike pattern because of the cementite. Bainite is stronger and harder than pearlite (although it is less strong and hard than martensite) but remains fairly ductile, especially at lower carbon concentrations.

A third nonequilibrium product can be formed by heating either pearlite or bainite to temperatures just below the eutectoid (typically around 700°C) for 18 to 24 hours. The resulting product, called *spheroidite*, is less hard and strong than pearlite but is far more ductile. Spheroidite gets its name from its microstructure in which the cementite particles transform from either the layers found in pearlite or the needles found in bainite into spheres of cementite suspended in an α-ferrite matrix, as shown in Figure 4-18.

Even the equilibrium structure of pearlite is affected by processing conditions. *Coarse pearlite*, consisting of thick alternating layers of cementite and α-ferrite, forms when steel is treated isothermally at temperatures just below the eutectoid. When lower-temperature treatments are used instead, the diffusion rate slows and the alternating layers become thinner. This product is called *fine pearlite*. Coarse pearlite is slightly more ductile than fine pearlite but less hard.

So far the discussion has treated steel as if it contained only carbon and iron. In many cases, other alloying elements are added to strengthen the solid solutions, add corrosion resistance, or increase the hardenability of the metal.

| *Hardenability* |
The ability of a material to undergo a martensitic transformation.

| *Bainite* |
Nonequilibrium product of steel with elongated cementite particles in a ferrite matrix.

| *Spheroidite* |
Nonequilibrium product of steel with cementite spheres suspended in a ferrite matrix.

| *Coarse Pearlite* |
Steel microstructure with thick alternating layers of cemenite and α-ferrite.

| *Fine Pearlite* |
Steel microstructure with thin alternating layers of cemenite and α-ferrite.

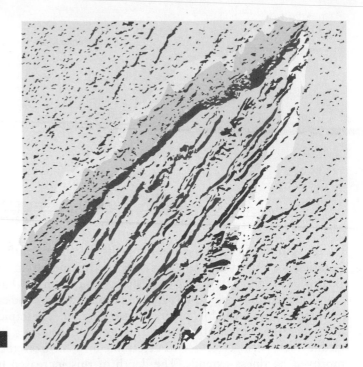

FIGURE 4-17 Microstructure of Bainite

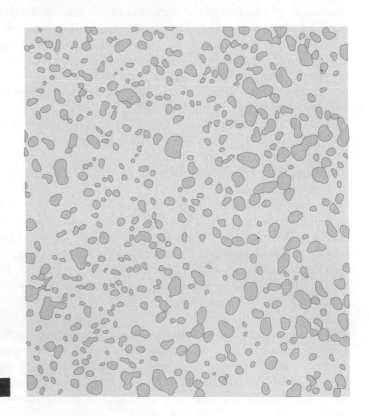

FIGURE 4-18 Microstructure of Spheroidite

Example 4-4

Two knights (Sir Kevin and Sir Robert) decide to fight a duel using their finest swords. Each knight takes his bainite sword with a eutectoid carbon concentration to the local blacksmith for sharpening. The blacksmith finds Sir Kevin to be a loathsome swine and decides to cheat in favor of Sir Robert. The blacksmith plans to convert one of the swords to martensite and the other to coarse pearlite, but he realizes that his tampering will be detected if he remelts the metal. How would the blacksmith convert the materials, and which knight will receive the pearlite sword and why?

Courtesy James Newell

SOLUTION

The blacksmith would convert Sir Robert's sword to martensite by heating it above 727°C for long enough to convert it to austenite, then plunge the sword into cold water to quench the structure without allowing diffusion. Sir Robert would get the martensite sword because it is harder and less ductile. Sir Kevin's sword would be converted to pearlite by heating it above 727°C for long enough to convert it to austenite, then allowing it to cool with isothermal treatments below the eutectoid temperature. We have not covered enough material yet to determine how long these treatments will take, but we will cover that topic later in the chapter.

These steels are known as *alloy steels* and may contain up to 50% of nonferrous elements.

The most common alloying elements are chromium and nickel. When at least 12% chromium is present in the alloy, the metal is classified as *stainless steel*. When chromium is added to the steel, most of it dissolves in the ferrite lattice while the rest joins the carbide phase of the cementite. Stainless steels tend to have a low carbon content, so relatively little of the chromium forms

| *Alloy Steels* |
Carbon–iron solid solutions with additional elements added to change properties.

| *Stainless Steel* |
Carbon–iron solid solutions with at least 12% chromium.

carbides. The chromium both stabilizes the lattice and essentially converts the steel to a chromium–iron binary alloy that remains in its α-ferrite form, which does not austenize. Even at elevated temperatures, the lattice remains in its BCC shape. The resultant lattice is more stable than the typical iron–carbon system. The chromium also oxidizes preferentially to the iron, creating a barrier that protects the steel from further oxidation. Stainless steels that contain chromium without nickel are called *ferretic stainless steels* and tend to be less expensive than those containing nickel.

When 12% to 17% chromium is added to molten steel containing between 0.15% and 1.0% carbon, the steel becomes capable of undergoing the martensitic transformation. These *martensitic stainless steels* are much stronger and harder than ferretic stainless steels but tend to be less resistant to corrosion.

Nickel is less prone to form carbides than iron, so it remains in the iron lattice when added to steel. Nickel has an FCC structure, so the addition of 7% to 20% nickel to steel helps the lattice maintain the FCC structure from austenite even when the steel is cooled to ambient temperatures. The FCC structure makes the steel more malleable and highly corrosion resistant. These high-nickel-content steels are called *austenitic stainless steels* because they retain the FCC structure of austenite.

For applications requiring greater strength and hardness, high-carbon steels (0.6% to 1.4%) are alloyed with materials that form carbides. Chromium, molybdenum, tungsten, and vanadium are common alloying elements. Because these alloys are hard, resist wear, and maintain a cutting edge, they are especially useful in industrial applications, including drill bits, dies, saws, and other tools. As a result, the high-carbon steel alloys are commonly referred to as *tool steels*.

Because steels can contain very different amounts of carbon along with any number of different alloying elements, a classification system was developed by the American Iron and Steel Institute (AISI) to efficiently identify the composition of the metal. The AISI/SAE system, summarized in Table 4-5, provides a simple reference to determine both the carbon content of the steel and the identities of primary alloying elements. The steel alloy is assigned a four-digit identification number. The first number identifies primary alloying elements. Carbon steel begins with 1, nickel steel with 2, and so forth. The second digit describes processing conditions in the case of carbon steel or the amount of the primary alloying element in the steel. The third and fourth digits represent the percentage

| Ferritic Stainless Steels |
Carbon-iron solid solutions with at least 12% chromium that contain no nickel.

| Martensitic Stainless Steels |
Carbon-iron solid solutions with at 12 to 17% chromium that can undergo the martensitic transformation.

| Austenitic Stainless Steels |
Carbon-iron solid solutions with at least 12% chromium that contain at least 7% nickel.

| Tool Steels |
Carbon-iron solid solutions with high carbon content that result in increased hardness and wear-resistance.

TABLE 4-5	AISI/SAE Designations for Steels	
Types	*First Digit*	*Second Digit*
Carbon steel	Always 1	Describes processing
Manganese steel	Always 1	Always 3
Nickel steel	Always 2	Percent nickel in steel
Nickel–chromium steel	Always 3	Percent nickel and chromium in steel
Molybdenum steel	Always 4	Percent molybdenum in steel
Chromium steel	Always 5	Percent chromium in steel
Chromium–vanadium steel	Always 6	Percent chromium and vanadium in steel
Tungsten–chromium steel	Always 7	Percent tungsten and chromium in steel
Silicon–manganese steel	Always 9	Percent silicon and manganese in steel
Triple-alloy steel (contains three alloys)	Either 4, 8, or 9	Percent remaining two alloys in steel

///

Example 4-5

Identify the compositions of 1095 and 5160 steels.

SOLUTION

1095 is a carbon steel (1) with no special processing (0) that contains 0.95% carbon (95). 5160 is a chromium steel (5), with approximately 1% chromium (1), and 0.60% carbon.

of carbon in the steel times 100. Because most steels contain less than 1% carbon, the first zero is omitted. For example, 1040 (rather than 10040) describes a basic carbon steel with 0.4% carbon; 6150 represents a chromium–vanadium steel with 1% chromium and/or vanadium and 0.5% carbon.

4.4 PHASE TRANSITIONS

All of the thermal transitions described in this chapter are driven by thermodynamics but require time. During the early stages of a phase transition, small nuclei of the new phase must form and remain stable long enough to begin to grow. As the grains of the new phase begin to grow from these nuclei, increasing amounts of the original or parent phase are consumed. The amount of transformation can be measured directly through microscopy or indirectly by a property that is influenced by the phase change. The *Avrami equation,*

$$y = 1 - \exp(-kt^n), \qquad (4.10)$$

is used to evaluate the fraction converted (y) to the new phase as a function of time (t). The constants k and n vary with both material type and temperature. Because the phase transformations are diffusion-related processes, the rate of transformation (r) varies with temperature and generally follows the Arrhenius equation,

$$r = Q_0 \exp\left(-\frac{E_A}{RT}\right), \qquad (4.11)$$

where Q_0 is a material-specific constant, E_A is the activation energy, R is the gas constant, and T is temperature.

For most metals, a curve comparing fractional conversion to the logarithm of time elapsed can be constructed, much like the one in Figure 4-19. The plot also indicates the time spent primarily on nucleation (where the slope of the line is small) and the onset of grain growth (where the slope increases sharply). Unfortunately, each such plot corresponds to the conversion rate at only one temperature. A series of curves, like those shown in Figure 4-20, would be required to fully characterize conversion at different temperatures.

Although these plots contain significant amounts of information, they are cumbersome to use in practice because a different plot is required for each temperature. Instead, the information from many such plots is combined into an *isothermal transformation diagram,* more commonly called a *T-T-T plot* (time-temperature-transformation), like the one shown in Figure 4-21. The diagram

| *Isothermal Transformation Diagram* |
Figure used to summarize the time needed to complete a specific phase transformation as a function of temperature for a given material.

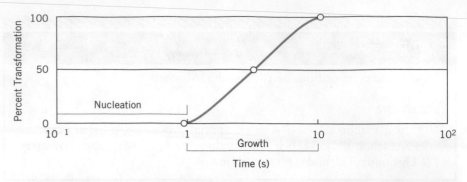

FIGURE 4-19   Percent Transformation versus Log (Time) for the Conversion of Austenite to Pearlite

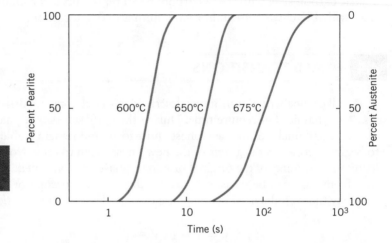

FIGURE 4-20   Percent Transformation versus Log (Time) for the Conversion of Austenite to Pearlite at Various Temperatures

From William D. Callister, Materials Science and Engineering, 6th edition. *Reprinted with permission of John Wiley & Sons, Inc.*

| Begin Curve |
Line on an isothermal transformation diagram representing the moment before the phase transformation starts.

| 50% Completion Curve |
Line on an isothermal transformation diagram indicating when half of the phase transformation has been completed.

| Completion Curve |
Line on an isothermal transformation diagram indicating when the phase transformation has been completed.

consists of three curves. The first represents 0% conversion and is called the *begin curve*. The dotted line in the center, the *50% completion curve*, identifies how much time is required for the transformation to reach the halfway point. Finally, the *completion curve* indicates how much time is required for the entire conversion.

Isothermal transformation diagrams are exceptionally valuable tools but do have limitations. Each plot corresponds to a single concentration. In the case of Figure 4-21, the graph applies only to the eutectoid concentration. The diagrams also apply only to isothermal transformations, in which the metal is held at a constant temperature throughout the entire phase transition.

Figure 4-21 also contains three isotherms labeled M (Start), M (50%), and M (90%). These lines represent the temperatures at which martensite begins for form (M [Start]), at which 50% of the austenite will have converted to martensite (M [50%]), and at which 90% of the austenite will have converted to martensite (M [90%]).

Figure 4-22 shows what happens to the microstructure during the process of the isothermal transition. Above 727°C, the entire microstructure would consist of grains of austenite. These grains would be stable and would not begin to transform. If the sample was cooled rapidly to 600°C, the austenite grains would no longer be stable. Thermodynamics would favor their conversion to ferrite and cementite (pearlite at equilibrium), but time is required for this to

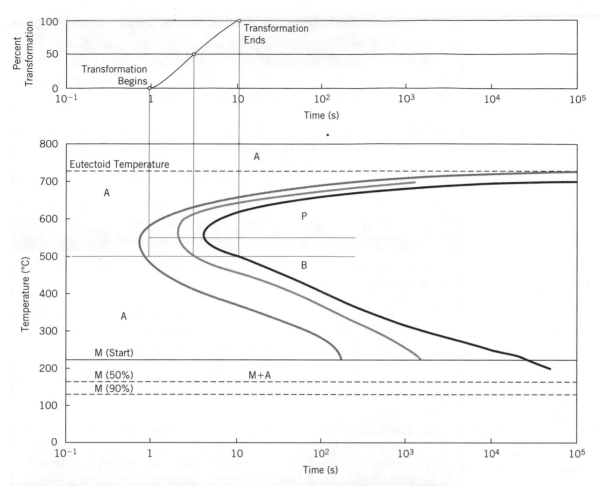

FIGURE 4-21 Isothermal Transformation Diagram for Austenite Converting to Pearlite at the Eutectoid Concentration

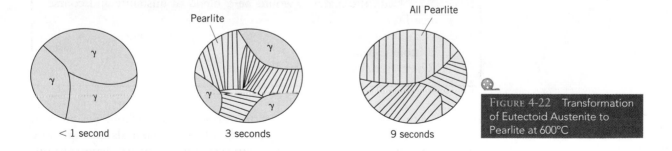

FIGURE 4-22 Transformation of Eutectoid Austenite to Pearlite at 600°C

occur. For the first 2 to 3 seconds, tiny nuclei of pearlite form spontaneously then begin to grow, consuming the parent (austenite) phase in the process. By about 6 seconds, the transformation is 50% completed, and the microstructure now consists of shrinking grains of austenite surrounded by growing grains of pearlite. Finally, by 9 seconds, the entire sample has converted to pearlite.

Isothermal transformation diagrams provide considerable insights into the phase transformations that occur in metals, but they apply only to metals maintained at a constant temperature. In most industrial processes, materials are

Example 4-6

How long would the blacksmith in Example 4-4 need to keep Sir Kevin's sword at 650°C to convert it completely to pearlite?

SOLUTION

Figure 4-21 shows that it takes approximately 90 seconds to reach the 100% completion curve at 650°C.

Example 4-7

For a eutectoid steel initially at 750°C, determine the resulting microstructure present if the steel is:
a. Plunged into cold water
b. Cooled rapidly to 650°C then held at temperature for 30 seconds
c. Cooled rapidly to 680°C then held at temperature for 10 seconds
d. Cooled rapidly to 680°C then held at temperature for 1 hour
e. Cooled rapidly to 680°C then held at temperature for 16 hours
f. Cooled rapidly to 500°C then held at temperature for 2 minutes

SOLUTION

a. The material would be quenched before any diffusion could take place, so the resulting microstructure would be that of martensite.
b. The material would be near the 50% completion line, so it would be a roughly equal blend of unstable austenite grains and growing pearlite grains.
c. Ten seconds is before the begin curve at 680°C, so the material would still be all austenite.
d. At 1 hour, the material would be a blend of austenite and coarse pearlite.
e. Sixteen hours is past the completion curve, so the material would be all coarse pearlite.
f. The material would be 100% bainite.

cooled continuously with the temperature dropping throughout the process. As a consequence, the final microstructures of most materials are far more complex because the transformations occur over a wide range of temperatures. In such cases, the cooling rate is a dominant factor in determining the final microstructure. Clearly, if eutectoid steel is cooled rapidly enough, it will form martensite. A slower cooling rate would result in the formation of pearlite instead. A continuous cooling transformation (CCT) diagram, like the one shown in Figure 4-23, is used to characterize the transformation.

The continuous cooling transformation diagram is a plot of temperature versus time with curved lines representing various cooling rates. The most important item on the plot is the curve representing the critical cooling rate. At any rate slower than this line, the austenite-to-pearlite transformation will

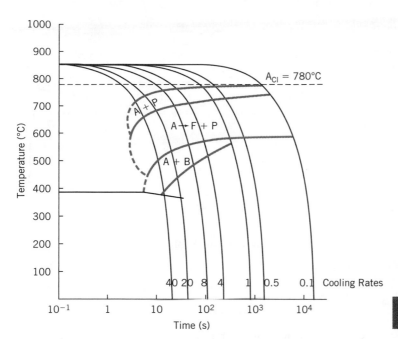

FIGURE 4-23   Continuous Cooling
Transformation Diagram for Eutectoid Steel

begin. At any faster rate, there is insufficient time for diffusion to occur and the material will become martensite. The M (Start), M (50%), and M (90%) isotherms occur at exactly the same temperatures as they did on the isothermal transformation diagrams. As with isothermal transformation diagrams, different CCT diagrams would be required for each alloy composition.

4.5   AGE HARDENING (PRECIPITATION HARDENING)

While considering the phase transformations associated with cooling and heating metals, it is worth discussing two other processes used to alter the properties of alloys. *Age hardening* (or precipitation hardening) uses changes in the solubility of solid solutions with temperature to foster the precipitation of fine particles of impurities. These particles strengthen malleable metals by serving as barriers to the propagation of dislocations through the matrix. Because the precipitated particles tend to be different in size from the metal atoms forming the lattice, the particles tend to distort the shape of the lattice.

The age-hardening process occurs in two stages: *solution heat treatment* and *precipitation heat treatment*. During solution heat treatment, the metal is heated and held at a temperature above the solvus line until one phase has totally dissolved into the other (e.g., heating to point T_1 in Figure 4-24). The β-phase dissolves completely into the α-phase. The alloy then is cooled rapidly to a temperature sufficient to prevent any significant diffusion (T_0 in Figure 4-24). For many alloys, this is room temperature. The resultant alloy contains a supersaturated solution of β dissolved in α.

In the second state, precipitation heat treatment, the temperature of the supersaturated alloy is raised to an elevated level but still below the solvus line (T_2 in Figure 4-24). At this temperature, the diffusion rate increases sufficiently to allow the β-phase to form as a fine precipitate. The rate at which precipitation occurs depends on diffusion rate (which itself is a function of temperature), alloy composition, and the relative solubilities of the α- and β-phases.

| *Age Hardening* |
Process that utilizes the temperature dependence of the solubilities of solid solutions to foster the precipitation of fine particles of impurities. Also called precipitation hardening.

| *Solution Heat Treatment* |
First step in age hardening that involves heating until one phase has completely dissolved in the other.

| *Precipitation Heat Treatment* |
Second stage in age hardening in which the diffusion rate increases enough to allow one phase to form a fine precipitate.

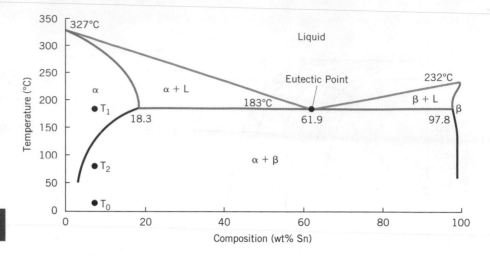

FIGURE 4-24 Illustration of Age Hardening

4.6 COPPER AND ITS ALLOYS

| *Chalcocite* |
The most common copper ore.

| *Chalcopyrite* |
Iron-containing mineral comprising about 25% of copper ores.

| *Blister Copper* |
Intermediate product during copper refining from which all iron has been removed.

| *Low-Alloy Coppers* |
Solid solutions containing at least 95% copper.

| *Brass* |
Alloy of copper and zinc.

| *Bronze* |
Alloy of copper and tin.

Copper ores are found in many different parts of the world. Approximately half the commercial copper ore supply is made up of *chalcocite* (Cu_2S), with *chalcopyrite* ($CuFeS_2$) accounting for an additional 25%. A high-temperature pyrometallurgical process is used to purify and concentrate the copper. The resultant liquid, called a copper matte, is blown with air (much like steel) to oxidize iron in the system. When most of the iron has been removed, the so-called *blister copper* is decanted and transferred to a refining furnace for final processing.

Pure copper has a distinctive red color and forms a green patina when oxidized, as shown in Figure 4-25. Copper is one of the few metals with significant commercial uses in a nonalloyed form. Copper is highly conductive (second only to pure silver among metals), corrosion resistant, and formable, with moderate density (8.94 g/cm^3), making it desirable for use in electronic wires and other construction applications. Copper cookware is highly prized because of its excellent and uniform heat conduction, but pure copper can become dangerous in contact with food, so almost all commercial copper cookware is lined with stainless steel or tin. Ingestion of copper can cause vomiting and cramps; consumption of as little as 27 grams can result in death. Chronic exposure often results in liver damage and inhibited growth.

Often small quantities of other metals are added to copper to increase its strength or hardness. These *low-alloy coppers* contain at least 95% copper and try to minimize the loss of conductivity while improving mechanical properties. Cadmium is a common additive in low-alloy coppers. The addition of 1% cadmium will significantly enhance the strength of the metal while sacrificing only 5% of conductivity.

Although frequently used as a pure metal or with small amounts of metal additives, copper can form 82 binary alloys. Two of the most common are *brass* (copper–zinc) and *bronze* (copper–tin). Brasses are strong, shiny, and more corrosion resistant than pure copper. The copper–zinc phase diagram is shown in Figure 4-26.

Below 35% zinc, the metal exists as a single phase, with zinc atoms in a solid solution with an α-copper lattice. Brasses in this range are strong and ductile and can be cold-worked readily. Increasing the zinc content increases the strength of the metal but decreases its corrosion resistance.

Above 35% zinc, a BCC β-phase dominates. If hot metal is cooled rapidly, the β-phase remains in place throughout the entire structure. At slower cooling rates, the α-phase precipitates out at grain boundaries, resulting in β-grains surrounded by α-precipitate. This microstructure, shown in Figure 4-27, is called the *Widmanstätten structure*, named after Count Alois von Beckh Widmanstätten, who discovered the structure in a meteorite in 1908.

Lead (up to 4%) often is added to brasses to enhance machinability and to fill pore spaces in the metal. Lead is essentially insoluble in copper and precipitates at the grain boundaries. As a result, the lead serves as a lubricant during machining.

Brasses are lustrous, corrosion resistant, and easily cast, making them ideal for decorative figures and architectural trim. Cast brasses also find use in plumbing fixtures, low-pressure valves, bearings, and gears.

Bronze alloys first appeared in ancient Sumeria around 3500 BC. Bronze is harder than most commercial alloys, corrosion resistant, and easily cast. Although the term *bronze* has become a generic for many hard copper alloys that resemble bronze, it originated with copper–tin systems. Most commercial bronzes contain around 10% tin. A copper–tin phase diagram is shown in Figure 4-28.

Bronze finds application in engine parts, bearings, bell making, and artistic sculptures. Bronze is especially well suited for casting in molds. When the molten metal is added to the mold, the metal expands to fill the entire volume. During cooling, the metal shrinks slightly, making it easy to remove but retaining the characteristic shape of the mold.

FIGURE 4-25
Recognizable Green Patina from the Oxidized Copper on the Statue of Liberty

PhotoDisc/Getty Images

| **Widmanstätten Structure** |
Microstructure present in brass in which β-phase grains surrounded by α-phase precipitate.

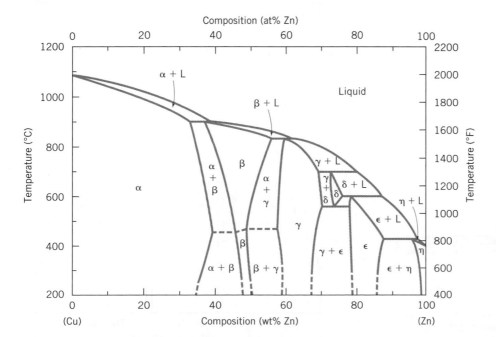

FIGURE 4-26 Phase Diagram for Brass (Copper–Zinc)

From T.B. Massalski, ed., Binary Alloy Phase Diagrams, 2nd edition. Vol. 3. ASM International, Materials Park, OH. Reprinted with permission of ASM International.

FIGURE 4-27   Widmanstätten Structure in Brass

Bronzes often include other elements, added in small quantities to improve specific properties. The addition of 1% to 3% silicon makes the bronze harder to cast but significantly improves its chemical resistance. Silicon bronzes are commonly used in chemical containers. Often up to 10% lead is added to bronze to soften the metal, making it easier to shape and enhancing its ability to hold a cutting edge. These **lead bronzes** are used often in artistic casting

| Lead Bronzes |
Copper–tin alloys that also contain up to 10% lead that is added to soften the metal.

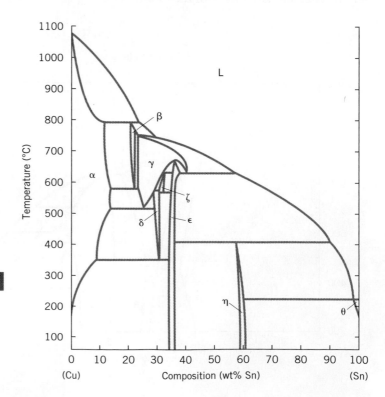

FIGURE 4-28   Copper–Tin Phase Diagram

From R. Hultgren and P.D. Desai, Selected Thermodynamic Values and Phase Diagrams for Copper and Some of Its Binary Alloys, *Incra Monograph I (New York: International Copper Research Association, Inc., 1971). Reprinted with permission of the International Copper Research Association.*

but are less strong and more brittle than normal bronze and are less useful for tools. Antimony frequently is added to bronze used for tools because it hardens the material and improves its ability to hold a cutting edge.

4.7 ALUMINUM AND ITS ALLOYS

Aluminum is the most common metallic element in the earth and has a density (2.70 g/cm³) approximately one-third that of steel. As a result, the tensile strength-to-weight ratio for aluminum is exceptional. Aluminum has a strong affinity for oxygen and is almost always found as oxides or silicates rather than in its native form. Nearly all commercial aluminum is produced from *bauxite*, which is a class of minerals rich in aluminum oxides. Bauxite must be treated to remove impurities and silicates before it can be converted to metallic aluminum.

| *Bauxite* |
Class of minerals rich in aluminum oxides that serves as the primary ore for aluminum production.

Aluminum is used extensively in aerospace applications, automobiles, and beverage cans and other packaging. The high ductility of aluminum allows it to be rolled into extremely thin foil. Aluminum forms an FCC lattice and is commonly alloyed with magnesium, copper, lithium, silicon, tin, manganese, and/or zinc.

Aluminum alloys are broadly classified as either *wrought* or *cast*. Wrought alloys are identified by a unique four-digit number. The first digit represents the primary alloying element, the second shows modifications, and the third and fourth show the decimal percentage of aluminum concentration. The major classes are summarized in Table 4-6. Casting alloys are distinguished from wrought alloys by the presence of a decimal point between the third and fourth digit.

| *Wrought* |
Shaped by plastic deformation.

| *Cast* |
Reshaped by being melted and poured into a mold.

TABLE 4-6 Aluminum Alloy Nomenclature		
Designation	*Alloying Elements*	*Purpose of Alloying Element*
Wrought		
1xxx	(>99% aluminum)	None
2xxx	Copper	Strength and machinability
3xxx	Manganese	Corrosion resistance and machinability
4xxx	Silicon or silicon and magnesium	Lowering of melting range
5xxx	Magnesium	Hardness and corrosion resistance
6xxx	Magnesium and silicon	Heat treatability and formability
7xxx	Magnesium and zinc	Stress corrosion resistance
8xxx	Lithium, tin, zircon, or boron	
Casting		
1xx.x	(>99% aluminum)	
2xx.x	Copper	
3xx.x	Silicon and copper or magnesium and copper	
4xx.x	Silicon	
5xx.x	Magnesium	
7xx.x	Magnesium and zinc	
8xx.x	Tin	

Aluminum alloys often are strengthened through a variety of cold-working techniques that increase strength but reduce corrosion resistance. Because strengthening is so common, most aluminum alloys carry a supplemental *temper designation* that indicates whether the material is strain hardened (H) or heat treated (T), also known as age hardening. Table 4-7 summarizes the temper designations for aluminum alloys.

| *Temper Designation* |
Nomenclature that shows whether an aluminum alloy was strain hardened or heat treated.

TABLE 4-7   Temper Designations for Aluminum Alloys

Symbol	Meaning
F	As fabricated
O	Annealed
H1x	Cold worked with x indicating the amount of cold work
H2x	Cold-worked then partially annealed
H3x	Cold-worked then stabilized at a low temperature to prevent age hardening
W	Solution treated
Tx	Age hardened with x providing additional detail on processing conditions

Example 4-8

For the following aluminum alloys, determine the primary alloying element, whether the alloy is cast or wrought, and whether it has been heat treated, strain hardened, or neither.

1199.H18   A380.0   4043

SOLUTION

1199 is a wrought alloy (no decimal after the third digit) with more than 99% aluminum (first digit is 1). It has been strain hardened by a factor of 8 but not annealed (from H18).

A380.0 is a cast alloy (decimal after the third digit) with manganese as the primary alloying element (starts with 3). It has been neither strain hardened nor heat treated (no temper designation).

4043 is a wrought alloy with silicon (or silicon and magnesium) as a primary alloying element. It has been neither strain hardened nor heat treated.

Example 4-9

Which of the alloys in Example 4-8 would be best suited for use in welding, and why?

SOLUTION

4043 would be the most useful in welding because the silicon lowers the melting range of the alloy.

What Limitations Do Metals Have?

4.8 CORROSION

The deterioration of metals in the environment costs companies billions of dollars each year. The free electrons that give metals their exceptional conductivity make the material particularly susceptible to chemical attack. This loss of material because of a chemical reaction with the environment is called *corrosion*. To fully understand corrosion, it is first necessary to review *electrochemistry*. During corrosion, metals transfer electrons to another material through a process called *oxidation*. The oxidation reaction occurs at a site called the *anode* and can be represented as

$$\text{Metal} \rightarrow \text{Metal ion}^{(n+)} + \text{n-lost electrons.} \tag{4.12}$$

The lost electrons must be picked up by another species through a process called *reduction*, which takes place at the *cathode*. The material gaining the electrons can be anything capable of receiving electrons but most commonly is an acid or another metal ion. The reduction reaction can be represented as

$$\text{Metal ion}^{(n+)} + \text{n-lost electrons} \rightarrow \text{Metal.} \tag{4.13}$$

Because the total number of electrons cannot change, oxidation and reduction reactions occur simultaneously. A specific example would be immersing magnesium (Mg) in a strong acid solution. The oxidation reaction,

$$\text{Mg} \rightarrow \text{Mg}^{2+} + 2e^-, \tag{4.14}$$

would represent the oxidation of the magnesium, while the reduction of the hydrogen ions from the acid would be given by

$$2H^+ + \ + 2e^- \rightarrow H_2. \tag{4.15}$$

The most famous electrochemical corrosion reaction is the rusting of iron. When metallic iron (Fe) is exposed to water with dissolved oxygen, a two-step reaction occurs. In the first stage, the iron is oxidized, as shown in Equation 4.16:

$$\text{Fe} \rightarrow \text{Fe}^{2+} + 2e^-. \tag{4.16}$$

This time, the iron ions react quickly to form $Fe(OH)_2$,

$$\text{Fe}^{2+} + 2OH^- \rightarrow \text{Fe(OH)}_2. \tag{4.17}$$

In the second stage of the reaction, the iron is oxidized again,

$$\text{Fe(OH)}_2 \rightarrow \text{Fe(OH)}_2^+ + e^-. \tag{4.18}$$

| *Corrosion* |
Loss of material because of a chemical reaction with the environment.

| *Electrochemistry* |
Branch of chemistry dealing with the transfer of electrons between an electrolyte and an electron conductor.

| *Oxidation* |
Chemical reaction in which a metal transfers electrons to another material.

| *Anode* |
The site at which oxidation occurs in an electrochemical reaction.

| *Reduction* |
Chemical reaction in which a material receives electrons transferred from a metal.

| *Cathode* |
The site at which reduction occurs in an electrochemical reaction.

The positive ion again reacts with a hydroxide ion from the water to form an insoluble compound, $Fe(OH)_3$, which is better known as **rust**.

$$Fe(OH)_2^+ + OH^- \rightarrow Fe(OH)_3. \qquad (4.19)$$

The previous examples involved a metal as the electron donor and hydrogen ions as electron receivers. However, metals can exchange electrons with other metals as well. The concept is clearly illustrated by the type of electrochemical cell shown in Figure 4-29. In an **electrochemical cell**, two solutions are separated by an impermeable barrier. On one side of the barrier, a piece of copper is placed in a solution containing copper ions (Cu^{2+}). On the other side, a piece of tin is placed in a solution containing tin ions (Sn^{2+}). If the two pieces of metal are connected by a wire, electrons will flow from the tin to the copper. A voltmeter placed on the wire would show a potential difference of 0.476 volts. During the process, the metallic tin would oxidize, generating more tin (Sn^{2+}) ions while the electrons would flow into the copper ion solution and reduce the copper ions (Cu^{2+}) to metallic copper, which would cause the tin to coat the surface of the copper piece. The net reaction is given by

$$Cu^{2+} + Sn \rightarrow Cu + Sn^{2+}. \qquad (4.20)$$

The copper served as the anode and gained metal while the tin served as a cathode and corroded. The process, called **electroplating**, is used to put thin layers or gold, silver, or copper on tableware but also has significant implications on corrosion.

The **galvanic series**, shown in Table 4-8, ranks different metals in order of their tendency to oxidize when connected with other metals in solutions with their ions. This table provides crucial information about which metal will serve as the anode and which will serve as the cathode.

The galvanic series distinguishes between passive and active forms of some alloys, including stainless steel. Passive alloys have become less anodic because of the formation of a thin oxide barrier on the surface of the metal that provides a hindrance to the diffusion of oxygen and inhibits additional corrosion. The spontaneous formation of this protective barrier is called **passivation**.

At least eight distinct forms of metallic corrosion can occur. All are impacted by environmental conditions, including temperature, submersion in liquid, acidity, and fluid velocity. More than one form of corrosion can occur simultaneously.

Uniform attack is the most common form of corrosion and is the easiest to design around. During uniform attack, the entire metallic surface is

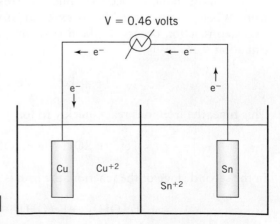

FIGURE 4-29 Electrochemical Cell with Copper and Tin

Table 4-8	Galvanic Series	
More Cathodic (less likely to oxidize)	Platinum
	Gold
	Graphite
	Titanium
	Silver
	Passive stainless steel
	Passive nickel
	Monel
	Copper–nickel alloys
	Bronzes
	Copper
	Brasses
	Active nickel
	Tin
	Lead
	Active stainless steel
	Cast iron
	Iron and steels
	Aluminum alloys
	Cadmium
	Aluminum
	Zinc
More Anodic (more likely to oxidize)	Magnesium and magnesium alloys

electrochemically attacked uniformly, and a residue is often left behind. Protective coatings often are used to protect the metallic surface from coming into contact with the corrosive environment.

Galvanic corrosion operates much like the electrochemical cell discussed previously. When dissimilar metals are electronically connected, the differences in their electrochemical potentials will lead to corrosion. The metal that is lower on the galvanic series will oxidize in favor of the more cathodic metal. When copper and steel tubing are joined in home water heaters, for example, the steel corrodes preferentially. There are several strategies to minimize galvanic corrosion, including selecting metals near each other in the galvanic series and preventing dissimilar metals from coming into electrical contact with each other. However, sometimes galvanic corrosion can be used to an advantage. The steel hulls of ships would corrode in seawater. However, small pieces of zinc are routinely attached to the hulls. The zinc preferentially corrodes and transfers electrons to the iron in the hull, protecting it from electrochemical attack. The zinc serves as a *sacrificial anode* and provides *cathodic protection* to the iron.

When small surface defects, including small holes or scratches, exist on a metal surface, corrosive materials tend to collect in these areas. The corrosive material in the hole oxidizes the metal, creating an even deeper pit and

| *Galvanic Corrosion* |
Loss of material resulting from the metal that is lower on the galvanic series oxidizing in favor of the more cathodic metal.

| *Sacrificial Anode* |
A metal low on the galvanic series used to oxidize and transfer electrons to a more important metal.

| *Cathodic Protection* |
Form of corrosion resistance provided by the use of a sacrificial anode.

| *Pitting* |
Form of corrosion
resulting from corrosive
material collecting in
small surface defects.

ultimately perforating the entire metal. This process, called *pitting*, is often difficult to detect because most of the surface of the metal is unaffected. The presence of chlorides (such as those found in salt water) worsens the pitting process by making the pits mildly acidic and autocatalyzing the corrosion reaction. Pitting is extremely difficult to prevent completely but can be reduced by polishing the surface of the metal to eliminate surface defects or by using a stainless steel with at least 2% molybdenum.

| *Crevice Corrosion* |
Loss of material resulting
from the trapping of stagnant
solutions against the metal.

Crevice corrosion is very similar to pitting and occurs anywhere that stagnant solutions can remain. It is most serious when it occurs under bolts, rivets, and gaskets. Even passive alloys are subject to crevice corrosion because the buildup of hydrogen ions in the stagnant films erodes the protective barriers. Crevice corrosion can be reduced by using welds instead of rivets to connect metals, by ensuring complete drainage to reduce the likelihood of forming stagnant films, by using Teflon® gaskets that will not absorb liquid, and by removing dirt and other deposits that promote the formation of stagnant liquids.

| *Intergranular
Corrosion* |
Loss of material
resulting from preferential
attack of corroding agents
at grain boundaries.

Intergranular corrosion, illustrated in Figure 4-30, occurs preferentially at grain boundaries. In some cases, the material at the grain boundaries becomes far more susceptible to corrosion than the rest of the metal. This preferential attack can result in macroscopic failure. Materials with precipitated phases at grain boundaries are particularly susceptible to intergranular corrosion. Austenitic stainless steels and stainless steel welds are particularly susceptible because at temperatures between 500°C and 800°C, chromium carbides ($Cr_{23}C_6$) precipitate out of the stainless steel lattice. When the chromium precipitates out of the lattice, a chromium-depleted region develops near the grain boundary. In this depleted region, the lattice-stabilizing properties of the chromium cease to exist and the region near the grain boundary becomes particularly susceptible to corrosive attack. The steel affected in this way has

| *Sensitized* |
Made more susceptible to
intergranular corrosion by
the localized depletion of a
precipitating element in the
region of a grain boundary.

been *sensitized*. Intergranular corrosion in austenitic stainless steels can be minimized by keeping the carbon content low to reduce the formation of carbides or by alloying the steel with another metal (such as titanium) with a greater affinity to form carbides than the chromium. In some cases, the chromium can be recovered from the carbide by heating the metal above 1000°C and rapidly quenching it.

| *Stress Corrosion* |
Loss of material resulting from
the combined influence of a
corrosive environment and
applied tensile stress.

Stress corrosion results from the combined influence of a corrosive environment and the application of a tensile stress. During stress corrosion, localized cracks form, as shown in Figure 4-31, and propagate until failure of the material results. Stress-corroded materials experience brittle failure even though the metal itself is usually ductile. Stress corrosion is a highly material-specific phenomenon. Most materials are subject to stress corrosion only when exposed to very specific corrosive materials. For example, lead is susceptible to stress corrosion only when exposed to lead acetate solutions, while brasses are susceptible when exposed to ammonia solutions. The only methods to avoid stress corrosion are to avoid exposing metals to corrosive solutions to

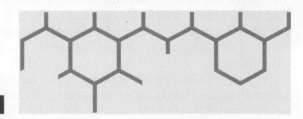

FIGURE 4-30   Schematic of Intergranular Corrosion

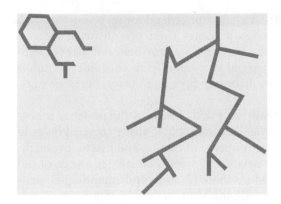

which they are susceptible or to reduce the magnitude of the applied stress. Reducing the applied stress is not as simple as it first appears, because residual stresses from uneven heating may suffice. A thermal annealing can be used to eliminate residual stresses.

Erosion corrosion results from the mechanical abrasion of a metal by a corrosive material. Essentially all metals are susceptible to erosion corrosion, but passive alloys are affected particularly because the mechanical abrasion wears away the oxide barrier that inhibits corrosion. The presence of bubbles or suspended particles accelerates the impact of erosion corrosion, and increased fluid velocity dramatically increases the corrosion. Erosion corrosion is most prevalent at places where flow becomes turbulent, including elbows, bends, or connection between pipes of different diameters. Erosion corrosion can be minimized by reducing the formation of turbulence, removing suspended solids, and reducing fluid velocity.

Selective leaching refers to the preferential elimination of one constituent of a metal alloy. When binary alloys experience corrosion, both metallic species are reduced and dissolved. However, the metal that is lower on the galvanic series tends to remain in the solution while replating the more cathodic material. The most common instance of selective leaching is the dezincification of brasses. The copper replates while the highly anodic zinc remains in solution. The result is a zinc-deficient copper lattice with significantly reduced mechanical properties. Alloy steels also tend to preferentially lose nickel by the same mechanism.

| *Erosion Corrosion* |
Loss of material resulting from the mechanical abrasion of a metal by a corrosive material.

| *Selective Leaching* |
The preferential elimination of one constituent of a metal alloy.

What Happens to Metals after Their Commercial Life?

4.9 RECYCLING OF METALS

Metals are the easiest class of materials to recycle. Pure metals can be remelted in a smelting furnace and recast into new products or alloyed with other metals. Recycling alloys is somewhat more challenging but operates in much the same fashion. In 2005 the United States recycled over 71 metric tons of metals (just over 50% of the total supply)[1], with nearly 80% coming from iron and steel (which are the most widely used industrial metals) and nearly 9% from aluminum. Metal for recycling (or *scrap metal*) is classified as *new scrap* when it comes from preconsumer sources, including stampings, trimming, filings, cuttings, and any material that failed to meet specifications. *Old scrap* is material recovered from consumer products that have completed their useful life, including cars, appliances, beverage cans, and commercial buildings.

Over 70% of all steel is eventually recycled. Much of the material is new scrap that comes from mill processing. This material is easily recycled because its exact composition is known. Improvements in casting and machining methods are reducing the availability of new scrap steel. The largest source of old scrap steel is junked automobiles. More than 12,000 car-dismantling companies operate in the United States, but significant quantities come from demolished buildings, recycled railroad tracks, and appliances.

Most steel mills melt scrap in basic-oxygen furnaces (BOFs) or in electric-arc furnaces (EAFs). The United States currently exports scrap steel to 44 different countries, with China purchasing the largest quantity.

Nearly 80% of all lead used industrially is recycled, in part because of the cost and difficulties in disposing it. Batteries are the most common source of recycled lead, accounting for more than 90% of the total. Nearly 91% of the total domestic demand for lead can be addressed from the lead smelted from recycled materials.

Aluminum recycling includes approximately 60% new scrap and 40% old scrap, with more than half of the new scrap coming from recycled beverage cans. In 2003, almost 50 billion aluminum cans were recycled, accounting for nearly half of the cans sold. Used beverage cans are shipped to smelting facilities where they are shredded and passed into a delacquering oven to remove any paints and residual moisture. The metal used in aluminum cans consists of two distinct aluminum–magnesium alloys. The alloy in the harder tops contains 4.5% magnesium, while the more formable sides contain only 1% magnesium. Some plants heat the aluminum to the temperature at which the harder alloy melts and sifts the remaining solids to separate the alloys. Others add the entire mixture in a smelting furnace to a mixture of molten metal with a low magnesium content. Because the recycled cans contain a net aluminum percentage between 1% and 4.5%, the excess fresh molten metal can have proportionally less magnesium and still form a new batch of the lower-magnesium alloy. The combined mixture is tested then cast into large ingots, which eventually are rolled into sheets to make the sides of cans.

Summary of Chapter 4

In this chapter we examined:

- The purposes and the differences of the four forming operations

- The principle of strain hardening, and how it affects mechanical properties of metals

- The three stages of annealing (recovery, recrystallization, grain growth) and their impact on hot working

- How to read and label a phase diagram

- The means to calculate phase compositions using either the lever rule or explicit mass balances

- The unique properties of carbon steel, including the role of microstructure and carbon content on both equilibrium and nonequilibrium structures

- The role of alloying elements in steel

- The diffusion process and its role in phase transformations

- The process of age hardening and its impact on properties

- How to use an isothermal transformation diagram (T-T-T plot) to determine the kinetics of phase transformations

- The unique properties of copper and its alloys

- The distinction between cast and wrought aluminum alloys

- The electrochemical nature of corrosion reactions and eight primary types of corrosion

- The recycling of industrial metals

Reference

[1] J. Papp, ed., *USGS 2005 Mineral Yearbook* (U.S. Department of the Interior, February 2007).

Key Terms

age hardening *p. 131*
alloys *p. 110*
alloy steels *p. 125*
annealing *p. 108*
anode *p. 137*
austenite *p. 119*
austenitic stainless steels *p. 126*
austenizing *p. 121*
bainite *p. 123*
bauxite *p. 135*

begin curve *p. 128*
binary eutectic *p. 116*
binary isomorphic *p. 111*
blister copper *p. 132*
brass *p. 132*
bronze *p. 132*
carbon steel *p. 118*
cast *p. 135*
cathode *p. 137*
cathodic protection *p. 139*

cementite *p. 119*
chalcocite *p. 132*
chalcopyrite *p. 132*
coarse pearlite *p. 123*
cold working *p. 106*
completion curve *p. 128*
corrosion *p. 137*
crevice corrosion *p. 140*
diffusion *p. 114*
diffusivity *p. 114*

Homework Problems

1. Recent evidence indicates that the reason the *Titanic* sank so quickly was that the steel used in its hull was very brittle. Describe the potential roles of microstructure, carbon content, and impurities on the brittleness of steel.

2. The accompanying partial phase diagram corresponds to hypothetical metals A and B that form an alloy with α and γ phases. Additionally, you know that a mixture of 70% A and 30% B melts completely to liquid at 1600°C.

   a. Draw the missing lines on the phase diagram and label the phases present in each region.

   b. Classify the system (e.g., ternary polymorphic).

   c. Calculate the mass fraction of material in the a phase at 1500°C if the total mass fraction of A is 40%.

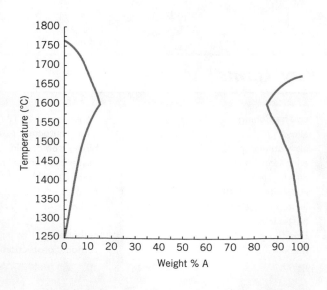

d. Determine the weight percent of A in the γ phase at 1500°C if the total mass fraction of A is 40%.

e. Label all of the following that appear on your phase diagram: solvus line, solidus line, liquidus line, eutectic point, eutectoid point, peritectic point.

f. Instead of cooling slowly, a sample with 70% A is rapidly quenched from 1600°C to room temperature. What effect would this have on the microstructure of the metal?

3. Coarse pearlite at the eutectoid concentration is heated to 600°C. Predict the mass fractions of ferrite and cementite present.

4. Take the steel in Problem 3 and heat it near 700°C for 20 hours. Then compare the microstructure and properties (hardness, ductility) of the steel that you produced with those of the steel with which you started.

5. A steel beam is made of coarse pearlite, and an architect decides that she would rather have bainite. Obviously, she does not wish to remelt the beam. Recommend a treatment process to convert the beam to bainite and explain (in terms of properties) why the architect might want bainite instead of pearlite or martensite.

6. Explain why aluminum cans cannot simply be melted (after removing the lacquer) and re-formed into new cans.

7. 500 grams of 1040 steel is cooled from 770°C to 500°C and allowed to remain at that temperature for several hours.

   a. Determine the amount and compositions of each of the phases present.
   b. Describe the microstructure.
   c. Can the T-T-T plot in Figure 4-23 be used to determine the time required to complete this phase transformation? Explain.

8. Using a stress-strain curve, explain the influence of cold working on the yield strength and ductility of steel.

9. A battleship sank because the hull of the ship corroded. The navy has asked you to suggest a method of corrosion prevention to make sure this does not happen to future vessels. What method(s) of corrosion prevention would you suggest? Why are other methods less suitable?

10. Explain the characteristics of erosion-corrosion and describe a situation where you might find this type of corrosion.

11. A copper–zinc alloy at its eutectoid concentration is cooled from 900°C to 770°C.

    a. Determine the weight percent of each phase and the weight percent of zinc present in each phase.
    b. Would the material likely have a uniform composition? Explain.

12. Explain the difference between the following terms:

    a. Hypoeutectoid and hypereutectoid
    b. Extruding and drawing
    c. Hardness and hardenability
    d. Eutectoid and euctectic
    e. Solidus and solvus lines

13. For each of the following treatments, predict the resulting microstructure if all materials began as euctectoid steel at 770°C.

    a. Cooled rapidly to 600°C, held for 2 minutes, then quenched.
    b. Cooled rapidly to 600°C, held for 1 second, then quenched.
    c. Cooled rapidly to 600°C, held for 5 seconds, then quenched.
    d. Cooled rapidly to 400°C, held for 90 seconds, then quenched
    e. Cooled rapidly to 400°C, held for 10 minutes, then heated to 700°C for 20 hours.

14. For each of the following treatments, predict the resulting microstructure if all materials began as euctectoid steel at 770°C.

    a. Cooled rapidly to 500°C, held for 1 minute, then quenched.
    b. Cooled rapidly to 700°C, held for 1 minute, then quenched.
    c. Cooled rapidly to 650°C, held for 30 seconds, then quenched.
    d. Cooled rapidly to 100°C, held for 90 seconds, then quenched
    e. Cooled rapidly to 550°C, held for 10 minutes, then heated to 700°C for 20 hours.

15. For each of the following treatments, predict the resulting microstructure if all materials began as euctectoid steel at 770°C.

    a. Cooled at a rate of 140°C per minute.
    b. Cooled at a rate of 45°C per minute.
    c. Cooled at a rate of 15°C per minute.

16. Given the following data for a hypothetical metal, calculate the time required to achieve 90% conversion from a γ-phase to an α-phase.

Fraction Converted	Time (sec)
0.25	200
0.50	280

17. Given the following data for a hypothetical metal, calculate the time required to achieve 90% conversion from a γ-phase to an α-phase.

Fraction Converted	Time (sec)
0.10	100
0.25	170

18. Why does bainite not form by continuous cooling of eutectoid steel?

19. Provide the most direct process to achieve the following transformations for eutectoid steel at room temperature:

    a. Convert spheroidite to bainite.
    b. Convert bainite to pearlite.
    c. Convert pearlite to martensite.
    d. Convert martensite to bainite.
    e. Convert coarse pearlite to fine pearlite.

20. Carbon steel with an initial tensile strength of 800 MPa and yield strength of 650 MPa is extruded with a force of 750 MPa. Describe the difference in properties and microstructure that result from performing the extrusion above or below the recrystallization temperature.

21. Based on the lead-tin phase diagram in Figure 4-11:

    a. What are the mass fractions of α and β present for a system containing 40wt% tin at 100°C?
    b. How much tin is present in each phase?

    c. If the material is heated, at what temperature does the first liquid form, and what is the composition of the liquid phase at that temperature?

22. Identify and classify all invariant points on the steel phase diagram.

23. What would the likely AISI/SAE designation be for the following alloys:

    a. Steel with 0.9% carbon and 5% molybdenum.
    b. Steel with 5% manganese and 0.2% carbon.
    c. Carbon steel with 0.4% carbon and no special treatment.
    d. Steel with 7% silicon and manganese with 0.4% carbon.

24. Identify and classify all invariant points on the brass phase diagram.

25. Explain why zinc is used as a sacrificial electrode in marine environments and how its presence protects the iron hulls of ships.

26. A solid solution containing 82% zinc and 18% copper is cooled from 1000°C to 20°C.

    a. When does the first solid form and what phase forms?
    b. What is the composition of the first solid?
    c. Will the entire solid formed in this region have the same composition? Explain.

27. Identify at least two specific aluminum alloys used in automobiles. Explain why the primary alloying element was chosen for each application.

28. Describe the roles of nickel and chromium in stainless steels.

29. Describe a physical situation in which each form of corrosion could take place, and suggest at least two methods of reducing its effect.

30. Discuss the advantages and disadvantages of an alloying element forming carbides in alloy steels.

31. Explain the operating principles of the four forming operations.

32. The final diameter of a cylindrical sample of metal that underwent 30% cold work is 7.5 cm. How big was the original sample?

33. Explain why segregation impacts mechanical properties.

34. Given the following series of percent transformation curves, construct a T-T-T plot for a hypothetical metal.

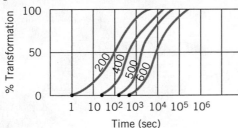

35. Identify the primary alloying elements, whether the material is cast or wrought, and any tempering performed for the following aluminum alloys:

a. 2017

b. 300.3

c. 1199 H14

5

Polymers

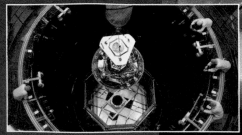

CONTENTS

Learning Objectives

By the end of this chapter, a student should be able to:

- Explain basic polymer terminology, including monomer, oligomer, copolymer, structural unit, and degree of polymerization.

- Identify the characteristics of thermoplastic and thermoset polymers.

- Explain the differences among random, block, alternating, and graft copolymers.

- Describe the basic structure and properties of major classes of polymers, including acrylics, polyamides (aramids and nylons), polyesters, polyolefins, high-volume thermoplastics, rayon, and elastomers.

- Understand what is meant by the glass transition temperature and its impact on polymer properties.

- Reproduce the correct series of reactions to generate a polymer of appropriate size via condensation or addition polymerization.

- Explain the differences among relative molecular mass, number average molecular weight, and weight average molecular weight.

- Explain the difference between primary and secondary bonding in polymers.

- Address the significance of constitutional issues, including the additive nature of secondary bonding, the impact of branching on physical properties, and the difference between irregular and ordered polymer networks.

- Identify asymmetric carbons in polymer chains.

- Distinguish between syndiotactic and isotactic dyads.

- Explain what is meant by polymer crystallinity and why different analytical techniques will yield different results.

- Understand the balance of attractive and repulsive forces that control conformation.

- Describe the difference in energy states among eclipsed, staggered, gauche, and trans-conformations.

- Explain the basic operation of polymer extrusion systems.

- Describe the operating principles of spinnerets, blown-film units, dies, and injection molding apparatus.

- Explain why none of the extrusion-based techniques will work for thermosets and how they are processed instead.

- Discuss the challenges associated with commercial recycling of polymeric materials.

What Are Polymers?

5.1 POLYMER TERMINOLOGY

| *Polymers* |
Covalently bonded chains of molecules with small monomer units repeated from end to end.

| *Monomers* |
Low-molecular-weight building blocks repeated in the polymer chain.

| *Oligomers* |
Small chains of bonded monomers whose properties would be altered by the addition of one more monomer unit.

The word *polymer* comes from the Greek roots "poly," meaning many, and "meros," meaning units or parts. *Polymers* are bonded chains of repeated units with covalent bonds from end to end. The building blocks that are repeated in the chain are called *monomers*. As monomers begin to connect together to form chains, they become *oligomers*. As more monomers add to the oligomer chain, it grows and eventually becomes a polymer when the addition of one more monomer unit would have no discernible effect on the properties of the chain.

A polymer might have 10,000 or 1,000,000 repeated monomer units in a chain, which makes it impractical to draw the entire polymer molecule. Instead, the polymer chain is identified by its *structural unit* (or repeat unit), which is the smallest part of the chain that repeats. The structural unit of polystyrene is shown in Figure 5-1.

The n represents the number of times the structural unit is repeated in the chain. Of course, polymers are not made one chain at a time. Large numbers of chains begin forming simultaneously, and different chains grow to different lengths. The number of repeat units is called the *degree of polymerization* of the polymer chain and is represented by the symbol $\overline{DP}_n$.

The covalently bonded atoms (usually carbon) that comprise the long, repeating center of the chain are called the *polymer backbone*. Atoms attached to the backbone are called *side groups* or substituents. Hydrogen is the most common side group, but methyl groups, benzene rings, hydroxide molecules, heteroatoms, or even other polymer chains can serve as side groups as well. When drawing a polymer chain, any side group not specifically shown is assumed to be hydrogen.

Polymers are classified based on their ability to be remelted and reshaped. *Thermoplastics* flow like viscous liquids when heated and continue to do so when reheated and recooled multiple times. Thermoplastics usually are mass produced as pellets that can be dyed, melted, and reshaped by end users. While the individual chains in thermoplastics have covalent bonds along their primary axis, the bonding between chains usually is limited to weak Van Der Waals interactions. In most cases there is no three-dimensional ordering between the chains and often little, if any, two-dimensional order. The apparently random relative positioning of adjacent chains often is described as "spaghetti on a plate," as shown in Figure 5-2. The lack of bonding between chains tends to reduce the tensile strength of thermoplastics but makes them relatively easy to recycle.

By contrast, when the chemicals that form *thermosets* are heated, they slowly undergo an irreversible chemical cross-linking reaction that bonds the chains together and causes the liquid to become an infusible solid mass. Once solidified, thermosets cannot be remelted or re-formed. For this reason, the polymerization reaction is performed in a mold or fiber spinneret so that the thermoset immediately takes on its final shape. The cross-linking between chains, illustrated in Figure 5-2, makes thermosets stronger and more resistant to chemical degradation than thermoplastics, but it also makes them difficult to recycle.

Decisions regarding whether to use thermosets or thermoplastics for specific applications create interesting ethical dilemmas. Cost and environmental impact tend to favor the use of thermoplastics. Imagine if grocery bags were made stronger using thermosets but could not be recycled and cost a quarter each. By contrast, aviation equipment, ballistic resistance materials, and many military supplies demand the higher performance offered by thermosets.

Many polymers are made of a single structural unit, repeated many times, but this is not the only possibility. When a polymer is formed from the polymerization of two or more monomers, it is called a *copolymer*. Four distinct classifications for copolymers made from two monomers (A and B) exist that distinguish how the monomers blend. These classifications are shown in Figure 5-3.

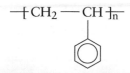

FIGURE 5-1 Structural Unit of Polystyrene

| **Structural Unit** |
Smallest repeating unit in a polymer. Also known as a repeat unit.

| **Degree of Polymerization** |
Number of repeat units in a polymer chain.

| **Polymer Backbone** |
Covalently bonded atoms, which are usually carbon, that comprise the center of the polymer chain.

| **Side Groups** |
Atoms attached to the polymer backbone. Also called substituents.

| **Thermoplastics** |
Polymers with low melting points due to the lack of covalent bonding between adjacent chains. Such polymers can be repeatedly melted and re-formed.

| **Thermosets** |
Polymers that cannot be repeatedly melted and re-formed because of strong covalent bonding between chains.

| **Copolymer** |
Polymer made up of two or more different monomers covalently bonded together.

| **Random Copolymers** |
Polymers comprised of two
or more different monomers,
which attach to the polymer
chain in no particular order
or pattern.

| **Alternating Copolymers** |
Polymers comprised of two
or more different monomer
units that attach to the chain
in an alternating pattern
(A-B-A-B-A-B).

| **Block Copolymers** |
Polymers comprised of two
or more different monomers
that attach to the chain in long
runs of one type of monomer,
followed by long runs of
another monomer
(AAAAAAABBBBBBBBAAAAA).

Random copolymers add either monomer in any order such that the probability of the next link the chain being monomer A or monomer B is not affected by the identity of the last monomer. By contrast, *alternating copolymers* always follow monomer A with monomer B and vice versa. The length of the chain may vary, but it will always follow the A-B-A-B pattern. *Block copolymers* involve long runs of monomer A, followed by long runs of monomer B, followed by

Thermoplastic

• Flows like viscous liquids when heated and
continue to do so when reheated and recooled
multiple times
• Weak Van Der Waal forces between chains
• Random relative positioning of adjacent
chains

Thermoset

• Cannot be remelted or reformed
• Strong cross-linking between chains due to
Van Der Waal forces
• Stronger and more resistant to chemical
degradation than thermoplastics, but also
difficult to recycle

FIGURE 5-2 Thermoplastic
versus Thermoset Polymers

Random: No particular pattern of monomers

Block: Long runs of one monomer, followed
by long runs of the other, followed by
more of the first

Alternating: One monomer is always followed
by the other and vice versa

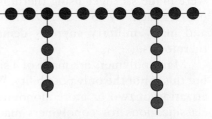

Graft: Chain of one monomer is connected
as a substituent to a chain of the
other monomer

FIGURE 5-3 Classes of
Copolymers

more monomer A. *Graft copolymers* result when a chain of one monomer (B) is connected as a substituent to a chain of the other monomer (A).

Polymer *blends* are formed by the mechanical mixing of polymers. Blends allow materials to experience a wider range of properties without going through the difficulty and expense of trying to synthesize a single polymer with the same combination of properties. Blends tend to exhibit properties between those of the original polymers. For example, the toughness of polystyrene can be enhanced by blending small amounts of a rubbery polymer that absorbs impact energy. Artificial surfaces for running tracks are a blend of polyurethane and rubber. Blends are far more difficult to recycle because of the problems in separating out the original polymers.

| *Graft Copolymers* |
Polymers in which one chain of a particular monomer is attached as a side chain to a chain of another type of monomer.

| *Blends* |
Two or more polymers mechanically mixed together but without covalent bonding between them.

5.2 TYPES OF POLYMERS

The naming of polymers is complex and sometimes confusing. Most naturally occurring polymers were named by either the source of the polymer (e.g., cellulose from plant cells) or the nature of the polymer (e.g., nucleic acids). The first synthetic polymers were named from the monomers that were used in their preparation. For example, molecules made from ethylene were called polyethylenes. Sometimes they were named for items present in the polymer chain, such as polyesters. The International Union of Pure and Applied Chemistry (IUPAC) has developed a system of nomenclature that uses strict rules to name polymers. Unfortunately, the IUPAC names and the common names both remain in use. Moreover, commercial polymers tend to be known by their IUPAC names, by a shortened abbreviation because the IUPAC names are often long and cumbersome, and by one or more commercial trade names. Fibers used for bullet-resistant vests often are made from poly-p-phenylene terephthalamide, which is abbreviated (PPTA) and sold under the trade names Kevlar® (DuPont) or Twaron® (Teijin). Figure 5-4 shows the structural unit of PPTA.

The complexity of the naming schemes makes it more useful to learn about the basic classes of commercial polymers. *Acrylic* fibers contain at least 85% polyacrylonitrile (PAN), shown in Figure 5-5. Acrylics are lightweight and durable, making them ideal for carpeting and clothing. Orlon® (DuPont) and Acrilan® (Solutia) are common commercial names for acrylic fibers. The most commercially significant acrylic fiber is polymethylmethacrylate (PMMA), shown in Figure 5-6. PMMA is transparent and does not shatter, making it

| *Acrylic* |
One type of polymer that contains at least 85% of polyacrylonitrile (PAN).

• Trade names: Kevlar, Twaron
• High tensile strength, lightweight, susceptible to ultraviolet degradation, nonconductive
• Used for bullet-resistant armor, sports equipment, fire-resistant clothes

FIGURE 5-4 Structural Unit of Poly-p-phenylene Terephthalamide (PPTA)

Courtesy James Newell

$$-\!\!\left[CH_2 - CH\right]_n$$
$$\qquad\qquad |$$
$$\qquad\qquad C\!\equiv\!N$$

- Trade names: Orlon, Acrilan
- Lightweight, durable
- Carbon-fiber precursor
- Used in tennis racquets, racing bikes, helmets

FIGURE 5-5   Structural Unit of Polyacrylonitrile (PAN)

David Morgan/iStockphoto (Mountain Bike); PhotoDisc/Getty Images (Helmet)

$$\qquad\quad CH_3 \qquad\qquad CH_3$$
$$\qquad\quad | \qquad\qquad\quad |$$
$$-\!\!\left[C - CH_2 - C - CH_2\right]_n$$
$$\qquad\quad | \qquad\qquad\quad |$$
$$\qquad\quad C\!=\!O \qquad\qquad C\!=\!O$$
$$\qquad\quad | \qquad\qquad\quad |$$
$$\qquad\quad OCH_3 \qquad\qquad OCH_3$$

FIGURE 5-6   Structural Unit of Polymethylmethacrylate (PMMA)

Courtesy James Newell

- Trade names: Plexiglas, Lucite
- Transparent, shatterproof, biocompatible
- Used in hockey rinks, taillights, implants

useful for taillights in cars and the protective plastic barriers that surround hockey rinks. PMMA is biologically compatible and is used for a variety of applications, including implants to repair bones, dentures, and hard contact lenses. PMMA is also a key ingredient in acrylic paint. PMMA is sold under the trade name Lucite® (Lucite).

Polyamides are polymers that contain amide ($CONH_2$) groups in the chain. Polyamides are classified further based on what the nitrogen atoms bond with in the polymer. If more than 85% of the amide groups are bonded to two aromatic rings, the polymer is considered an *aramid*. Poly-p-phenylene benzobisoxazole (PBO) fiber, shown in Figure 5-7, is an aramid fiber sold under the commercial name Zylon® (Toyobo). Aramids display exceptional tensile strength and heat resistance, making them ideal for use in ballistic materials, ropes, and as reinforcement fibers in composites.

Nylon fibers are polyamides with less than 85% of the amide groups bonded to aromatic rings. Nylon fibers are durable and very strong, although typically

| *Polyamides* |
Polymers that contain amide (—N—) groups in the chain.

| *Aramid* |
Polymer in which more than 85% of the amide groups are bonded to two aromatic rings.

| *Nylon* |
Type of polyamide in which less than 85% of the amide groups are bonded to aromatic rings.

FIGURE 5-7   Structural Unit of Poly-p-phenylene Benzobisoxazole (PBO)

Courtesy James Newell

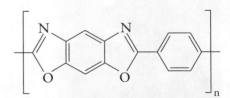

- Trade name: Zylon
- Transparent, shatterproof, biocompatible
- Used in hockey rinks, taillights, implants

not as strong as aramids. They are also easily dyed and stretched. Nylon was invented in 1935 by Wallace Carothers of DuPont and made its first impact in toothbrush bristles in 1938. By 1940 nylon stockings became a marketing success. Nylon was also used to replace silk in parachutes. The name *nylon* was never trademarked and now refers to a variety of materials, but the most commonly used is Nylon 6,6, shown in Figure 5-8.

Nylon 6,6 is sold under a variety of trade names, including Capron® (Allied Signal), Dartek® (DuPont), Primamid® (Prima Plastics), and Celanese® (Ticona). The polar amide groups in the fiber backbone enable adjacent chains to experience hydrogen bonding with each other, significantly improving the strength and crystallinity of the fibers.

Polyesters are long-chain polymers containing at least 85% of an ester (C—O—C) of a substituted aromatic carboxylic acid. Polyester fibers are strong and can be dyed or made transparent. The most common polyester, polyethylene terephthalate (PET), is used in clear beverage bottles and carpet fibers. PET, shown in Figure 5-9, makes up 5% of all commercial polymers. Polyesters resist staining and shrinking and have found widespread use in clothing. Dacron® (Invista) and Mylar® (DuPont Teijin) are brand names for other polyester products.

Polyolefins are oil-like polymers that contain only hydrogen and aliphatic (nonaromatic) carbon in the polymer. Polyethylene (PE) and polypropylene (PP) are the two most common polyolefins. Polyethylene is the simplest of

| *Polyesters* |
Long-chain polymers that contain at least 85% of an ester of a substituted aromatic carboxylic acid. These fibers are strong and can be dyed or made transparent.

| *Polyolefins* |
Polymers that contain only hydrogen and aliphatic carbon.

- Trade names: Capron, Dartek, Primamid, Celanese
- Durable, strong, easily dyed, stretched
- Used for clothing, rope, toothbrush bristles

FIGURE 5-8   Structural Unit of Nylon 6,6

Courtesy James Newell

- Trade names: Dacron, Mylar
- Strong, easily dyed, transparent, stain resistant
- Used for clothing, carpeting, plastic bottles

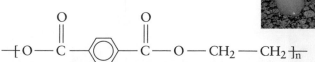

FIGURE 5-9   Structural Unit of Polyethylene Terephthalamide (PET)

Courtesy James Newell

all polymers with a simple carbon chain completely saturated with hydrogen atoms, as shown in Figure 5-10. Polyethylene is inexpensive, easily manufactured, and resistant to chemicals.

The properties of polyethylene change significantly depending on the nature of the chain. Under high-temperature and high-pressure conditions, some of the hydrogen atoms are knocked off and replaced by other polyethylene chains. The formation of side chains along the polymer backbone in called *branching*. The presence of branches disrupts the interactions between adjacent chains, lowering the tensile strength, melting point, stiffness, crystallinity, and density of the polymer.

Low-density polyethylene (LDPE) contains many branches that tend to be relatively long. The resulting polymer has low crystallinity and lower strength but much greater flexibility. LDPE melts at low temperature and is easily processed, making it ideal for high-volume applications where strength is not critical. Many plastic wraps, toys, produce bags, and squeeze bottles are made from LDPE.

High-density polyethylene (HDPE) is chemically identical to LDPE but is more linear with fewer branches. The resultant polymer is stronger and more crystalline. Milk bottles, trash bags, chemical storage tanks, and plastic cups are made from HDPE. Brand names for HDPE resins include Paxon® (Exxon-Mobil Chemical) and Unival® (Dow).

Polypropylene (PP) is similar to polyethylene, but a methyl group ($-CH_3$) replaces one of the four hydrogens in the structural unit, as shown in Figure 5-11. The resulting polymer is much more rigid than polyethylene but also harder and more abrasion resistant. Polypropylene also melts at a much higher temperature than PE but still is inexpensive to produce. This combination of properties makes polypropylene ideal for furniture, dishwasher-safe utensils and plates, coffee makers, drinking straws, DVD containers, and medical devices that must be sterilized.

| **Branching** |
Formation of side chains along the polymer backbone.

FIGURE 5-10 Structural Unit of Polyethylene (PE)

Courtesy James Newell

- Trade names: Paxon, Unival
- Simplest polymeric structure
- Inexpensive
- Used for plastic bags, toys, chemical storage

$$-\!\!\left[\,CH_2 - CH_2\,\right]_n$$

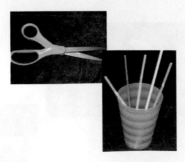

FIGURE 5-11 Structural Unit of Polypropylene (PP)

Courtesy James Newell

- Trade names: Ektar, Fortilene
- Rigid, inexpensive, abrasion resistant, high melting point
- Used for drinking straws, cooking utensils, TV cabinets, stadium seats

$$-\!\!\left[\,CH_2 - CH\,\right]_n$$
$$|$$
$$CH_3$$

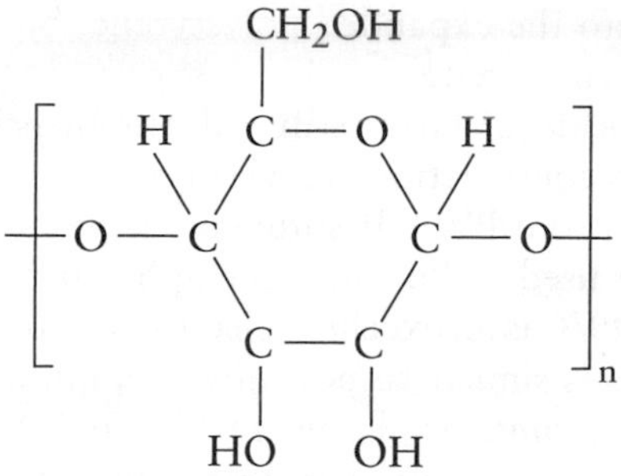

$$CH_2OH$$

[Structural diagram of rayon]

- Trade names: Bemberg, Galaxy, Danufil, Viloft
- Light, absorbs water, comfortable, soft, smooth
- Used for clothing, home furnishings

FIGURE 5-12   Structural Unit of Rayon

Courtesy James Newell

Rayon was the first synthetic polymer fiber developed. The *viscose process*, in which cellulose from wood or cotton is treated with alkali then extruded through a spinneret, dates back to 1892. Rayon initially was called *artificial silk*, and is still referred to as *regenerated cellulose*. Rayon is lightweight and absorbs water well, making it useful for clothing and home furnishings, including blankets and sheets. The structure of rayon is shown in Figure 5-12.

Polyvinylidene chloride (PVDC) polymers are much better known by the trade name Saran® (Dow). PVDC, shown in Figure 5-13, is produced in a thin film with tightly bound molecules that create a barrier to moisture and oxygen. As such, the material makes an ideal wrap for foods.

Polystyrene and polyvinyl chloride are the other *high-volume thermoplastics (HVTPs)* that this book will consider. Polystyrene (PS) is chemically similar to polyethylene, but one of the hydrogen atoms is replaced by an aromatic ring, as shown in Figure 5-14. Polystyrene is rigid, provides excellent thermal insulation, and can be made either transparent or in a wide variety of colors, leading to its use in disposable coffee cups and CD jewel cases. The most common use of polystyrene is as expanded polystyrene, which consists of 5% polystyrene and 95% air. Dow Chemical manufactured expanded polystyrene under the trade name Styrofoam®. Expanded polystyrene can be molded into a variety of shapes easily; for that reason, most custom packing materials and packing peanuts are made from expanded polystyrene. Initially, environmentally damaging chlorofluorocarbons (CFCs) were used in the process

| *Rayon* |
Lightweight polymer that absorbs water well; the first synthetic polymer ever constructed.

| *Viscose Process* |
Technique used to make rayon, which involves treating cellulose from wood or cotton with alkali and extruding it through a spinneret.

| *High Volume Thermoplastics (HVTPs)* |
Simple polymeric materials produced as pellets in large quantities.

- Trade names: Saran, Diofan, Ixan
- Barrier to moisture and oxygen, hard
- Used for food wrap, common constituent of many copolymers

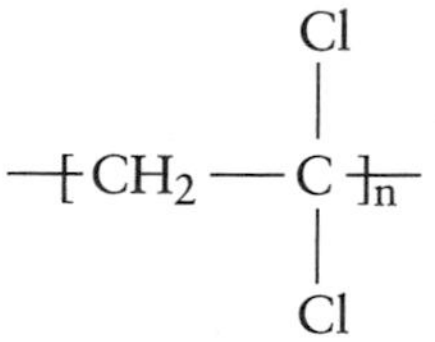

$$-\left(CH_2 - \underset{\underset{Cl}{|}}{\overset{\overset{Cl}{|}}{C}}\right)_n-$$

FIGURE 5-13   Structural Unit of Polyvinylidene Chloride (PVDC)

Courtesy James Newell

- Trade names: Novacor, Styrofoam
- Inexpensive, stiff, thermally insulating
- Used for packaging material, CD jewel cases, disposable cups for hot beverages

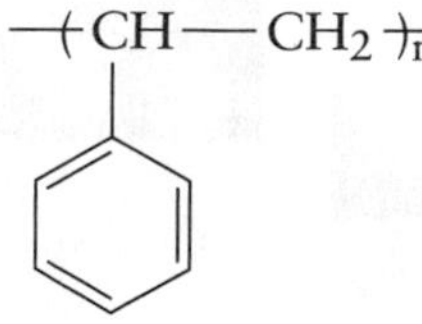

$$-\left(CH - CH_2\right)_n-$$

[benzene ring]

FIGURE 5-14   Structural Unit of Polystyrene (PS)

Courtesy James Newell

of blowing air into the expanded polystyrene, but now all manufacturers use more benign blowing agents.

Polyvinyl chloride (PVC) is a stiff and extremely inert polymer that has found widespread use in construction. Vinyl siding on houses, piping, and cooking oil bottles are made from PVC. If softening agents called *plasticizers* are added, PVC can also be used in flooring, car upholstery, medical devices, and some electrical cables. PVC is an excellent electrical insulator and is weather resistant. The polymer also is similar to polyethylene, but a chlorine atom replaces one of the hydrogen atoms, as shown in Figure 5-15. The chlorine atoms also make the polymer flame resistant. In recent years concerns have been raised about the widespread use of PVC and its role in dioxin contamination when landfilled. Chlorinated dioxins form during the production or combustion of chlorine-containing organic compounds. Significant advances have occurred in reducing the production of dioxins during PVC manufacture, but house fires, backyard burning of trash, and fires in municipal landfills containing discarded PVC create significant quantities of dioxins, often well above the safe limit. Entire towns including Love Canal, New York, and Times Beach, Missouri, have experienced significant problems from dioxin contamination. The U.S. Environmental Protection Agency (EPA) lists chlorinated dioxins as a serious threat to public health.

Elastomers are broadly defined as polymers that can stretch by 200% or more then return to their original length when the stress is released. Although there are many types of elastomers, the class can be split into the polyurethanes and the aliphatic thermosets (or rubbers).

Polyurethanes are polymers with urethane linkages, like the one shown in Figure 5-16, in their backbone. The broad definition results in a wide variety of materials falling under the polyurethane classification. Many polyurethanes are used to make foam, including the foam cushions on most chairs and couches. In a liquid form, polyurethane is used as a paint or water sealant. Perhaps the most famous polyurethane application is Spandex, a flexible polymer fiber that easily stretches and regains its shape. Spandex fibers contain at least 85% polyurethane and have unique properties. They can stretch over 500% and recover their shape and are resistant to sweat, body oils, and detergents. The

| **Elastomers** |
| Polymers that can stretch by 200% or more and still return to their original length when the stress is released. |

| **Polyurethanes** |
| Broad category of polymers that includes all polymers containing urethane linkages. |

FIGURE 5-15   Structural Unit of Polyvinyl Chloride (PVC)

Courtesy James Newell

- Trade names: Geon, Viclon
- Low cost, chemical and moisture resistant, outdoor stability
- Used for pipes, upholstery, flooring, electrical cables, medical devices

$$-\!\!\left(\!CH\!-\!CH_2\!\right)_{\!n}$$
$$|$$
$$Cl$$

FIGURE 5-16   Urethane Linkage in a Polymer (Polyurethane)

Courtesy James Newell

- Trade names: Esthane, Spandex Pellanthane
- Low cost, chemical and moisture resistant, outdoor stability
- Used for foam cushions, active wear, scooter wheels, coatings

first Spandex applications replaced rubber in women's bra panels and straps, but the market has grown to a wide range of active wear. Spandex derives its properties from the polymeric fibers that have a blend of rigid regions and flexible regions, as shown in Figure 5-17. Spandex is sold under the brand names Lycra® (Invista) and Dorlasten® (Bayer).

Natural rubber has been produced for centuries by native South Americans from the sap of the rubber tree (*Hevea braziliensis*). After Europeans arrived in the sixteenth century, rubber found commercial uses in shoes, waterproof coating, and a variety of other products. However, the natural rubbers tended to soften and creep at high temperatures. Charles Goodyear added sulfur to the rubber at an elevated temperature and found that chemical cross-linkages formed between adjacent polymer chains in a process that became known as *vulcanization*. This strengthened the material substantially without significantly damaging its elastic properties. The polymeric chain in natural rubber consists primarily of isoprene, shown in Figure 5-18.

The vulcanization can occur because the polymer backbone still contains double bonds. Sulfur atoms form primary bonds between the carbon atoms in adjacent polymer chains. A maximum of two cross-linkages per structural unit are possible, although stearic issues make the actual number significantly lower.

The location of the methyl group on the polymer backbone plays a significant role in the properties of the polymer. When most of the structural units have the methyl group on the same side (the cis-formation), the material is very elastic and can be softened readily. The cis-formation is dominant in natural rubber. However, when the methyl groups tend to occur on opposite sides (the trans-formation), the polymer becomes much harder and less elastic. The trans-polyisoprene, sometimes called *balata*, is used on soles of shoes and was used in the center of baseballs during World War II, but was quickly replaced when home run production dropped.

Polybutadiene is a synthetic elastomer developed during the buildup to World War II, when U.S. supplies of natural rubber were threatened. The chemical structure of polybutadiene is similar to polyisoprene, but the methyl group is replaced by a hydrogen atom, as shown in Figure 5-19. The missing methyl group reduces the tensile strength, stiffness, and resistance to solvents. However, polybutadiene is much cheaper than polyisoprene and adheres well

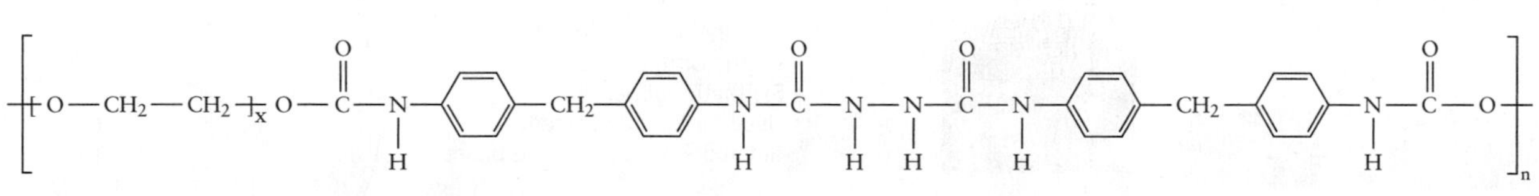

| *Vulcanization* |
Process by which chemical cross-linkages can form between adjacent polymer chains, strengthening the material without significantly damaging its elastic properties.

FIGURE 5-17 Structural Unit of Spandex (x can be any length but 40 to 50 is common)

• Other names: natural rubber, isoprene rubber
• Used for additives in tires and shoe soles

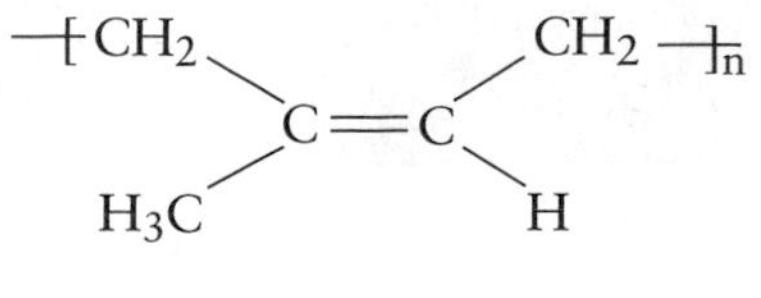

FIGURE 5-18 Structure of Polyisoprene

to metals. Nearly 70% of the polybutadiene produced is used in tires. Most of the rest is used as tougheners in blends with other polymers.

Polychloroprene is a synthetic rubber developed by Wallace Carothers at DuPont. The structure of polychlorprene is similar to polybutadiene, but one of the hyrodgen atoms on the double-bonded carbons is replaced by a chlorine atom, as shown in Figure 5-20. The large chlorine atom improves the resistance to oils, strength, stiffness, flame resistance, and thermal stability of the polymer. Bonding between polymer chains is accomplished by a vulcanization process in which metal oxides form oxygen connectors between adjacent chains. DuPont initially named the polymer "Duprene" but today sells polychloroprene under the trade name Neoprene®.

Polychloropene is used in wet suits, boots, latex gloves, molded foams, cables, and as raw materials for adhesives.

The various polymers described in this section exhibit a wide variety of properties; however, an even larger range can be achieved by blending polymers. Blended polymers are suitable materials for numerous applications, but selecting the best polymer for a specific task is not trivial. Table 5-1 lists key physical and mechanical properties for the polymers discussed in this section.

Although the properties in Table 5-1 are familiar (tensile strength, melting point, etc.), the *glass transition temperature* (T_g) is a physical property that is unique to polymers and some ceramics. At low temperatures, polymers behave much like glassy solids, where the only movement comes from molecular vibrations. As temperature increases, polymers with low molecular weights tend to melt like other materials, but high-molecular-weight polymers develop a cooperative movement prior to melting. Between this temperature and the melting point, polymer chains flex and uncoil, behaving more like a rubbery material than either a glass or liquid.

In many ways, the polymer can be viewed as a spring connected to two posts. Below the glass transition temperature, the only movement is at the molecular level and is unnoticeable. Above the glass transition temperature, someone stretches the spring, which now flexes and vibrates. The explosion that destroyed the space shuttle *Challenger* resulted from a rubber o-ring that became glassy when the air temperature dropped below the glass transition

| *Glass Transition Temperature* |
Second-order thermodynamic transition in which the onset of large-scale chain mobility occurs in polymers. Below T_g, the polymer is glasslike and brittle. Above T_g, the polymer becomes rubbery and flexible.

FIGURE 5-19   Structural Unit of Polybutadiene

Courtesy James Newell

- Manufacturers: Goodyear, Bayer, Bridgestone
- Synthetic rubber
- Used for tires and treads for automobiles, trucks, and buses

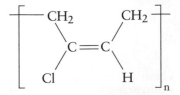

FIGURE 5-20   Structural Unit of Polychloroprene

Courtesy James Newell

- Trade names: Neoprene, Duprene
- Similar to polybutadiene but addition of chlorine improves stiffness and stability
- Used for latex gloves, boots, wet suits, adhesives

TABLE 5-1   Physical Properties of Commercial Polymers

Polymer	Category	T_m (°C)	T_g (°C)	Density (g/cm³)	Tensile Strength (MPa)	Tensile Modulus (MPa)
PMMA	Acrylic	265–285	105	1.19	55–76	2400–3400
PBO	Aramid	n/a	n/a	1.54	5800	180,000
Nylon 6,6	Nylon	255	n/a	1.14	90	3400
PET	Polyester	245–265	80	1.4	172	4275
LDPE	Polyolefin	110	n/a	0.92	10.3	166
HDPE	Polyolefin	130–137	n/a	0.94–0.97	19–30	800–1400
PP	Polyolefin	164	−20	0.903	35.5	1380
Viscose	Rayon	n/a	n/a	1.5	28–47	9.7
PVDC	HVTP	160	−4	1.17	34.5	517
PS	HVTP	180	74–110	1.04	46	2890
PVC	HVTP	175	81	1.39	55	2800
Polyisoprene	Elastomer	40	−63	0.970	17–25	1.3
Polybutadiene	Elastomer	n/a	−110 to −95	1.01	18–30	1.3
Polychloroprene	Elastomer	n/a	−45	1.32	25–38	0.52

Values from Polymer Handbook, 4th edition, *J. Brandrup, E. Immergut, and E. Grulke, eds. (Hoboken, NJ: John Wiley & Sons, 1999).*

temperature of the polymer. Instead of forming a tight seal, the now-glassy air ring allowed gases to pass by, and the shuttle exploded. We consider the glass transition temperature more closely when we examine crystallinity in polymers.

| *Cracking* |
Process of breaking large organic hydrocarbons into smaller molecules.

How Are Polymer Chains Formed?

Petroleum is the primary raw material for most polymers. When crude oil is distilled, the valuable petroleum products are removed, leaving behind less valuable, higher-molecular-weight compounds that are heated over a catalyst. As a result, the large hydrocarbons are broken into smaller molecules through a process called *cracking*. These smaller molecules serve as the initial building blocks for the monomers that are converted to polymers. One of the most common monomers, ethylene, is also obtained from natural gas.

Because of the cost of crude oil and the environmental impact of refining it, significant effort is being placed into finding commercially viable alternative sources for polymer precursors. Corn, barley, soybeans, and wood have all been identified as renewable sources of polymer precursors, but large-scale commercial production is currently hindered by high costs and low yields.

Whatever the source of the precursor materials, most polymers are made through one of two reaction schemes: *addition polymerization* and *condensation polymerization*.

| *Addition Polymerization* |
One of the two most common reaction schemes used to create polymers, involving three steps: initiation, propagation, and termination. Also called chain growth polymerization and free-radical polymerization.

| *Condensation Polymerization* |
Formation of a polymer that occurs when two potentially reactive end groups on a polymer react to form a new covalent bond between the polymer chains. This reaction also forms a by-product, which is typically water.

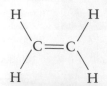

FIGURE 5-21 Vinyl Monomer (Ethylene)

| *Vinyl Monomer* |
Double-bonded organic molecule used to begin addition polymerization.

| *Initiation* |
First step in the process of polymerization, during which a free radical is formed.

Addition polymerization, which is also known as *free-radical polymerization* or *chain growth polymerization*, begins with a *vinyl monomer*, like the one shown in Figure 5-21. The groups that surround the double-bonded carbon can be different, and the identity of the monomer changes accordingly. If all four atoms are hydrogen, the vinyl monomer is ethylene. If one hydrogen atom is replaced by a benzene ring, the monomer is styrene. More than one hydrogen atom can be replaced with other atoms. A vinyl monomer with one chlorine atom would be vinyl chloride; two chlorine atoms would make the monomer vinylidene.

Regardless of the specific vinyl monomer, all addition polymerization reactions occur in three steps:

1. Initiation
2. Propagation
3. Termination

Figure 5-22 illustrates the complete reaction using a styrene monomer. The *initiation* step can be induced by heat, radiation (including visible light), or

Initiation

$$HO-OH + Light \longrightarrow 2HO\cdot$$

Free Radical Reacts with First Monomer

$$HO\cdot + \underset{\underset{H}{|}}{\overset{\overset{H}{|}}{C}} = \underset{\underset{H}{|}}{\overset{\overset{H}{|}}{C}} \longrightarrow HO - \underset{\underset{H}{|}}{\overset{\overset{H}{|}}{C}} - \underset{\underset{H}{|}}{\overset{\overset{H}{|}}{C}}\cdot$$

Propagation (Repeats Until Termination)

$$HO - [\underset{\underset{H}{|}}{\overset{\overset{H}{|}}{C}} - \underset{\underset{H}{|}}{\overset{\overset{H}{|}}{C}}]\cdot_m + \underset{\underset{H}{|}}{\overset{\overset{H}{|}}{C}} = \underset{\underset{H}{|}}{\overset{\overset{H}{|}}{C}} \longrightarrow HO - [\underset{\underset{H}{|}}{\overset{\overset{H}{|}}{C}} - \underset{\underset{H}{|}}{\overset{\overset{H}{|}}{C}}]\cdot_{m+1}$$

Propagation (Repeats Until Termination)

Primary Termination

$$HO - \left[\underset{\underset{H}{|}}{\overset{\overset{H}{|}}{C}} - \overset{\overset{H}{|}}{\underset{\underset{\bigcirc}{|}}{C}} \right]_M \cdot + \cdot OH \longrightarrow HO - \left[\underset{\underset{H}{|}}{\overset{\overset{H}{|}}{C}} - \overset{\overset{H}{|}}{\underset{\underset{\bigcirc}{|}}{C}} \right]_M OH$$

Mutual Termination

$$HO - \left[\underset{\underset{H}{|}}{\overset{\overset{H}{|}}{C}} - \overset{\overset{H}{|}}{\underset{\underset{\bigcirc}{|}}{C}} \right]_M \cdot + \cdot \left[\underset{\underset{H}{|}}{\overset{\overset{H}{|}}{C}} - \overset{\overset{H}{|}}{\underset{\underset{\bigcirc}{|}}{C}} \right]_N OH \longrightarrow HO - \left[\underset{\underset{H}{|}}{\overset{\overset{H}{|}}{C}} - \overset{\overset{H}{|}}{\underset{\underset{\bigcirc}{|}}{C}} \right]_{M+N} OH$$

FIGURE 5-22 Reaction Sequence for the Addition Polymerization of Styrene

the addition of a chemical (generally hydrogen peroxide). The purpose of the initiator is to induce the formation of a highly reactive unpaired electron called a *free radical*. Once formed, the free radical attacks the double bond, forming a new bond and transferring the free radical to the end of the chain. If we consider the chemical initiation of the addition polymerization of styrene using hydrogen peroxide, the H—O—O—H spontaneously decomposes into a pair of O—H· radicals, which are then able to attack the double bond of a styrene molecule.

During the second stage, *propagation*, the newly formed styrene radical can attack the double bond in another styrene molecule, again breaking the double bond, transferring the radical to the end, and adding one more molecule to the growing chain.

The new polystyrene radical can continue attacking the double bonds in styrene monomers, each time growing by one monomer length. The polymer continues to grow as long as it reacts with more monomer.

The final stage, *termination*, occurs when two free radicals react with each other and end the polymerization reaction. Two distinct types of free radicals are present in the system: growing polystyrene chains and unreacted initiator radicals. When a growing polymer radical with M monomer units added reacts with another growing radical with N monomer units, *mutual termination* occurs, resulting in a completed polymer chain with M + N units.

Alternatively, the growing polystyrene radical of length M can combine with a primary free radical from the hydrogen peroxide resulting in *primary termination*. In this case, a finished polymer with M monomer units results.

Three points about addition polymerization must be considered:

1. The free radical may form on either the substituted or the unsubstituted side of the vinyl monomer (except when ethylene is the monomer). Thus, the new monomer may add head to head only, head to tail only, or a blend of the two, depending on the relative stabilities of the free radicals.

2. Although the polymer may have thousands (or millions) of identical monomer units added, there will be only two *end groups*. In this case, both will be —OH groups. The end groups have little significance on the mechanical properties of most polymers but can be used to determine the number of chains formed through titration.

3. Many different chains are reacting at the same time, and whether they react with another vinyl monomer or with a free radical is probabilistic. Therefore, a distribution of chain lengths will form, as shown in Figure 5-23. Controlling that distribution is a major challenge for polymer scientists and engineers.

| *Free Radical* |
Molecule containing a highly reactive unpaired electron.

| *Propagation* |
Second stage of the polymerization process during which the polymer chain begins to grow as monomers are added to the chain.

| *Termination* |
Final step in the polymerization process, which causes the elongation of the polymer chain to come to an end.

| *Mutual Termination* |
One of the two different types of termination in the polymerization process. During this type of termination, the free radicals from two different polymer chains join to end the propagation process.

| *Primary Termination* |
Last step in the polymerization process, which occurs when the free radical of a polymer chain joins with the free radical on an end group.

| *End Groups* |
Two substituents found at both ends of a polymer chain, which have little to no effect on mechanical properties.

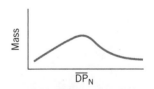

FIGURE 5-23 Distribution of Chain Lengths from Addition Polymerization

| *Step-Growth Polymerization* |
Formation of a polymer that occurs when two potentially reactive end groups on a polymer react to form a new covalent bond between the polymer chains. This reaction also forms a by-product, which is typically water. Also known as condensation polymerization.

5.4 CONDENSATION POLYMERIZATION

The alternative to addition polymerization is condensation polymerization, which sometimes is referred to as *step-growth polymerization*. Unlike addition polymerization, condensation polymerization does not require sequential steps or any initiation. In condensation, potentially reactive functional groups on the ends of molecules react. A new covalent bond forms between the functional groups, and a small molecule (usually water) is formed as a by-product. *Functional groups* are specific arrangements of atoms that cause an organic compound to behave in predictable ways. Most chemical reactions occur at functional groups. Figure 5-24 shows the functional groups frequently found in polymers. The most common condensation polymerizations occur between an acid and an alcohol, such as the reaction between terephthalic acid and ethylene glycol to form PET and water, as shown in Figure 5-25.

The polymer still contains potentially active functional groups on each end that are capable of reacting with another acid, alcohol, or with another growing chain. The polymer continues to grow until finally providing a polymer of the form

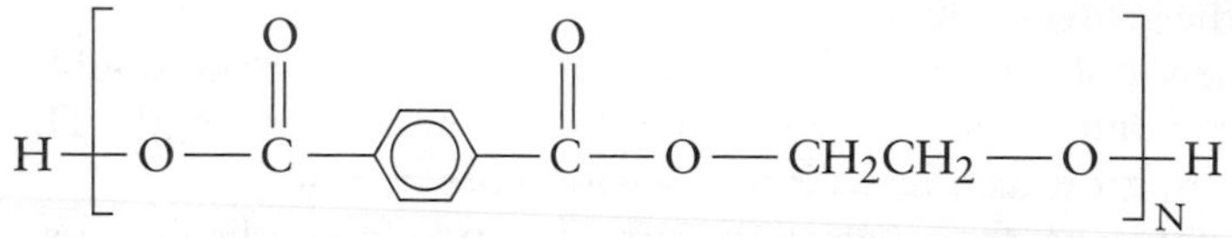

along with $2(n - 1)$ water molecules. PET is called a *homopolymer* because it has a single repeat unit. The functional group from the glycol will react only with an acid, while the functional group from the acid will react only with an alcohol. This ensures that the monomer always will add A-B-A-B-A-B. Theoretically, condensation polymerization could continue until all of the available monomer has reacted to form a single giant polymer chain. In reality, the polymerization reaction reaches some equilibrium, and, as the chain grows larger, stearic hindrance inhibits further growth. Again, a distribution of chain lengths will result from the polymerization process. In some cases the reaction can be terminated by adding a material with only one functional group. This termination is called *quenching* the reaction. In addition to acids and alcohols, acids and amines also experience condensation polymerizations. Nylon 6,6 is formed by the condensation reaction of adipic acid and hexamethylene diamine, for example.

Name	Structure	Name	Structure
Acid		Epoxy	
Alcohol	—C—O—H	Ester	
Aldehyde		Ether	—C—O—C—
Amide		Isocyanate	—N=C=O
Amine		Ketone	
Aromatic			

FIGURE 5-24 Functional Groups Found in Polymers

164   Chapter 5 | Polymers

Terephthalic Acid   Ethylene Glycol

Yields

Polyethylene Terephthalate (PET) and Water

5.5 IMPORTANCE OF MOLECULAR WEIGHT DISTRIBUTIONS

Because both types of polymerization reactions result in multiple chains of different lengths, each chain will have a very different molecular weight. Any molecular weight used to represent a polymer sample will have to represent an average of the wide range of chain lengths. To avoid confusion, *relative molecular mass* (**RMM**) is defined as

$$RMM = \frac{m}{1.0001},$$ (5.1)

where m is the mass of any given polymer chain and 1.0001 is 1/12 of the mass of a carbon-12 atom. Ultimately, RMM is just the molecular-weight definition used for traditional nonpolymeric materials. Consider a polyethylene chain with $\overline{DP}_n = 10,000$. The structural unit contains two carbon atoms, each with a molecular mass of 12, and four hydrogen atoms, each with a molecular weight of 1. Thus, the relative molecular mass of the chain is given by

$$RMM = 10,000 * [(2 * 12) + (4 * 1)] = 280,000$$

The 280,000 applies only to that single, specific chain. Larger chains in the same batch will have higher RMMs, while shorter chains will have lower RMMs. Tensile strength and other mechanical properties vary with molecular weight. For example, Figure 5-26 shows the relationship between tensile strength and molecular weight for polyethylene.

Suppose RMM was plotted against mass for a given sample, as shown in Figure 5-27. M_i represents the molecular mass of a given fraction (i). The weight of the ith fraction (W_i) is given by the equation

$$W_i = n_i M_i,$$ (5.2)

where n_i represents the number of molecules in the ith fraction. The total weight (W) of the polymer chains is defined as

$$W = \sum_{i=1}^{n} W_i.$$ (5.3)

| **Relative Molecular Mass (RMM)** |
Term used to represent the average molecular weight of a sample containing a wide range of polymer chain lengths. This term is used to avoid confusion between the number average molecular weight and the weight average molecular weight.

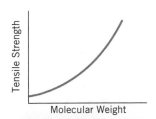

Tensile Strength

Molecular Weight

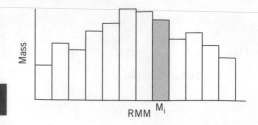

FIGURE 5-27   Mass versus RMM for a Hypothetical Polymer

A sample containing 10 moles of polymer chains with a relative molecular mass of 500 would have a weight of 5000 g:

$$W_i = 10 \text{ mol} * 500 \text{ g/mol} = 5000 \text{ g.}$$

The ability to distinguish between the number of molecules in a given fraction and the mass of molecules in a given fraction provides the opportunity to define different average molecular weights for the polymer. The **number average molecular weight** ($\overline{M}_n$) is the simplest and most direct averaging. $\overline{M}_n$ is obtained by dividing the mass of the specimen by the total number of moles present:

$$\overline{M}_n = \frac{\sum n_i M_i}{\sum n_i} = \frac{\sum W_i}{\sum n_i} = \frac{W}{\sum n_i} \tag{5.4}$$

Using this average, all molecules contribute equally to the averaging. Properties that depend on the total number of molecules, regardless of their size, correlate well with number average molecular weight.

The **weight average molecular weight** ($\overline{M}_w$) normalizes based on the weight of the individual fractions rather than just the number of molecules. As a result, large chains have a greater impact than small chains. The weight average molecular weight is defined as

$$\overline{M}_W = \frac{\sum W_i M_i}{\sum W_i} = \frac{\sum W_i M_i}{W} = \frac{\sum n_i M_i^2}{\sum n_i M_i}. \tag{5.5}$$

$\overline{M}_W$ is a more appropriate average when large molecules dominate behavior, such as viscosity and toughness. The M_i^2 term in the numerator of Equation 5.5 makes large molecules very important. The difference between the two averages will become clearer in Example 5-1.

Both number average and weight average molecular weights ignore the insignificant effect of the two end groups per chain. Two points about these two molecular weights should be noted:

1. $\overline{M}_W$ is always greater than $\overline{M}_n$ for real polymers.
2. The ratio of $\overline{M}_w/\overline{M}_n$ provides information about the breadth of the distribution of molecular weights.

In general, polymer chains with higher molecular weights will be stronger, more creep resistant, and tougher. The physical properties of polymers depend on constitution, configuration, and conformation of the chains, which are the topics of the next sections.

Example 5-1

Consider a blend made by adding 1 gram of C_5H_{12} to 1 gram of a large paraffin wax, $C_{100}H_{202}$. Determine the number average and weight average molecular weights for the blend.

SOLUTION

The RMM for C_5H_{12} is $(5 * 12) + (12 * 1) = 72$ g/mol. Since 1 gram of C_5H_{12} is provided, we can determine the number of moles of C_5H_{12} by dividing by the RMM.

$$1 \text{ g } C_5H_{12}/72 \text{ g/mol} = 1.39 \times 10^{-2} \text{ mol of } C_5H_{12}.$$

Similarly, the RMM for $C_{100}H_{202}$ is

$$(100 * 12) + (202 * 1) = 1402 \text{ g/mol}$$

and

$$1 \text{ g } C_{100}H_{202}/1402 \text{ g/mol} = 7.13 \times 10^{-4} \text{ mol of } C_{100}H_{202}.$$

From Equation 5.4,

$$\overline{M}_n = \frac{W}{\sum n_i} = \frac{1 \text{ g} + 1 \text{ g}}{(1.39 \times 10^{-2} \text{ mol}) + (7.13 \times 10^{-4} \text{ mol})}$$

$$= 143 \text{ g/mol}$$

From Equation 5.5,

$$\overline{M}_w = \frac{\sum W_i M_i}{\sum W_i} = \frac{(1 \text{ g})(72 \text{ g/mol}) + (1 \text{ g})(1402 \text{ g/mol})}{1 \text{ g} + 1 \text{ g}}$$

$$= 737 \text{ g/mol}$$

Because many more moles of the smaller hydrocarbon were present, the number average molecular weight is closer to the RMM of the C_5H_{12}. However, the mass of the $C_{100}H_{202}$ was much greater, so the mass average molecular weight wound up being five times higher than the number average.

What Influences the Properties of Polymers?

5.6 CONSTITUTION

Constitution includes all of the issues related to bonding, including primary and secondary bonding, branching, formation of networks, and end groups. *Primary bonding* forms the polymer backbone, including side groups, and is always covalent. All bonds must be saturated, so all main-chain carbons must have four bonds; nitrogen, three bonds; and oxygen, two bonds. *Functionality* refers to the number of different bonds a molecule has formed. All members of the backbone must have a functionality of at least 2. *Secondary bonding* refers to the bonding between adjacent

| *Constitution* |
All issues related to bonding in polymers including primary and secondary bonding, branching, formation of networks, and end groups.

| *Primary Bonding* |
Covalent bonding of the polymer backbone and side groups.

| *Functionality* |
Number of bonds a molecule has formed.

| *Secondary Bonding* |
Highly distance-dependent bonding between adjacent polymer chains; usually includes hydrogen bonding, dipoles, and Van der Waals forces.

polymer chains. In most cases secondary bonding results from a combination of three sources:

1. Van Der Waals forces
2. Dipolar attraction
3. Hydrogen bonding

As Table 5-2 shows, all secondary bonding is highly dependent on distance. The chains must be close together for secondary bonding effects to matter. On average, hydrogen bonding (if present) tends to be stronger than Van Der Waals forces.

Even though primary bonds are substantially stronger than secondary bonds, secondary bonding plays a key role in the mechanical properties of polymers. Although individual secondary bonds are weak, secondary bonding is an additive function of chain overlap. To move the chain, all of the secondary bonds must be broken. If each structural unit gets 8.4 kJ/mol of secondary bonding from Van Der Waals forces and 100 structural units are in close enough proximity to neighboring chains to experience secondary bonding, then 840 kJ/mol of energy is required to overcome the secondary bonds. The additive nature of secondary bonds explains why high-molecular-weight polymer chains tend to be stronger than low-molecular-weight chains of the same material. Longer chains have more opportunity to form entanglements and develop the spatial regularity needed for secondary bonding, as illustrated in Figure 5-28.

Branching, as explained in the discussion of polyolefins, involves another polymer chain taking the place of a side group along the main chain. The atom at which the two chains connect is called the *branch point*, as shown in Figure 5-29. Branches are different from side groups because they are constitutionally identical to the main chain. There is no way to distinguish the main chain from the branch, although the shorter chain often is designated the branch. The presence of branches reduces the likelihood of forming entanglements and developing significant secondary bonding. As a result, highly branched polymers are less strong and less tough, but branching does make the polymer easier to melt, more susceptible to solvents, and more degradable.

TABLE 5-2	Bond Energies and Lengths for Common Polymeric Bonds		
Bond Type	*Bond Classification*	*Bond Length (nm)*	*Bond Energy (kJ/mol)*
C—C	Covalent—Primary	0.154	347
C—H	Covalent—Primary	0.110	414
C—N	Covalent—Primary	0.147	305
C—O	Covalent—Primary	0.146	360
C—Cl	Covalent—Primary	0.177	339
Hydrogen Bonding	Secondary	0.24–0.32	12.5–29
VDW and Dipoles	Secondary	0.3–0.5	8.4 for VDW only to 42 for highly polar

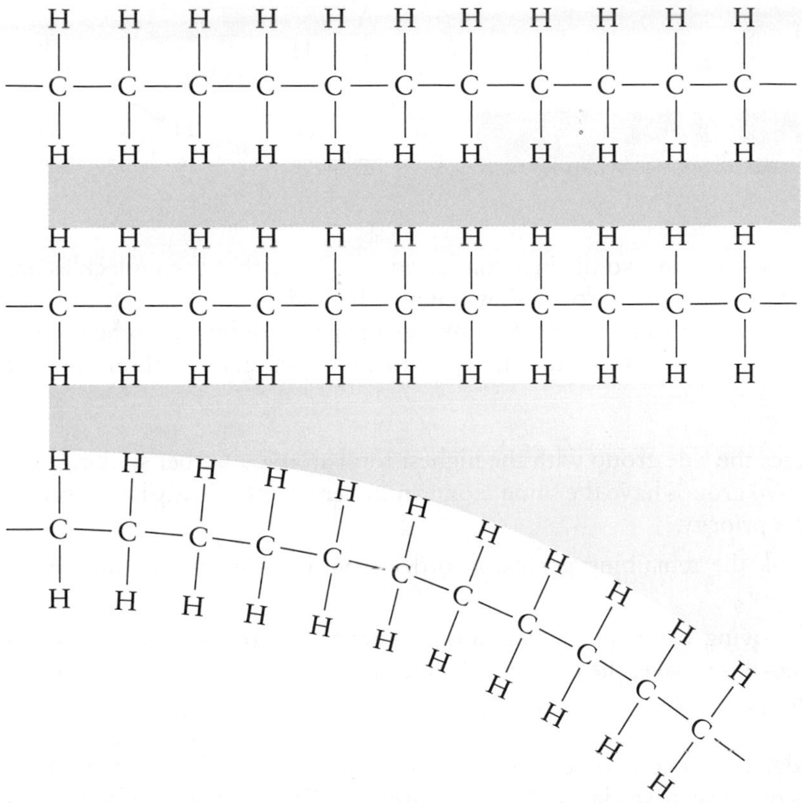

FIGURE 5-28   Van der Waals Interactions in Long Chains

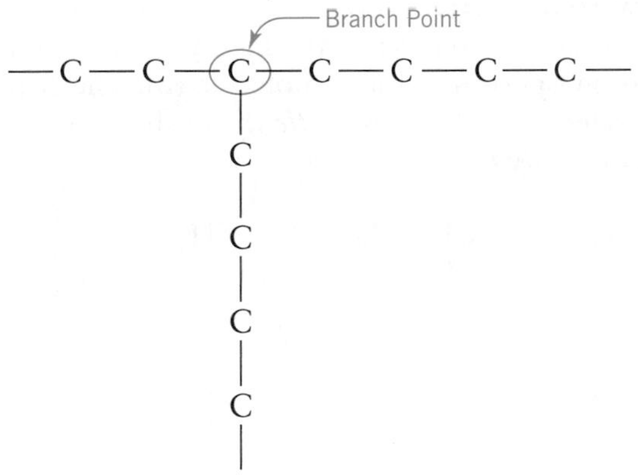

FIGURE 5-29   Branch Point of a Polymer

5.7   CONFIGURATION

Configuration is the spatial arrangement of substituents around the main chain carbon atom that can be altered only by the breaking of bonds. A carbon atom is capable of multiple configurations if and only if it is asymmetric (i.e., it has four different substituents). Any carbon atom with double bonds or repeated substituents (e.g., two hydrogen atoms) is not asymmetric and cannot have multiple configurations. Figure 5-30 shows an

| *Configuration* |
Spatial arrangement of substituents around the main chain carbon atom that can be altered only by the breaking of bonds.

FIGURE 5-30 Asymmetric Carbon Atom in Two Distinct Configurations

S-Configuration   R-Configuration

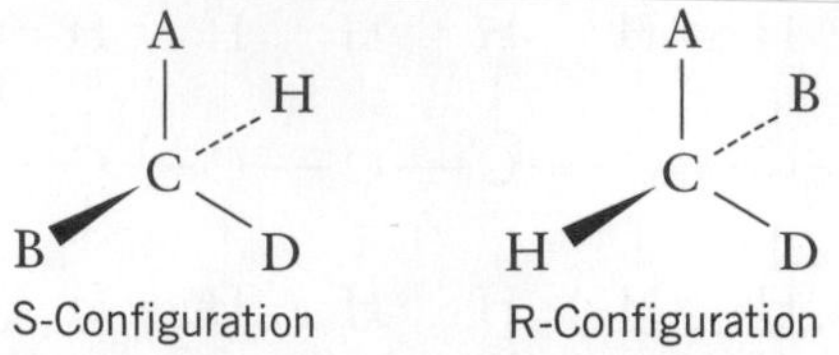

asymmetric carbon in two distinct configurations. Note that the molecules are mirror images and that no rotation will make them identical.

Rules exist to distinguish between mirror images in polymers to help identify which molecule is present. Starting with the main chain carbon atom at the center,

1. Place the side group with the highest total atomic number sticking out. If two groups have the same atomic number, the physically larger group gets priority.

2. Rank the remaining groups in order from highest atomic number to lowest.

3. If moving from the highest atomic number to the lowest requires a clockwise move, the carbon is labeled (R). If the move is counterclockwise, the carbon is labeled (S).

Consider the next two segments from polypropylene chains. They may appear identical at first glance, but they are not. The relative configurations of the asymmetric carbons are different, and there is no way that the second segment can rotate to match the first perfectly. The relative configuration of adjacent asymmetric carbons is called *tacticity*.

If we look at a flat projection of the two segments, the methyl group is on the same side of molecule (a) and on opposite sides of molecule (b). The two carbon atoms in molecule (a) are referred to as an *isotactic dyad*; the two carbons in molecule (b) are a *syndiotactic dyad*.

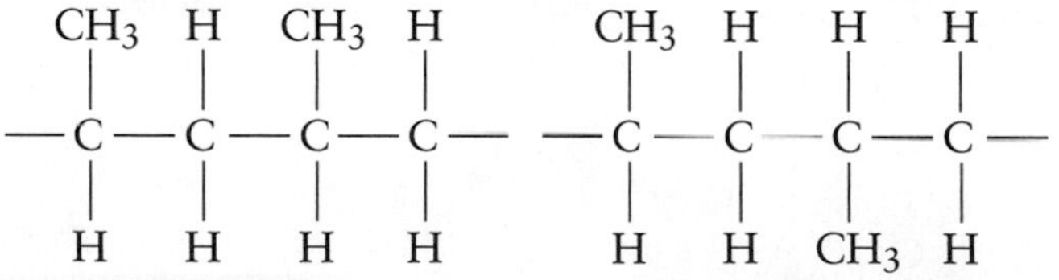

| **Tacticity** |
Relative configuration of adjacent asymmetric carbons.

| **Isotactic Dyad** |
Configuration of a substituent in a polymer, in which the substituent is located on the same side of the polymer chain in all repeating units.

| **Syndiotactic Dyad** |
Configuration of a polymer in which the substituent is located on opposite sides of the molecule in each repeating unit.

| **Atactic** |
Term used to describe a polymer that contains significant numbers of both syndiotactic and isotactic dyads.

(a) Isotactic Dyad   (b) Syndiotactic Dyad

A polymer can have all syndiotactic dyads, all isotactic dyads, or a blend of each. If both syndiotactic dyads and isotactic dyads are present in significant numbers, the polymer is said to be *atactic*. Some chemical reactions involving polymers are affected by the different electron clouds associated with syndiotactic and isotactic configurations, but the property affected most directly by tacticity is crystallinity.

Unlike metals or ceramics, polymers do not form Bravais lattices. However, it is energetically favorable for some polymers to form regions of two-dimensional or three-dimensional order. To form crystalline regions, polymers need a regular structure. Both isotactic and syndiotactic alignments are regular, but an atactic chain does not have spatial regularity. As a result, both isotactic

Example 5-2

Determine if there are any asymmetric carbons in the polymers shown. If so, determine whether they are (R) or (S) carbons. (*Note:* DP_{n1} and DP_{n2} represent the rest of the polymer chain in that direction.)

(a)
$$DP_{n_1} - \underset{\underset{H}{|}}{\overset{\overset{H}{|}}{C}} - DP_{n_2}$$

(b)
$$DP_{n_1} - \overset{\overset{O}{\|}}{C} - DP_{n_2}$$

(c)
$$+\overset{\overset{O}{\|}}{C} - \underset{}{\bigcirc} - \overset{\overset{O}{\|}}{C}OCH_2CH_2O+$$

(d)
$$DP_{n_1} - \underset{\underset{H}{|}}{\overset{\overset{CH_3}{|}}{C}} - DP_{n_2} \qquad DP_{n_1} > DP_{n_2}$$

(e)
$$DP_{n_1} - \underset{\underset{H}{|}}{\overset{\overset{CH_3}{|}}{C}} - DP_{n_2} \qquad DP_{n_1} = DP_{n_2}$$

SOLUTION

a. No, there are no asymmetric carbons. The two hydrogen atoms are identical so the carbon has only three unique substitutents. It is not asymmetric.
b. No, there are no asymmetric carbons. The double bond leaves only three substituents.
c. No, there are no asymmetric carbons.
d. With four different substituent groups, the carbon is asymmetric. DP_{n1} is the largest substituent. $DP_{n2} > CH_3 > H$, so we would have to travel counterclockwise to go in order. Thus, the carbon atom is (S).
e. If $DP_{n1} = DP_{n2}$, the carbon has two identical substituents and is not asymmetrical.

and syndiotactic polypropylene chains will crystallize, but atactic chains will not. Bulky side groups restrict the formation of crystallites in polymers. The aromatic ring in polystyrene makes it difficult to crystallize. However, hydrogen bonding between chains helps fix the polymers in place and promotes crystallinity. At best, polymers are semicrystalline.

Crystallinity in polymers is more complicated than in metals. In many ways, crystalline regions are like defects in an amorphous continuum, as shown in Figure 5-31. Parts of chains form these ordered regions, which would be like pieces of fruit suspended in Jell-O®. These small regions of order are effectively crystallites in polymer.

Unlike the clearly defined crystals in metals and ceramics, polymers have no large-scale lattices. Instead, there are small regions of order that may be only a few chains wide. With no clear lattice structure, we need to ask: How crystalline does it need to be to count? The answer is that it depends on how the crystallinity is measured, because different analytical techniques will provide quite different estimates for polymer crystallinity.

As discussed in Chapter 2, X-ray diffraction is an exceptional technique for studying crystal systems. However, around 10 layer planes are needed to generate the positive reinforcement needed to obtain coherent diffraction, which implies that any crystallites that are less than 10 layer planes thick will be ignored entirely by X-ray diffraction.

An alternative method for estimating crystallinity is to apply a simple mixing rule based on densities. The crystallinity (α) could be estimated as a ratio of the differences between the sample and an amorphous polymer (a) to that of a perfect crystal (c) to the amorphous polymer:

$$\alpha = \frac{\rho - \rho_a}{\rho_c - \rho_a}.\qquad(5.6)$$

The value of ρ_a is obtained from quenching studies, while the value of ρ_c comes from X-ray diffraction on a pure single crystal. This technique will count any order, even two small chain segments that happen to be close to each other, as crystalline. Thus the relative density method tends to overestimate crystallinity.

A third approach involves calorimetry. The energy required to heat the sample to its melting point is recorded and compared to that required to melt a pure crystal sample. The crystallinity (α) is estimated as:

$$\alpha = \frac{\Delta H_m}{\Delta H_{m,c}},\qquad(5.7)$$

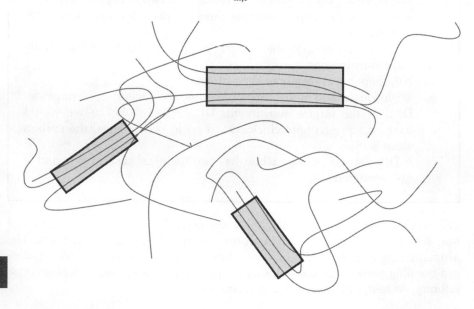

FIGURE 5-31 Schematic of Polymer Crystallinity

where ΔH_m is the enthalpy of melting of the sample and $\Delta H_{m,c}$ is the enthalpy of melting of a pure crystal sample. This technique would work, but more of the sample crystallizes during the process while its temperature rises between the glass transition temperature and the melting point. The testing technique itself alters the sample and, as a result, will overestimate crystallinity.

The variation in estimation between the techniques can be substantial. The estimate of crystallinity in a drawn polyethylene sample that tends to form small crystallites might be as low as 2% from X-ray diffraction and as high as 20% from relative densities. There is no single answer to how crystalline is crystalline in polymers.

5.8 | CONFORMATION

*C*onformation refers to the spatial geometry of the main chain and substituents that can be changed by rotation and flexural motion. Conformation is a function of molecular energetics. Attractive forces (bond energies) are favorable and will make molecules want to be closer together. Repulsive forces are unfavorable and will make molecules tend to stay apart. Five total forces combine to influence conformation:

$$E_{Tot} = E_B + E_{ee} + E_{nn} + E_{kin} + E_{ne}, \tag{5.8}$$

where E_{Tot} is the total energy, E_B is bond energy (attractive), E_{ee} is the energy from the interaction of electron clouds (repulsive), E_{nn} is the energy from the interaction between nuclei (repulsive), E_{kin} is a repulsive kinetic energy term associated with bonding, and E_{ne} is the energy from the interaction between the nucleus and electron cloud (attractive).

Because the electron clouds of adjacent molecules come in much closer contact than the nuclei, E_{ee} is much greater than E_{nn}. Although the bond energy (E_B) is often quite high, it is independent of spatial geometry, so conformation is really controlled by three terms: E_{ee}, E_{kin}, and E_{ne}. If the repulsive forces are greater than the attractive forces, the most stable geometry will have substituents as far apart as possible. Repulsive forces are almost always greater than attractive forces, except when the presence of heteroatoms in the substitutent groups allows for hydrogen bonding or other significant interaction between adjacent groups.

Let us consider an ethane (C_2H_6) molecule looking down the carbon chain with the six hydrogen atoms free to rotate. The sum of the attractive forces for ethane is 19.7 kcal/mol while the repulsive forces total 22.4 kcal/mol. Because the repulsive forces are stronger than the attractive forces, the most stable conformation for ethane would have all of the hydrogen atoms separated by 60 degrees, as shown in Figure 5-32.

Consider the rotation of the three hydrogen atoms attached to the first carbon atom. At 0° rotation, the hydrogen atoms on the adjacent carbon atoms are completely aligned. Because the repulsive forces are stronger than the attractive ones, this alignment is at a higher energy state and is less favorable. As the hydrogen atoms begin to rotate, there is less interaction, and the energies become more favorable until they are as far apart as possible at 60° offset. Beyond 60°, the electron clouds around the hydrogen begin to interact again. Because all of the substituents are the same, the energies cycle symmetrically through the entire 360° rotation, as shown in Figure 5-33.

| Conformation |
Refers to the spatial geometry of the main chain and substituents that can be changed by rotation and flexural motion.

FIGURE 5-32 Ethane Molecule in the Most Stable (Trans) Conformation

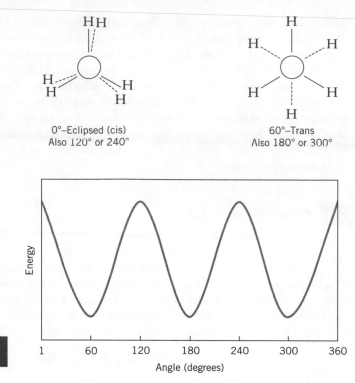

0°–Eclipsed (cis)
Also 120° or 240°

60°–Trans
Also 180° or 300°

Energy

1   60   120   180   240   300   360
Angle (degrees)

The issue of conformation becomes more important when the substituents are different. Instead of ethane, consider a molecule of butane (C_4H_{10}). Again, the repulsive forces are greater than the attractive forces, but this time each of the two central carbon atoms has a methyl group attached, as shown in Table 5-3.

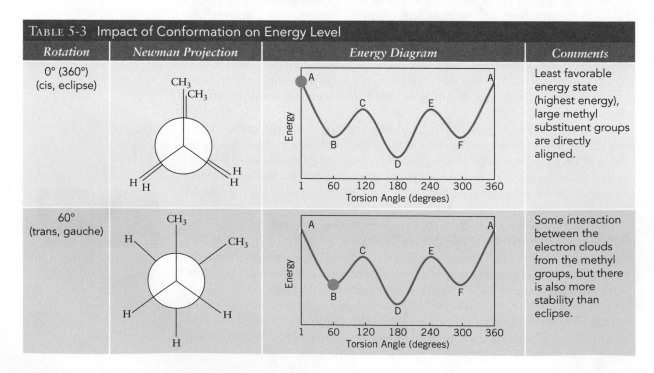

TABLE 5-3   Impact of Conformation on Energy Level

Rotation	Newman Projection	Energy Diagram	Comments
0° (360°) (cis, eclipse)	CH₃ CH₃ H H H H	A B C D E F	Least favorable energy state (highest energy), large methyl substituent groups are directly aligned.
60° (trans, gauche)	CH₃ H CH₃ H H H	A B C D E F	Some interaction between the electron clouds from the methyl groups, but there is also more stability than eclipse.

TABLE 5-3 *(continued)*

Rotation	Newman Projection	Energy Diagram	Comments
120° (cis, staggered)			Electron cloud from the —CH₃ group begins to interact with the electron cloud of the substituent hydrogen atoms.
180° (trans, anti)			Optimal energy state, methyl groups are as far apart as possible.
240° (cis, staggered)			Same as 120°.
300° (trans, gauche)			Same as 60°.

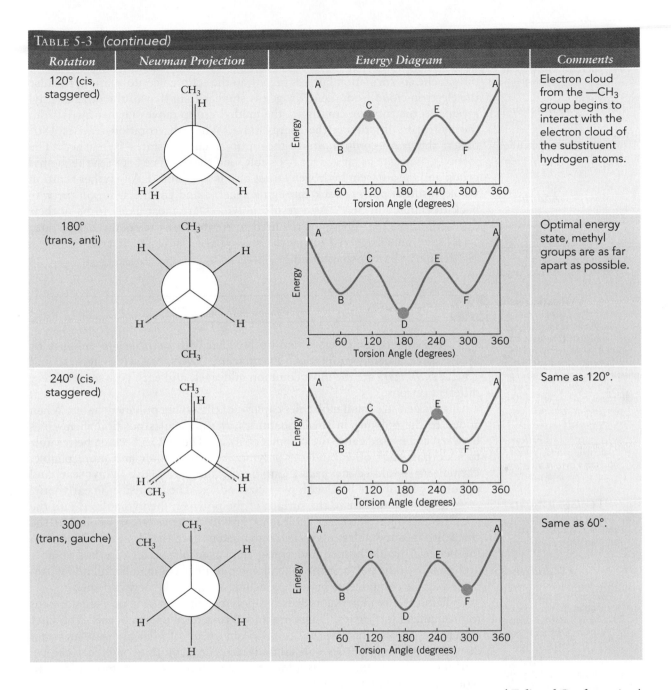

The least favorable energy state for this molecule occurs in the *eclipsed conformation* or cis-conformation when the large methyl substituent groups are directly aligned. The closeness of the large groups results in substantial repulsion. The energy state improves as the methyl group begins to rotate away until it reaches 60° offset. At this point, there is still some interaction between the electron clouds from the methyl groups, but there is also more stability. The 60° offset between the largest groups is called the *gauche conformation*. As the rotation continues, the electron cloud from the methyl group begins to interact with the electron cloud of the substituent hydrogen atoms, resulting in

| Staggered Conformation |
Arrangement of the largest
substituents where the
substituents are offset
by 120 degrees.

| Trans-Conformation |
Conformation in which the
largest substituents are
offset by 180 degrees. This
conformation is typically
the most favorable one.

| Additives |
Molecules added to a polymer
to enhance or alter specific
properties or molecules
added to concrete for
purposes other than altering
a specific property.

| Plasticizers |
Additives that cause swelling,
which allows the polymer
chains to slide past one
another more easily, making
the polymer softer and more
pliable. Also used to decrease
the viscosity of cement paste
to make it easier to flow the
concrete into its final form.

| Fillers |
Additives whose primary
purpose is to reduce the
cost of the final product.

| Coloring Agents |
Pigments or dyes that change
the way light is absorbed
or reflected by a polymer.

| Stabilizers |
Additives that improve a
polymer's resistance to
variables that can cause
bonds to rupture, such
as heat and light.

| Dyes |
Additives dissolved directly
into the polymer, causing the
polymer to change color.

| Pigments |
Coloring agents that do not
dissolve into the polymer.

higher energies and less stability. When the rotation reaches 120°, the methyl group is in its closest contact with the hydrogen, and the molecule is said to be in a *staggered conformation*. The energy state is significantly higher than in the gauche conformation but less high than the eclipsed conformation, because the electron cloud from the hydrogen is smaller than that of the other methyl group. As the rotation continues, the methyl group moves farther away from the hydrogen, improving the energy state. When the rotation reaches 180° offset, the *trans-conformation*—the optimal energy state—is reached. The methyl groups are as far apart as possible, and the distance between the methyl group and the adjacent hydrogen atoms also is maximized. Any futher rotation would result in passing back through staggered and gauche again on the way back to eclipsed at 360°. As a result, polymers are most likely to be in their trans-conformation. Energy is required to rotate over the energy barrier into a less favorable staggered or gauche state. Larger substituent groups have a higher energy barrier to rotation.

5.9 ADDITIVES

Commercial polymers often are blended with *additives* to enhance or alter specific properties. *Plasticizers, fillers, coloring agents*, and *stabilizers* are the most common additives, and each type serves a very different purpose.

Plasticizers are small molecules capable of dissolving polymer chains. When added to the polymer in small quantities, the molecules position themselves between chains and cause the polymer to swell. The chains become better able to slide past each other, and the polymer becomes softer and more pliable. Plasticizers should be inexpensive, nontoxic, and nonvolatile. Vinyl seats and upholstery in cars are PVC with plasticizer added. The plasticizer in early vinyl seats tended to migrate to the surface of the polymer, then volatilized into the air. The escaping plasticizer molecules provided the new car smell, but the smell also meant that less plasticizer remained to keep the PVC soft. Over time, the material would harden and crack. Improved plasticizers are less volatile and remain in the polymer for much longer, but car companies still add extra, more volatile components because consumers expect the new car smell.

Fillers are any material added to a polymer that provide no enhancement to mechanical properties. Fillers are added to reduce product costs. The filler material costs less than the polymer, so any blend of filler and polymer that does not affect properties substantially will be a more economical choice. Carbon black, for example, is added to the elastomers in automobile tires to reduce costs and as a coloring agent.

Coloring agents change the way light is absorbed or reflected by the polymer. *Dyes* are dissolved directly into the polymer and are often organic molecules. Dyed polymers may be clear or translucent. By contrast, *pigments* do not dissolve into the polymer and make the product opaque. Carbon black is the most common pigment used in industry.

Stabilizers are materials added to polymers to improve their resistance to heat and light. Heat or ultraviolet light cause some bonds to rupture. As a result, the polymer loses strength, becomes more brittle, and often discolors. Susceptibility to degradation varies with the polymer. Polyethylene experiences decay quickly, but PMMA remains impervious to ultraviolet light for years. Stabilizers slow the degradation by serving as free-radical acceptors that protect the integrity of the polymer chain. Without the addition of stabilizers, PVC molecules would tend to lose chlorine atoms to form hydrochloric acid (HCl) and leave a free radical behind. In the presence of oxygen, the unstable chain forms ketones, and the mechanical properties of the polymer change substantially. Metallic salts commonly are used as stabilizers.

How Are Polymers Processed into Commercial Products?

5.10 POLYMER PROCESSING

Thermoplastic polymers are produced in enormous quantities as small pellets. Purchasers melt the pellets and convert them into fibers, films, or shaped parts. The principle for all of these operations focuses around the use of an *extruder* to melt the pellets and force them into a shaping device. Most people are familiar with extruders even if the do not recognize the name. Pasta makers, glue guns, meat grinders, and most Play-Doh® accessories are simple extruders. The components of the extrusion apparatus used in polymers include a *hopper* that stores a quantity of the pellets and feeds them directly into a chamber (called the *barrel*) with a heated, turning screw. The heated screw melts the pellets into liquid polymer, then pushes the melt forward. Unwanted volatiles including water and solvents are vented out. A schematic of an extruder is shown in Figure 5-34.

Extrusion is a continuous process. As long as the hopper has pellets and the motor continues to turn the screw, the process can continue. Extrusion produces large volumes of polymeric materials. The motor controls the drive speed, which can be changed as processing needs change. Thermal bands that surround the barrel provide the heat.

The purpose of the extruder is to provide a molten polymer that is pumped into its desired shape and allowed to cool. In the simplest applications, the molten polymer is pushed through a series of mesh screens that filter out any unmelted particles, dirt, or other solid contaminants. The filters, called a *screen pack*, eventually become clogged and must be be replaced. The molten polymer then enters a shaping tool called a *die* that is mounted on the end of the extruder. The die usually is made of stainless steel and tapers inward to form the polymer into a simple shape, such as a rod, pipe, or tube. Once it emerges from the die, the shaped polymer is cooled with water to help it solidify into its final form. Often the polymer is passed through a series of sizing plates during the cooling process to ensure that the part assumes the proper shape.

Coated wires are made in the same fashion, except the die is offset so that a bare wire can be fed through it as well. The molten polymer coats the wire in the die. The exit hole to the die has a diameter slightly larger than the wire, which controls the thickness of the coating, as shown in Figure 5-35.

| *Extruder* |
Device used in the processing of polymers that melts polymer pellets and feeds them continuously through a shaping device.

| *Hopper* |
Part of the extrusion apparatus that holds a large quantity of polymer pellets as they are fed into the barrel.

| *Barrel* |
Piece of the extrusion apparatus that contains a heated screw which is used to melt the polymer and force the polymer forward into the next chamber.

| *Screen Pack* |
Piece of the extrusion apparatus that is used as a filter to separate unmelted particles, dirt, and other solid contaminants from the molten polymer.

| *Die* |
Part of the polymer-processing apparatus through which the polymer is pushed, causing the polymer to form a simple shape, such as a rod or tube.

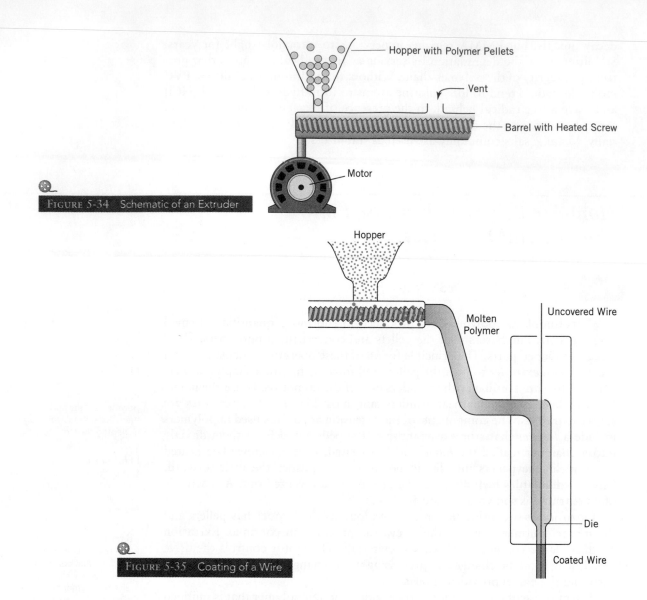

Hopper with Polymer Pellets

Vent

Barrel with Heated Screw

Motor

FIGURE 5-34 Schematic of an Extruder

Hopper

Molten Polymer

Uncovered Wire

Die

Coated Wire

FIGURE 5-35 Coating of a Wire

| *Spinneret* |
Circular, stationary block with small holes through which molten polymer can flow to take the shape of a fiber.

| *Tow* |
Large spool that is used to wind solidified polymer fibers after they have been pushed through the spinneret.

| *Melt Spinning* |
Process of pushing polymers through a spinneret and winding the solidified fibers onto a tow, which imposes a shear stress on the fibers upstream as they emerge from the spinneret.

There are many variations in extruder design, including some that use a pair of screws operating together to melt and pump the polymer. However, all polymeric extrusion processes operate on the general principle just described.

When the goal of the extrusion is to produce thin fibers, the die is replaced by a *spinneret*. The word *spinneret* is really a misnomer because the part remains stationary during the fiber spinning process. A spinneret is a circular die plate with a series of small holes (usually 0.005 inches or less in diameter) in it, as shown in Figure 5-36.

When the molten extruded polymer is pushed to the top of the spinneret, gravity draws the material through the holes in small strings. As the polymer passes through the air below the spinneret, it cools and solidifies. A revolving waywinder winds the solidified fibers into a large removable spool called a *tow*, as shown in Figure 5-37. The waywinder also imposes a shear stress on the solidifying fibers upstream as they emerge from the spinneret. This shear helps reduce the fiber diameter and improve the strength of the fiber. The overall process is called *melt-spinning*. Although the holes in most spinnerets are

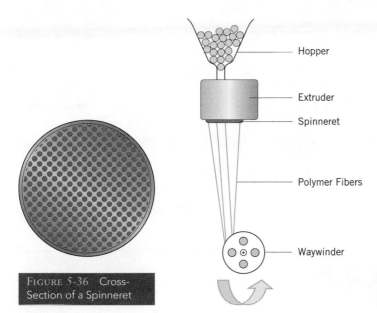

FIGURE 5-36 Cross-Section of a Spinneret

Hopper

Extruder

Spinneret

Polymer Fibers

Waywinder

FIGURE 5-37 Melt-Spinning Process

round, many other shapes can be made including y-shaped holes to form the trilobal fibers used in carpeting.

The thin films used in coatings, garbage bags, and food wraps also begin with an extrusion process. When the molten polymer emerges from the extruder, it passes into a tubular die. Air is forced in from beneath the die and flows upward through the polymer, forming a bubble. The bubble expands and cools, spanning a larger area with a progressively thinner layer of polymer surrounding the air. The stretching also causes the molecules to move closer and develop a more crystalline orientation. The point at which the bubble develops this more-oriented formation is called the *freeze line* and can be identified by a loss of clarity. When the bubble reaches the collapsing frame, the material is pinched together to maintain the bubble, then wound onto a roll very similar to fiber spinning, as shown in Figure 5-38.

The final polymer processing process that we will examine is *injection molding*. Unlike the extrusion systems that have been discussed, injection molding can make parts with complex shapes. The injection-molding process begins the same way as extrusion processes, with pellets in a hopper feeding into a barrel that contains a heated screw. However, the screw in an injection-molding machine has a reciprocating motion that melts the polymer through a blend of heat and mechanical action. The molten polymer collects in a small reservoir at the end of the barrel. When the appropriate weight of molten polymer (called the *shot size*) has accumulated in the reservoir, the end of the barrel opens and the screw pushes all of the liquid forward into a mold. The screw then retracts and the polymer is allowed to cool in the mold. Once cooled, the mold is opened and the part is removed. The design of the mold controls the shape of the product.

The shot size varies from a few ounces to as much 40 pounds. Early injection-molding systems had problems with premature hardening of the polymer and incomplete flows, but modern techniques have controlled these problems. Figure 5-39 shows a schematic on injection molding.

All of the techniques discussed so far have involved the melting of thermoplastic pellets. Thermoset polymers cannot be formed from any of these

| *Freeze Line* |
Term associated with blown-film apparatus, which indicates the point at which the molecules develop a more crystalline orientation around the bubble of air.

| *Injection Molding* |
Type of polymer processing that is similar to extrusion but can be used to develop parts with complex shapes rapidly.

| *Shot Size* |
Specified weight of a polymer that is injected into the mold at the end of the barrel during the injection molding process.

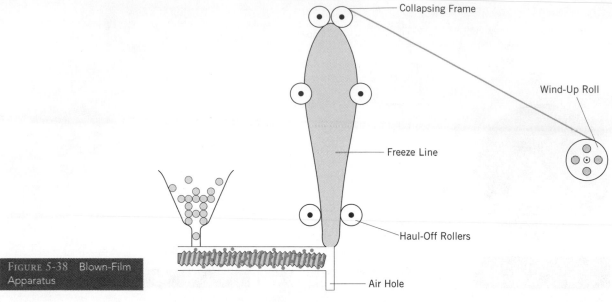

Collapsing Frame

Wind-Up Roll

Freeze Line

Haul-Off Rollers

Air Hole

FIGURE 5-38 Blown-Film Apparatus

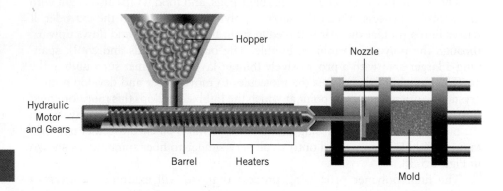

Hopper

Nozzle

Hydraulic
Motor
and Gears

Barrel   Heaters

Mold

FIGURE 5-39 Injection-Molding Apparatus

techniques because the cross-linking between chains prevents them from melting. Instead, the polymerization of thermosets must occur in their final shape. Often the polymerization reactions are carried out in molds. The commercially available "creepy crawlers" for children involve mixing reagents in spider-shape molds and putting them in the oven to form a solid thermoset polymeric creature. Some thermosets will cross-link at room temperature and require only time; others need to be heated.

| *Solution Spinning* |
Process used to make thermoset fibers by performing the polymerization reaction in a solvent as the material flows through a spinneret and into a quenching bath.

Thermoset fibers are produced through a technique called *solution spinning*, illustrated in Figure 5-40. The reagents for the polymerization are mixed in a solvent (often a concentrated acid solution) and passed through a spinneret. The polymerization reaction occurs during the spinning process. The newly formed fibers emerge from the spinneret, pass through an air gap that helps realign the molecules, then enter a quenching bath where the solvent is removed. The fibers emerge from the quenching bath and are wound onto a tow, much like a thermoplastic material. Many high-performance polymers, including PPTA and PBO, are made by a solution-spinning process.

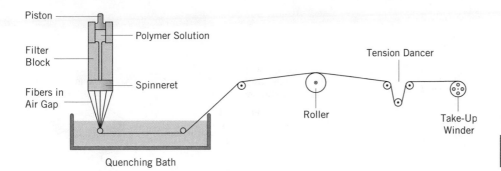

Piston

Filter
Block

Fibers in
Air Gap

Polymer Solution

Spinneret

Quenching Bath

Tension Dancer

Roller

Take-Up
Winder

What Happens to Polymers When They Are Discarded?

After polymers are converted into commercial products and sold to consumers, they are used for a period of time, then discarded. Until the 1990s, almost 90% of polymeric materials in commercial use wound up landfills, with about 10% incinerated. In 1991, 74 billion tons of plastic materials were discarded in the United States with less than 2% recycled.

European laws mandate almost total recycling of most materials, including thermoplastic polymers, and have targeted complete recycling by 2010. The United States lags behind in these efforts, both because of recycling costs that would be passed on to consumers and the difficulties in collecting, transporting, and sorting the waste from so vast a country. Figure 5-41 shows the life cycle of materials used in the manufacture of many polymers.

5.11 RECYCLING OF POLYMERS

Technically, polymer recycling includes both plant recycle and postconsumer waste. Plant recycle involves the regrinding and remelting of scrap polymer that never left the plant in a finished product. Postconsumer waste includes all polymeric materials that were discarded after leaving the plant.

The primary difficulty arises from the diversity of polymeric materials in use. When glass and aluminum cans are recycled, they can be converted into essentially the same products repeatedly, but that rarely happens with polymers. Most commercial polymers include coloring agents, plasticizers, and other additives that must be removed for any practical recycling activity. A second difficulty arises in sorting. A bottle made from PVC looks very much like a bottle made from PET, but chemically they are very different. They cannot be mixed together without significantly altering the properties of the new material.

To facilitate the sorting process, the Society of the Plastics Industry developed an identification code that is placed on most commercial polymer products. Table 5-4 summarizes these codes.

The numbering system does not imply that all of these polymers are recycled. In most communities, PET and HDPE are accepted for recycling. In some cases, LDPE is recycled, but few commercial companies have found it economical to

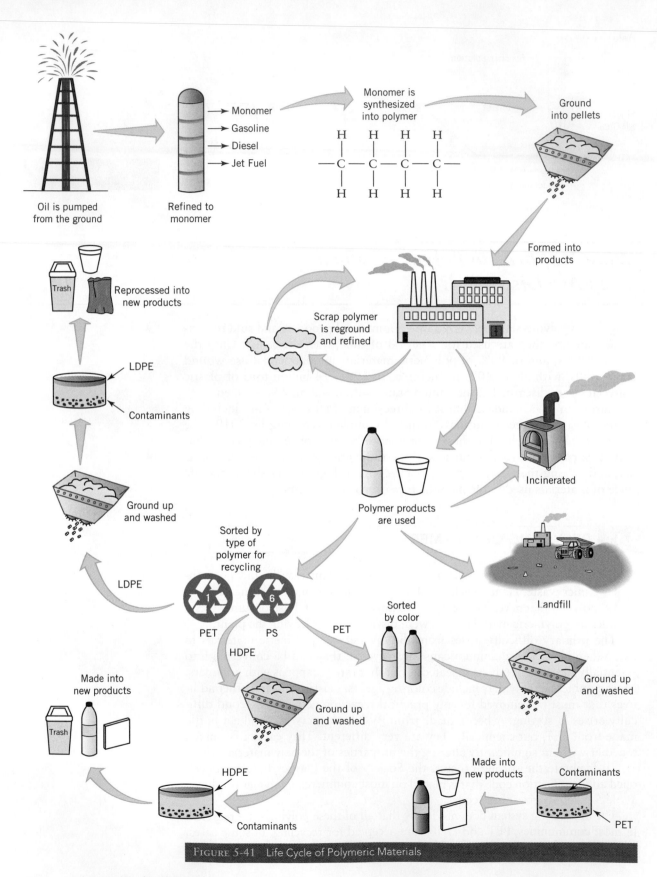

FIGURE 5-41   Life Cycle of Polymeric Materials

TABLE 5-4   Recycling Symbols for Common Plastics

Symbol	Material	Common Uses
1 PET	Polyethylene terephthalate (PET)	Soda bottles Carpet fibers
2 HDPE	High-density polyethylene (HDPE)	Milk bottles Shampoo bottles Plastic bags Hard plastic cups Sports bottles
3 PVC	Polyvinyl chloride (PVC)	Oil bottles Piping Castings
4 LDPE	Low-density polyethylene (LDPE)	Plastic grocery bags Shrink wrap Plastic lumber
5 PP	Polypropylene (PP)	Drinking straws Bottle tops Plastic furniture
6 PS	Polystyrene (PS)	Packaging Beverage cups Meat packaging

recycle other polymers. Most other polymers and polymer blends continue to wind up in landfills.

PET is one of the easier polymers to recycle. Beverage bottles are sorted by color, then ground into pellets and washed. PET is denser than water, so it sinks to the bottom during washing while the HDPE and any residue from the label float to the top. The clean pellets are harvested and find many uses, including carpet fibers, new bottles, and filling for pillows. The limiting factors for PET recycling tend to be the need to sort by hand and the cost of shipping the bulky empty soda bottles to the recycler.

HDPE also must be sorted by hand. Clear HDPE items are ground into small flakes and washed. The polyethylene floats while many contaminants sink. The recycled HDPE is then either dyed or not and reprocessed into a variety of products. Bottles containing materials for human consumption do not use recycled materials. Colored HDPE items are mixed together, ground into flakes, and dyed black.

Many recycling centers do not accept LDPE, but those that do operate much like HDPE recyclers. Trash bags, tubing, and plastic lumber often contain some recycled LDPE.

PVC is difficult to recycle, in part because it is seldom used alone. Generally, commercial PVC has been treated with antioxidants, coloring agents, plasticizers, and additives to make it more resistant to ultraviolet light. PVC also requires more process energy to manufacture than any of the major thermoplastics.

Plastics that are not recycled wind up back in the ecosystem. Many of the polymers themselves are generally benign, but the colorants and plasticizers in them may contain toxins including lead and cadmium. Some studies attribute as much as 28% of all toxic cadmium contamination in municipal waste to discarded plastics. For some polymers, incineration provides an energy-conscious alternative because the heat given off during the incineration can be used to generate steam. Hydrocarbons burn well and (with suitable control and scrubbing) generate only carbon dioxide and water as by-products. However, inorganics and other pollutants accumulate in the incinerator ash. Tests have revealed levels of furan, dioxin, cadmium, and lead in incinerator ash that exceed federal guidelines.

| *Design for Recycling (DFR)* |
Effort to consider the life cycle consequences when designing a material or product.

Significant research is focused on making the recycling of polymers more commercially viable. Several systems of automatically sorting recycled polymers are being developed and tested. There is also a significant thrust to *design for recycling (DFR)*, which essentially means to consider life cycle in product design. Product designers are encouraged to use recycled polymers when practical and to use a single polymer instead of a blend where practical.

In addition to the difficulties in sorting and removing additives, used polymers invariably experience some degradation from exposure to the environment. Heat and ultraviolet light damage the polymer, and while stabilizers slow the progression, they cannot completely stop it. Oxidation also results in broken bonds and lowered mechanical properties. Entropy considerations also result in more tangled and less flexible chains over time. There is no way to regain the original properties of the polymer completely. As a result, recycled polymers provide only a fraction of the total chains in new applications.

Summary of Chapter 5

In this chapter we examined:

- The basic terminology and nomenclature of polymers

- The characteristics and applications of the classes of commercial polymers

- The mechanisms of addition and condensation polymerization

- The types of bonding found in polymers and their impact on properties

- How to determine tacticity in polymers and its impact on crystallinity

- The energy barriers that govern conformation

- How polymers are processed into useful materials through extrusion, blown films, melt spinning, injection molding, and solution spinning

- Issues impacting the recycling and/or disposal of polymeric materials

Key Terms

acrylic *p. 153*
addition polymerization *p. 161*
additives *p. 176*
alternating copolymers *p. 152*
aramid *p. 154*
atactic *p. 170*
barrel *p. 177*
blends *p. 153*
block copolymers *p. 152*
branching *p. 156*
coloring agents *p. 176*
condensation polymerization
  p. 161
configuration *p. 169*
conformation *p. 173*
constitution *p. 167*
copolymer *p. 151*
cracking *p. 161*
degree of polymerization *p. 151*
design for recycling (DFR) *p. 184*
die *p. 177*
dyes *p. 176*
eclipsed conformation *p. 175*
elastomers *p. 158*
end groups *p. 163*
extruder *p. 177*
fillers *p. 176*
free radical *p. 163*
freeze line *p. 179*

functional groups *p. 164*
functionality *p. 167*
gauche conformation *p. 175*
glass transition temperature *p. 160*
graft copolymers *p. 153*
high-volume thermoplastics
  (HVTPs) *p. 157*
homopolymer *p. 164*
hopper *p. 177*
initiation *p. 162*
injection molding *p. 179*
isotactic dyad *p. 170*
melt spinning *p. 178*
monomers *p. 150*
mutual termination *p. 163*
number average molecular
  weight *p. 166*
nylon *p. 154*
oligomers *p. 150*
pigments *p. 176*
plasticizers *p. 176*
polyamides *p. 154*
polyesters *p. 155*
polymers *p. 150*
polymer backbone *p. 151*
polyolefins *p. 155*
polyurethanes *p. 158*
primary bonding *p. 167*
primary termination *p. 163*

propagation *p. 163*
quenching *p. 164*
random copolymers *p. 152*
rayon *p. 157*
relative molecular mass (RMM)
  p. 165
screen pack *p. 177*
secondary bonding *p. 167*
shot size *p. 179*
side groups *p. 151*
solution spinning *p. 180*
spinneret *p. 178*
stabilizers *p. 176*
staggered conformation *p. 176*
step-growth polymerization
  p. 163
structural unit *p. 151*
syndiotactic dyad *p. 170*
tacticity *p. 170*
termination *p. 163*
thermoplastics *p. 151*
thermosets *p. 151*
tow *p. 178*
trans-conformation *p. 176*
vinyl monomer *p. 162*
viscose process *p. 157*
vulcanization *p. 159*
weight average molecular weight
  p. 166

Homework Problems

1. Show two different pathways to form syndiotactic polyvinyl chloride with a degree of polymerization of four.

2. A blend contains 45wt% polystyrene and 55wt% polypropylene. What is the mole fraction of polystyrene in the blend?

3. Show the reactions needed to form Nylon 6,6 with a degree of polymerization of three.

4. Determine the average degree of polymerization for a PMMA with a relative molecular mass of 120,000.

5. Calculate the number average molecular weight and weight average molecular weight for a polymer with the listed mass fractions:

RMM Range	Mass Fraction
0–2500	0.03
2500–5000	0.12
5000–7500	0.08
7500–10,000	0.24
10,000–12,500	0.22
12,500–15,000	0.15
15,000–17,500	0.11
17,500–20,000	0.05

6. Calculate the number average molecular weight and weight average molecular weight for a polymer with the listed mass fractions:

RMM Range	Mass Fraction
0–5000	0.01
5000–10,000	0.10
10,000–15,000	0.12
15,000–20,000	0.14
20,000–25,000	0.18
25,000–30,000	0.25
30,000–35,000	0.13
35,000–40,000	0.07

7. How would the stress-strain curve change for a polymer if it were tensile-tested above its glass transition temperature and below its glass transition temperature?

8. Which of the polymers in Table 5-1 are capable of forming isotactic and syndiotactic dyads?

9. Show the reactions needed to generate a polymer with a degree of polymerization of four from this monomer:

$$H_2C = CH(C_3H_3O_2)$$

10. Rank these polymers in order of their ability to crystallize. Explain the basis for your order: PE, PP, PVC, PVDC.

11. For a given polymer consisting of 1000 chains, the number average molecular weight is 4000 and the quantity $\sum n_i M_i^2 = 1.96 \times 10^{11}$. Determine the weight average molecular weight for the sample.

12. Explain the difference between a copolymer and a polymer blend.

13. For polymethylmethacrylate,
    a. Describe the influence of primary and secondary bonding, including the interactions between molecules in a chain and between adjacent chains.
    b. Are there any tacticity issues? If so, draw and label the relevant structural isomers.
    c. Show the series of reactions that lead to the formation of PMMA.

14. Show which combinations of functional groups in Figure 5-27 are capable of undergoing condensation reactions with each other and what the reaction products would be.

15. Explain the difference between the melting temperature and glass transition temperature of a polymer.

16. A student claims that he can prove that the crystallinity of a specific polymer sample is 27%. What is wrong with this statement?

17. Find 10 items around your residence that are made from polymers. Identify the primary polymer used in each item and draw its structural unit.

18. Explain why aramids tend to be stronger than other polyamides.

19. Discuss the primary barriers to establishing commercial recycling of other thermoplastic polymers.

20. Explain why it is more difficult for the carbon chain to rotate in a polystyrene molecule than it would be in a polyethylene molecule.

21. Why is it more difficult to make long chains from condensation polymerization than from addition polymerization?

22. Why is forming a narrow molecular weight distribution important?

23. How would copolymerization affect crystallinity?

24. For the following applications, suggest a possible polymeric material. Describe its suitability based on mechanical properties, cost, and recycleability:
    a. Flyswatter
    b. Child's picnic table
    c. Padding

25. Provide a qualitative life cycle analysis for plastic grocery bags.

26. Why can't thermoset polymers be injection molded?

27. Explain the difference between injection molding and extrusion through a die.

28. Why is controlling temperature so important in extrusion and injection molding?

29. What is the greatest number of theoretical cross-linkages possible by vulcanizing 100 g of polyisoprene chains?

30. Explain why the terephthalic acid used in PET formation always reacts with the ethylene glycol end of the growing chain and never with the terephthalic acid end.

31. What influence should branching have on recycling polymers?

32. Explain the constitutional, configurational, and conformational issues present in polystyrene. (Begin by defining the terms and then explain how they apply to polystyrene—e.g., what possible configurations can the molecule take, etc.)

6

Ceramics and Carbon Materials

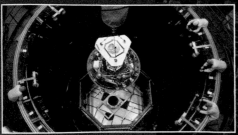

CONTENTS

Learning Objectives

By the end of this chapter, a student should be able to:

- Define a ceramic material and identify physical and mechanical characteristics common to ceramics.

- Calculate the coordination number for a ceramic lattice.

- Identify and describe the primary crystal systems common in ceramics.

- Understand the role of octahedral and tetrahedral interstitial sites in the crystal structure of ceramics.

- Describe the seven basic classes of ceramic materials.

- Explain the role of abrasives in industry.

- Describe the operating principles behind sandpaper and other abrasive material processes.

- Explain the relevance of the silicon phase diagram to glassmaking.

- Explain the float glass process.

- Identify the microstructures and properties associated with glasses.

- Describe the role of the SiO_4^{4-} tetrahedron in glasses.

- Describe the differences between sintering and vitrification.

- Explain the fundamentals of cement chemistry.

- List the primary constituents in Portland cement and define their role in the hydration process.

- Describe the role of cement in industrial design.

- Distinguish between the different classes of Portland cement.

- Define whitewares and explain the firing and glazing processes.

- Discuss the role of refractories in industry.

- Distinguish between the primary classes of refractory bricks.

- Define structural clay products and explain their use in industry.

- Describe the ongoing developments in advanced ceramics.

- Explain why graphite looks like a polymer but behaves like a ceramic.

- Explain how diamonds get their unique mechanical properties.

- Describe how carbon fibers are made and their use in industry.

- Describe carbon nanotubes and their potential uses in industry.

What Are Ceramic Materials?

6.1 CRYSTAL STRUCTURES IN CERAMICS

| *Ceramics* |
Compounds that contain metallic atoms bonded to nonmetallic atoms such as oxygen, carbon, or nitrogen.

When most people hear the word *ceramics*, they tend to think of fired clay statues or dinnerware. In fact, the word *ceramic* is derived from the Greek word *keramos*, meaning pottery. However, ceramics encompass a much larger range of material types and uses.

As discussed in Chapter 1, ceramic materials are compounds that contain metallic atoms bonded to nonmetallic items, most commonly oxygen, nitrogen, or carbon. This broad definition results in an enormous range of material properties. Most ceramics have ionic bonding, but many contain a mixture of ionic and covalent bonds. These bonds provide most ceramic materials with hardness, wear resistance, and chemical stability. Most ceramic materials are immune to corrosion, because essentially they are already corroded.

When the bonding is primarily ionic, the resulting lattice consists of metallic cations and nonmetallic anions. The presence of these ions makes the ceramic lattice far more complicated than that of a pure metal. First, the net charge of the system must balance. If titanium ions (Ti^{4+}) are in a lattice with oxide ions (O^{2-}), then there must be twice as many oxide ions present to balance the charge of the titanium ions.

Cations tend to be smaller than anions, so the ratio of the cation radius (r_c) to the anion radius (r_a) is less than 1. Larger cations, shown in Figure 6-1(b) can contact more anions than smaller cations, as shown in Figure 6-1(a). Each cation would prefer to be in contact with as many anions as possible, while each anion would prefer to contact as many cations as possible. The *coordination number* of the lattice is defined as the number of anions that each cation contacts, and it is controlled by the atomic radii and geometry. Table 6-1 summarizes the minimum ratio of atomic radii necessary to achieve a stable lattice for each coordination number. Table 6-2 summarizes ionic radii for cations and anions.

| *Coordination Number* |
The number of anions in contact with each cation in a ceramic lattice.

These coordination numbers impact the types of crystal structures that ceramics form. These structures are classified roughly by the number of different cations and anions present in each lattice. The simplest structures, containing one cation for every anion, are classified as AX systems, where A represents the cation and X the anion. When the charges of the cation and anion

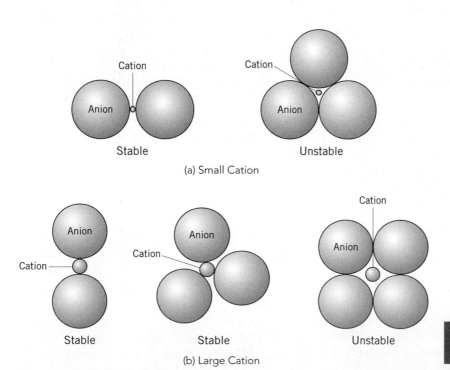

(a) Small Cation

(b) Large Cation

FIGURE 6-1   Illustration of the Influence of Ionic Radii Ratios on Stable Coordination Numbers

TABLE 6-1 Summary of Ionic Radii Ratios for Stable Coordination Numbers

Coordination Number	r_c/r_a Value
2	<.155
3	.155–225
4	.225–414
6	.414–732
8	.732–1.0

TABLE 6-2 Ionic Radii

Atomic Species	Ion	Ionic Radius (nm)	Atomic Species	Ion	Ionic Radius (nm)
Actinium	Ac^{3+}	0.118	Nickel	Ni^{2+}	0.078
Aluminum	Al^{3+}	0.057	Niobium	Nb^{4+}	0.074
Antimony	Sb^{3+}	0.09		Nb^{5+}	0.069
Arsenic	As^{3+}	0.069	Nitrogen	N^{5+}	0.01–0.02
	As^{5+}	~0.04	Osmium	Os^{4+}	0.067
Astatine	At^{7+}	0.062	Oxygen	O^{2-}	0.132
Barium	Ba^{2+}	0.143	Palladium	Pd^{2+}	0.05
Beryllium	Be^{2+}	0.054	Phosphorus	P^{5+}	0.03–0.04
Bismuth	Bi^{3+}	0.12	Platinum	Pt^{2+}	0.052
Boron	B^{3+}	0.02		Pt^{4+}	0.055
Bromine	Br^-	0.196	Polonium	Po^{6+}	0.067
Cadmium	Cd^{2+}	0.103	Potassium	K^+	0.133
Calcium	Ca^{2+}	0.106	Praseodymium	Pr^{3+}	0.116
Carbon	C^{4+}	<0.02		Pr^{4+}	0.1
Cerium	Ce^{3+}	0.118	Promethium	Pm^{3+}	0.106
	Ce^{4+}	0.102	Radium	Ra^+	0.152
Cesium	Cs^+	0.165	Rhenium	Re^{4+}	0.072
Chlorine	Cl^-	0.181	Rhodium	Rh^{3+}	0.068
Chromium	Cr^{3+}	0.064		Rh^{4+}	0.065
	Cr^{6+}	0.03–0.04	Rubidium	Rb^+	0.149
Cobalt	Co^{2+}	0.082	Ruthenium	Ru^{4+}	0.065
	Co^{3+}	0.065	Samarium	Sm^{3+}	0.113
Copper	Cu^+	0.096	Scandium	Sc^{2+}	0.083
	Cu^{2+}	0.072	Selenium	Se^{2-}	0.191
Dysprosium	Dy^{3+}	0.107		Se^{6+}	0.03–0.04
Erbium	Er^{3+}	0.104	Silicon	Si^{4-}	0.198
Europium	Eu^{3+}	0.113		Si^{4+}	0.039
Fluorine	F^-	0.133	Silver	Ag^+	0.113

TABLE 6-2 (continued)

Atomic Species	Ion	Ionic Radius (nm)	Atomic Species	Ion	Ionic Radius (nm)
Francium	Fr^+	0.18	Sodium	Na^+	0.102
Gadolinium	Gd^{3+}	0.111	Strontium	Sr^{2+}	0.127
Gallium	Ga^{3+}	0.062	Sulfur	S^{2-}	0.174
Germanium	Ge^{4+}	0.044		S^{6+}	0.034
Gold	Au^+	0.137	Tantalum	Ta^{5+}	0.068
Hafnium	Hf^{4+}	0.084	Tellurium	Te^{2-}	0.211
Holmium	Ho^{3+}	0.105		Te^{4+}	0.089
Hydrogen	H^-	0.154	Terbium	Tb^{3+}	0.109
In	In^{3+}	0.092		Tb^{4+}	0.089
Iodine	I^-	0.22	Thallium	Tl^+	0.149
	I^{5+}	0.094		Tl^{3+}	0.106
Iridium	Ir^{4+}	0.066	Thorium	Th^{4+}	0.11
Iron	Fe^{2+}	0.087	Thulium	Tm^{3+}	0.104
	Fe^{3+}	0.067	Tin	Sn^{4-}	0.215
Lanthanum	La^{3+}	0.122		Sn^{4+}	0.074
Lead	Pb^{4-}	0.215	Titanium	Ti^{2+}	0.076
	Pb^{2+}	0.132		Ti^{3+}	0.069
	Pb^{4+}	0.084		Ti^{4+}	0.064
Lithium	Li^1	0.078	Tungsten	W^{4+}	0.068
Lutetium	Lu^{3+}	0.099		W^{6+}	0.065
Magnesium	Mg^{2+}	0.078	Uranium	U^{4+}	0.105
Manganese	Mn^{2+}	0.091	Vanadium	V^{3+}	0.065
	Mn^{3+}	0.07		V^{4+}	0.061
	Mn^{4+}	0.052		V^{5+}	~0.04
Mercury	Hg^{2+}	0.112	Ytterbium	Yb^{3+}	0.1
Molybdenum	Mo^{4+}	0.068	Yttrium	Y^{3+}	0.106
	Mo^{5+}	0.065	Zinc	Zn^{2+}	0.083
Neodymium	Nd^{3+}	0.115	Zirconium	Zr^{4+}	0.087

are different, the lattice is classified as an A_mX_p system, where m $\neq$ p. Finally, when a ceramic has two cation species with different charges, the system is classified as an $A_mB_nX_p$ system.

The interstitial voids between atoms in the lattice become very important when considering ceramic structures because atoms or ions can occupy these openings. Two distinct types of interstitial spaces exist in the FCC (face-centered cubic) lattices common to many ceramics. Four interstitial vacancies occur at the points at which the three atoms in a plane come in contact with a fourth atom in an adjacent plane, as shown in Figure 6-2. These are called *tetrahedral positions* because lines drawn from the center of the atoms form

| *Tetrahedral Positions* |
Four interstitial sites present in lattices that form a tetrahedron when lines are drawn from the center of the sites.

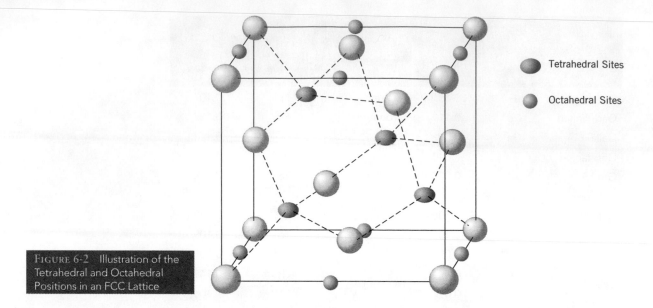

Tetrahedral Sites

Octahedral Sites

FIGURE 6-2 Illustration of the Tetrahedral and Octahedral Positions in an FCC Lattice

| *Octahedral Positions* |
Intersitial spaces between six atoms in a lattice that form an octahedron.

| *Sodium Chloride Structure* |
Lattice system in which anions fill the face and corner sites in a FCC lattice while an equal number of cations occupy the interstitial regions.

a tetrahedron. Similarly, there are eight *octahedral positions* where six atoms are equally far apart from the interstitial space. In general, the large anions fill the close-packed planes while the smaller anions occupy the interstitial sites.

One of the most common AX-systems, the *sodium chloride structure* shown in Figure 6-3, is highly ionic and involves equal numbers of cations and anions. The Cl^- anions occupy the corner and face sites in the FCC lattice structure, and Na^+ cations occupy the interstitial regions. Because each cation contacts six anions, the coordination number of the lattice is 6. Iron oxide (FeO) and manganese sulfide (MnS) are other common ceramics with the sodium chloride structure.

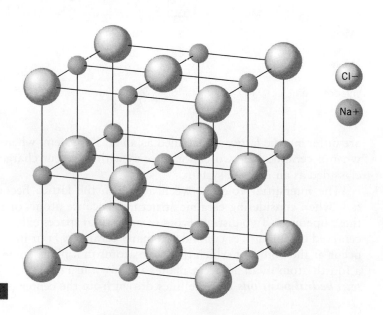

Cl−

Na+

FIGURE 6-3 Sodium Chloride Structure

Example 6-1

Verify that a coordination number of 6 is valid for a ceramic consisting of Na^+ and Cl^- ions.

SOLUTION

From Table 6-2, the ionic radius of a sodium ion is 0.102 nm and that of a chloride ion is 0.181 nm. The cation to anion ratio becomes:

$$\frac{r_c}{r_a} = \frac{r_{Na^+}}{r_{Cl^-}} = \frac{0.102 \text{ nm}}{0.181 \text{ nm}} = 0.56$$

Table 6-1 shows that the range of ionic radius ratios for a coordination number of 6 spans from 0.414 to 0.732. Because 0.56 falls within this range, 6 is the correct coordination number.

The *zinc-blend structure* is another AX system, but this time each unit cell contains the equivalent of four anions and four cations, as shown in Figure 6-4. Two equivalent forms of the structure exist. In one, all corner and face positions in the FCC lattice are occupied by anions while all of the cations occupy interstitial tetrahedral positions. An exactly equivalent structure results if the cations and anions are reversed. As a result, each anion is bonded to four cations, resulting in a coordination number of 4. Along with zinc species, silicon carbide lattice forms a zinc-blend structure.

For systems where the cation and anion have different charges, an A_mX_p system results like the one found in the *calcium fluoride (CaF₂) structure*, shown in Figure 6-5. The CaF₂ structure consists of an FCC lattice with calcium ions occupying the lattice sites and F^- ions in the tetrahedral sites. The unique feature of the calcium fluoride structure is that the octahedral sites in

| *Zinc-Blend Structure* |
FCC-lattice system with equal number of cations and ions in which each anion is bonded to four identical cations.

| *Calcium Fluoride Structure* |
FCC-lattice system with cations occupying lattice sites, anions in the tetrahedral sites, and octahedral sites left vacant.

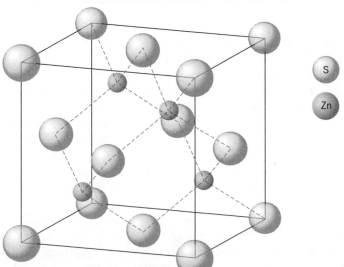

S

Zn

FIGURE 6-4 Zinc-Blend Structure

6.1 | Crystal Structures in Ceramics   195

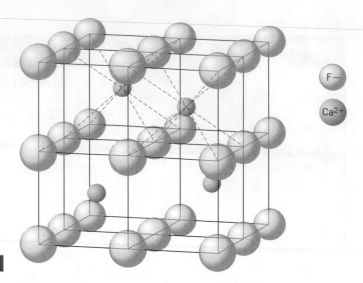

FIGURE 6-5 Calcium Fluoride Structure

F−

Ca2+

///

Example 6-2

Calculate the coordination number for uranium oxide.

SOLUTION

From Table 6-2, the ionic radius of a U^{4+} ion is 0.105 nm and that of an oxide ion is 0.132 nm. The cation to anion ratio becomes:

$$\frac{r_c}{r_a} = \frac{r_{U^{4+}}}{r_{O^{2-}}} = \frac{0.105 \text{ nm}}{0.132 \text{ nm}} = 0.795$$

According to Table 6-1, the ionic radius ratio of 0.795 corresponds to a coordination number of 8.

the lattice remain unoccupied. In the case of uranium oxide (UO_2), these unoccupied spaces may be used to trap fissionable by-products.

The *perovskite structure*, shown in Figure 6-6, provides an example of an $A_m B_n X_p$ system. In this structure, two anions with potentially different charges exist in the same ceramic as a single anion species. Perovskites, such as barium titanate ($BaTiO_3$) and calcium titanate ($CaTiO_3$), generally contain a metallic cation capable of holding a larger charge like titanium. In these systems, the nontitanium cation occupies the corner sites of an FCC lattice with the oxide anions occupying the faces sites of the lattice. The Ti^{4+} cation fills the octahedral interstitial site. Materials with the perovskite structure display piezoelectric properties, which are discussed in Chapter 8.

The *spinel structure*, shown in Figure 6-7, is another example of an $A_m B_n X_p$ system. In this case, magnesium aluminate ($MgAl_2O_4$) forms an FCC lattice with the oxide anions in the corner and face sites. The Mg^{2+} cations fill the four tetrahedral sites, while the eight octahedral sites are occupied by Al^{3+}

| *Perovskite Structure* |
FCC-lattice system with two species of cation, one occupying corner sites and the other occupying the octahedral sites. An anion species occupies the face sites in the lattice.

| *Spinel Structure* |
FCC-lattice system with two species of cation, one occupying the tetrahedral sites and the other occupying the octahedral sites. An anion species occupies the face and corner sites in the lattice.

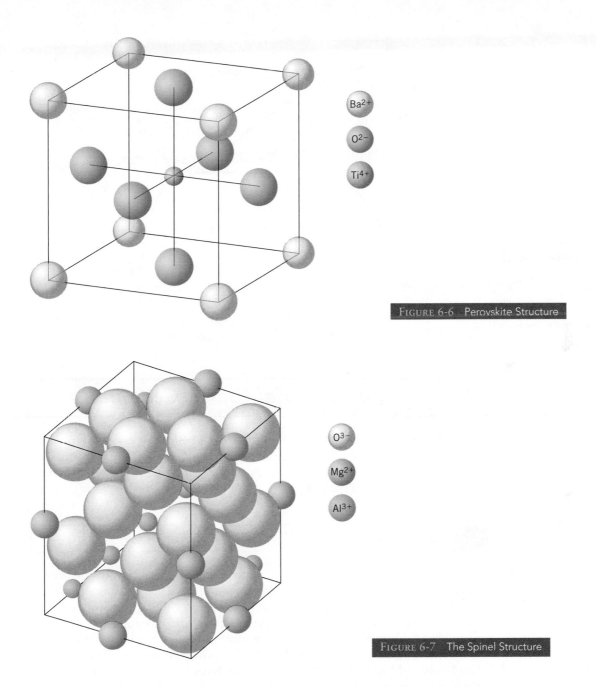

Ba^{2+}

O^{2-}

Ti^{4+}

FIGURE 6-6   Perovskite Structure

O^{3-}

Mg^{2+}

Al^{3+}

FIGURE 6-7   The Spinel Structure

cations. The spinel structure has special significance because it has ferromagnetic properties.

Ceramic systems are represented in phase diagrams just like metals. In general, ceramic phase diagrams involve blends of oxides and/or silicates. Figure 6-8 shows a typical ceramic phase diagram for magnesium oxide–alumina. The same principles of interpreting the phase diagram that applied to metals apply to ceramic diagrams as well, including the calculation of phase composition from tie lines and the identification of solidus lines, liquidus lines, solvus lines, and all invariant points.

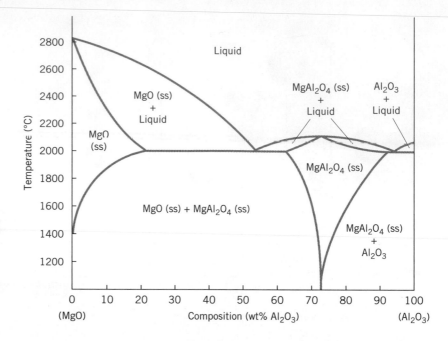

FIGURE 6-8 MgO–Al$_2$O$_3$ Phase Diagram

From B. Hallstedt, "Thermodynamic Assessment of the System MgO–Al$_2$O$_3$," Journal of the American Ceramic Society, Vol. 75, No. 6 (1992): 1502. Reprinted with permission of the American Ceramic Society

What Are the Industrial Uses of Ceramics?

The American Ceramic Society classifies ceramic materials into seven distinct groups: *abrasives, glasses, cements, refractories, structural clay products, whitewares,* and *advanced ceramics.* Each category possesses distinct properties and challenges.

6.2 ABRASIVES

| Abrasives |
Materials used to wear away other materials.

Abrasives are used to wear away other materials through processes including grinding, sanding, lapping, and pressure blasting. The abrasive particle acts like a cutting instrument, ripping away part of the softer material. For this reason, hardness is the most important characteristic of an abrasive material, but the abrasive must also resist fracture, so toughness is a concern. Many ceramics make ideal abrasives because of their blend of toughness and hardness, coupled with enough resistance to heating to endure the elevated temperatures generated by friction during abrasive processes.

The most familiar commercial abrasive product is *sandpaper,* in which a resin is used to affix coarse particles to a backing. Sandpaper dates back to thirteenth-century China, where bits of seashells were bonded with natural gum to bits of parchment or shark skin. The first U.S. patent for sandpaper dates back to 1834 and involved securing pieces of glass to paper. Modern sandpapers have evolved beyond glass, and some no longer use paper as a backing. Instead, Mylar, cotton, and rayon backings are commonly used.

Four distinct classes of abrasives are used in sandpaper: alumina, garnet, silicon carbide, and "ceramic." Alumina (Al$_2$O$_3$) is by far the most common sandpaper abrasive and possesses many beneficial properties. Alumina is dense

(3.97 g/cm^3), tough (9 on the Moh hardness scale), insoluble in water, and has a high melting point (2288K). Alumina forms a *corundum structure*, shown in Figure 6-9, in which the Al^{3+} ions occupy the octahedral sites in the hexagonal-close-packed (HCP) lattice. Because the system must remain electrically neutral, there can be only two aluminum ions for every three oxygen ions. As a result, the aluminum ions can occupy only two-thirds of the octahedral sides, and the lattice becomes slightly distorted. The close packing of the aluminum and oxygen ions leads to the exceptional hardness of the material. Many gemstones are alumina with specific impurities, including rubies (Cr^{3+} provides the red color) and sapphire (Fe^{2+} and Ti^{4+} provide the blue color).

| *Corundum Structure* |
HCP-lattice system with cations occupying two-thirds of the octahedral sites.

Alumina particles can be formed in a variety of shapes and sizes and are *friable*, so that they form fragments with new sharp edges when they break under stress from heat and pressure. The abrasive aluminum oxide particles on the sandpaper are self-renewing.

| *Friable* |
Forming sharp edges when broken under stress.

Garnet [A$_3$B$_2$(SiO$_4$)$_3$] actually refers to a variety of materials classified as aluminum garnets, chromium garnets, or iron garnets, depending on the identity of the B-metal in its structure. Garnet is a softer material than alumina and is not friable, making it less suitable for the removal of large amounts of material. The sharp garnet edges smooth out during the sanding process, producing a much smoother surface than aluminum oxide and sealing off the grain of the wood (if wood is the material being sanded). As a result, garnet sandpaper typically is used on soft woods and for final finishing. More than 100,000 tons of garnet sandpaper are produced each year and are recognizable by their distinctive orange color.

Silicon carbide (SiC) is harder than alumina and is friable, although most wood is not hard enough to fracture the surface of the particles. The combination of hardness and resistance to high temperatures makes SiC ideal for abrading metals, plaster, and fiberglass.

Ceramic sandpapers are the hardest and most expensive. Ceramic particles are deposited on the backing using a *sol-gel* process in which metal salts are forced into a colloidal suspension, called a sol, which is then placed into a mold. Through a series of drying and heat treatments, the sol converts into a wet, solid

| *Sol-gel* |
Material formed by forming a colloidal suspension of metal salts then drying them in a mold into a wet, solid gel.

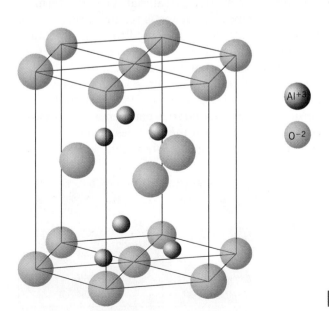

Al^{+3}

O^{-2}

FIGURE 6-9 Corundum Structure of Aluminum Oxide

Grit	Grade	Uses
40–60	Coarse	Stripping or large-scale sanding
80–120	Medium	Smoothing
150–180	Fine	Final wood sanding
220–240	Very Fine	Sanding between staining coats
280–320	Extra Fine	Mark removal
360–600	Super Fine	Removal of surface scratches

TABLE 6-3   Sandpaper Grit Sizes

gel. The gel is dried further, forming an extremely hard ceramic. Sol-gel sandpapers are used to shape and level wood, and are typically found on belt sanders.

Whatever the abrasive agent, sandpapers are classified by grit, which is the number of abrasive particles per square inch. Table 6-3 summarizes the grit sizes, grades, and uses for commercial sandpaper.

Sandpaper is classified as a coated abrasive because the particles are bonded to a flexible backing, but other forms of abrasives are used in commercial processes and products. Lapping abrasives are extremely fine particles that are used for fine polishing. Typically, lapping abrasives are sold as either a dry powder or a paste that is applied with water to a rotating wheel. The item to be polished is moved gently back and fourth against the covered wheel. Diamond dust is the hardest lapping abrasive. Bonded abrasives fuse hard ceramic particles with a binding material that enables them to be pressed into useful shapes, including grinding wheels, cylinders, blocks, and cones. The primary technique used to form bonded abrasives is *powder pressing*.

| *Powder Pressing* |
The formation of a solid material by the compacting of fine particles under pressure.

Unlike polymers and metals, most ceramics cannot be shaped by melting the material and pouring it into a mold. Instead, ceramic particles are mixed with a binding agent (usually water), which lubricates the particles during compaction, and are forced into a shape by pressure. In the simplest form, *uniaxial powder pressing,* the ceramic powder is compressed in a metal die by pressure in a single direction. As a result, the compacted material takes the form of the die and platens. When a more complicated shape is required, the ceramic material is placed in a rubber chamber, and water is used to apply hydraulic pressure uniformly in all directions. This technique is called *isotactic powder pressing* and is significantly more expensive than uniaxial pressing.

$

Although pressure has fused the particles together, a sintering process is needed to improve the strength of the pressed object. During sintering, the pressed object is heated, causing the grain boundaries between adjacent particles to coalesce. The interstitial region between the particles begins to shrink and gradually develop into small, spherical pores, as shown in Figure 6-10.

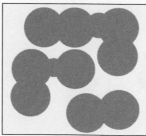

FIGURE 6-10
Microstructural Changes
Resulting from Sintering

*G*lasses differ from most solid materials because they do not exhibit a crystalline structure. Although glassy materials can be made from a number of metallic oxides, *silica glass* is by far the most common. Ironically, noncrystalline silica glass is made from a crystalline material, silicon dioxide (SiO_2), that occurs in a variety of allotropic forms depending on temperature and formation path. The phase diagram for SiO_2 is shown in Figure 6-11. The most common crystal structure for SiO_2 is quartz (or α-quartz), a trigonal structure. When quartz is heated to 573°C, its transforms into a hexagonal lattice and becomes β-quartz (or high quartz). As temperature is increased to 867°C, the SiO_2 transforms again, this time into a hexagonal lattice called *tridymite*, which further converts to cubic *cristobalite* at 1713°C.

Although there is nothing particularly unusual about the heating of quartz to form liquid SiO_2, the behavior of the material when cooling is quite different. As the SiO_2 melt cools, the phase diagram indicates that cristobalite should form. However, the extremely high viscosity of the molten SiO_2 prevents the tetrahedral SiO_2 molecules from rearranging into a crystalline structure. As the liquid continues to cool, it forms an amorphous glass instead of the thermodynamically favored cristobalite crystal structure.

Unlike metals, glass does not transform directly from a liquid to a rigid solid. Instead, it undergoes a transition from a highly viscous liquid to a semisolid that becomes rigid only when the glass transition temperature is reached. Just as in polymers, the glass transition temperature represents the point at which large-scale molecular motion becomes possible. When a glass is heated above the glass transition temperature, it develops large-scale motion through a process called *vitrification*.

The basic unit of glass is a SiO_4^{4-} tetrahedron, like the one shown in Figure 6-12. In this arrangement, each silicon atom sits in the center of a tetrahedron with an oxygen atom in each corner. The bonds between the silicon and oxygen atoms contain both ionic and covalent characteristics. A *loose network* forms like the one shown in Figure 6-13 with the tetrahedrons touching corner to corner with no long-range order.

To reduce the viscosity of these loose networks, metal oxides called *network modifiers* often are added to the mixture. Common network modifiers include Na_2O, K_2O, and MgO. The oxygen atoms from these oxides break up the loose network by occupying a tetrahedral site. Other oxides may be added that cannot form their own network but can join an existing loose network of SiO_4^{4-} tetrahedrons. These compounds are called *intermediate oxides* and generally are used to impart special properties to the glass. For example, aluminosilicate glasses can withstand higher temperatures, while the addition of lead oxide increases the index of refraction, making the glass more lustrous and ideal for crystal glassware, windows, or jewelry. Lead-glass wine decanters, like the one shown in Figure 6-14, were popular until the late 1980s, when it was discovered that alcoholic beverages could leach out some lead from the glass.

Most glasses are comprised of a loose network of SiO_4^{4-} tetrahedrons, network modifiers, and intermediate oxides. *Soda–lime glass*, the most common formulation, consists of 72% SiO_2, 14% soda (N_2O), 7.9% lime (CaO), 1.8% alumina (Al_2O_3), 1% lithium oxide (Li_2O), 1% zinc oxide (ZnO), and less than

| *Glasses* |
Inorganic solids that exist in a rigid, but noncrystalline form.

| *Silica Glass* |
Noncrystalline solid formed from the cooling of molten silicon dioxide (SiO_2).

| *Tridymite* |
High-temperature polymorph of silicon dioxide (SiO_2) that exhibits a hexagonal lattice.

| *Cristobalite* |
High-temperature polymorph of silicon dioxide (SiO_2) that exhibits a cubic lattice.

| *Vitrification* |
Heating process in which a glassy solid develops large-scale motion.

| *Loose Network* |
Molecular arrangement in glasses in which there is no long-range order but the adjacent SiO_4^{4-} tetrahedrons share a corner oxygen atom.

| *Network Modifiers* |
Additives used to reduce the viscosity of loose networks in glasses.

| *Intermediate Oxides* |
Additives used to impart special properties to glasses.

| *Soda–Lime Glass* |
Most common glass composition that includes silicon dioxide (72%), soda (14%), and lime (7.9%) as primary constituents.

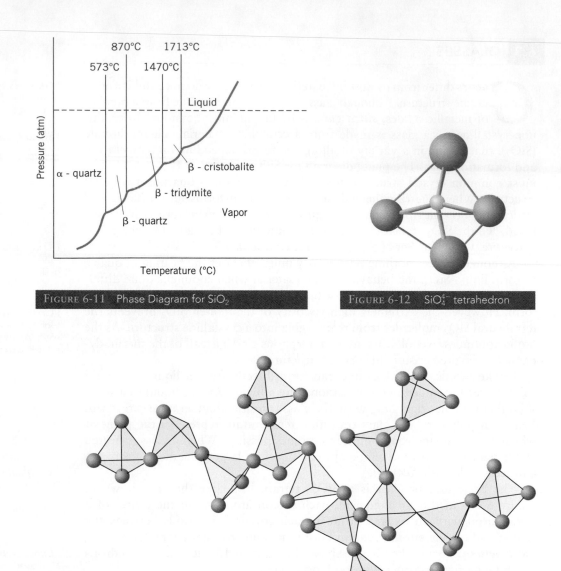

FIGURE 6-11   Phase Diagram for SiO₂

FIGURE 6-12   SiO₄⁴⁻ tetrahedron

FIGURE 6-13   Loose Network of
SiO₄⁴⁻ Tetrahedrons

1% each of barium oxide (BaO), potassium oxide (K_2O), and antimony oxide (Sb_2O_3).

The other reason for including specific additives to glasses is to alter their color. When free of impurities, silica glasses are colorless. By adding specific metallic oxide pigments to the mix, glassmakers can create a transparent or translucent glass with striking colors. Table 6-4 summarizes the additives for specific glasses. Most striking on this list is uranium oxide, shown in Figure 6-15, which produces a green/yellow glass (often called Vaseline glass) that glows in the dark. In the 1940s the use of uranium oxide was banned in glassmaking, as the government wanted to control access to uranium, and due to concerns about the health of glassworkers. By the late 1950s these restrictions were eliminated,

but the high cost of uranium oxide makes its use prohibitively expensive for most glassware.

More than 80% of all commercial glass is produced through the *float-glass process*, shown in Figure 6-16, which was developed by Sir Alastair Pilkington in 1959. In this process, the constituents for glass are melted in a furnace at 1500°C. A fine ribbon of glass is drawn from the furnace and floats on top of a pool of molten tin. The viscous glass and the molten tin are completely immiscible, so a perfectly smooth surface forms between the glass and tin. When the surface of the glass has hardened sufficiently, rollers pull the glass through an annealing furnace called a *Lehr*. While in the Lehr, the glass is slowly cooled to remove any residual stresses, and any coatings are applied.

| *Float-Glass Process* |
Glass production technique in which a fine ribbon of glass is drawn from a furnace and floated on the surface of a pool of molten tin.

| *Lehr* |
Annealing furnace used in glass manufacture.

TABLE 6-4   Additives for Colored Glasses

Color	Additive(s)
Amber	Manganese oxides
Black	Manganese, copper, and iron oxides
Light blue	Copper oxides
Dark blue	Cobalt oxides
Green	Iron oxides
Green-yellow	Uranium oxides
Red	Selenium oxides
White	Tin oxides
Yellow	Lead and antimony

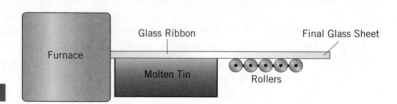

Furnace

Glass Ribbon

Final Glass Sheet

Molten Tin

Rollers

FIGURE 6-16 Float Glass Process

| Cement |
Any material capable of
binding things together.

| Hydraulic Cements |
Binding materials that require
water to form a solid.

| Nonhydraulic Cements |
Binding materials that
form a solid without the
need for water.

| Portland Cement |
Most common hydraulic
cement made from pulverizing
sintered calcium silicates.

| Concrete |
Most important commercial
particulate composite,
which consists of a blend of
gravel or crushed stone and
Portland cement.

6.4 CEMENTS

Although *cement* is actually a generic term referring to any material capable of binding things together, from a materials perspective, the term signifies either *hydraulic cements*, which require water to form a solid, or *nonhydraulic cements*, which form solids without the need for water.

The most common hydraulic cement, *Portland cement*, is made by pulverizing nodules of sintered calcium silicates. Because these calcium silicates are abundant in limestone, chalk, shell deposits, and shale, Portland cement is one of the least expensive construction materials available. The most common use of Portland cement is as the matrix material for the nearly ubiquitous building material *concrete*, a particulate composite that is discussed in Chapter 7.

Portland cement can be produced through either a dry or wet process, with the dry process requiring more energy. In either process, a source of calcium carbonate is ground and blended with quartz (SiO_2) and a clay or silt that provides Fe_2O_3 and Al_2O_3 to the mixture. Because of the prevalence of different iron, aluminum, calcium oxides, and silicates, a standard system of abbreviated nomenclature has been developed. This shorthand notation

Abbreviation	Compound
A	Al_2O_3
C	CaO
F	Fe_2O_3
H	H_2O
S	SiO
$\overline{S}$	SO_3

TABLE 6-5 Shorthand Nomenclature for Constituents of Portland Cement

is summarized in Table 6-5 and is used throughout this section. When more than one compound exists, the ratios are expressed as subscripts, such that a compound with 3 moles of CaO for each mole of Al_2O_3 would be abbreviated as C_3A.

The mixed particles are fed to a rotary kiln, beginning a four-stage process that includes *evaporation-dehydration*, *calcination*, *clinkering*, and cooling. During the evaporation-dehydration stage, the mixture is heated between 250°C and 450°C to drive off all free water. As the heating continues to around 600°C, any water bound to the oxides and silicates is removed. By 900°C, the calcination process begins, in which the calcium carbonate is converted to calcium oxide and carbon dioxide is liberated, as shown in Equation 6.1:

$$CaCO_3 \rightarrow CaO + CO_2 \qquad (6.1)$$

In the same temperature range, the calcium oxide (C) reacts with the aluminum oxide (A) and the ferric oxide (F) to form tetracalcium aluminoferrite (C_4AF) and tricalcium aluminate (C_3A), as shown in Equations 6.2 and 6.3:

$$4C + A + F \rightarrow C_4AF \qquad (6.2)$$
$$3C + A \rightarrow C_3A \qquad (6.3)$$

As the temperature of the kiln increases to around 1450°C, the clinkering phase begins, and the remaining calcium oxide reacts with the silicates from the quartz to form dicalcium silicate (C_2S) and tricalcium silicate (C_3S), as shown in Equations 6.4 and 6.5:

$$2C + S \rightarrow C_2S \qquad (6.4)$$
$$3C + S \rightarrow C_3S \qquad (6.5)$$

The product that emerges from the kiln is called *clinker* and consists of a distribution of particles that average about 10 mm in diameter. The clinker passes through a cooling phase, then is sent to a ball mill where it is ground along with 5% powdered gypsum ($C\overline{S}H_2$) until the resulting mixture achieves an average particle size of approximately 10 μm, with a range of around 1 μm to 100 μm. Table 6-6 summarizes the typical composition of Portland cement.

| *Evaporation-Dehydration* |
First stage in the processing of Portland cement in which all free water is driven off.

| *Calcination* |
Second stage in the processing of Portland cement in which calcium carbonate is converted to calcium oxide.

| *Clinkering* |
Third stage in the processing of Portland cement in which calcium silicates form.

Table 6-6 Composition of Portland Cement

Compound	Abbreviation	Mass Percent
Tricalcium silicate	C_3S	55
Dicalcium silicate	C_2S	20
Tricalcium aluminate	C_3A	10
Tetracalcium aluminoferrite	C_4AF	8
Gypsum	$C\bar{S}H_2$	5

The exact composition of the Portland cement can be varied to make it better suited to specific applications. Table 6-7 summarizes the different types of Portland cement and the applications for which they are suited. Types I and II account for more than 90% of all Portland cements.

The Portland cement particles are stored in dry conditions until needed. When the time arrives to build the cement structure, the cement particles are mixed with water to form a *cement paste* and begin a series of hydration reactions that develop the final properties of the solid cement. The calcium silicates (C_3S and C_2S) make up three-quarters of the total mass and provide most of the strength in the cement. The calcium silicates undergo highly exothermic reactions with water to form calcium silicate hydrates (C-S-H) and calcium hydroxide (CH), as summarized in Equations 6.6 and 6.7:

| **Cement Paste** |
Mixture of cement particles and water.

$$2C_3S + 7H \rightarrow C_3S_2H_8 + 3CH \qquad (6.6)$$

$$2C_2S + 5H \rightarrow C_3S_2H_8 + CH \qquad (6.7)$$

The calcium silicate hydrates are amorphous particles that are extremely small and include a variety of compositions, so the C-S-H designation used does not imply an exact ratio between the constituents.

The calcium silicate reactions really occur in five distinct stages. Stage 1 occurs during the first minutes after water is mixed with the cement. Calcium and hydroxide ions are released from the C_3S, resulting in heat generation and a rapid rise in pH. CH and C-S-H also begin to crystallize in Stage 1. After approximately 15 minutes, the cement enters a dormant period (Stage 2) during which

Table 6-7 Types of Portland Cement

Type	Special Properties
I	General use
II	Some sulfate resistance
III	Gains strength quickly
IV	Low heat of hydration (important in large structures)
V	High sulfate resistance
IA	Type I with air-entraining agent blended in
IIA	Type II with air-entraining agent blended in
IIIA	Type III with air-entraining agent blended in

the reaction slows. A coating of C-S-H develops on the surface of the cement, which creates a diffusion barrier for the water. As the thickness increases, the reaction rate becomes increasingly controlled by the diffusion. After 2–4 hours, a critical mass of ions is achieved and the reaction rate (Stage 3) accelerates. During Stage 3, both C_3S and the less reactive C_2S hydrate rapidly. After about 8 hours, the rate of reaction decelerates and diffusion completely controls all rates (Stage 4). Finally, a steady state (Stage 5) occurs in which the hydration is essentially independent of temperature.

At the same time, the tricalcium aluminate (C_3A) also undergoes a hydration reaction. Without the gypsum, the C_3A would react rapidly with the water, causing premature setting and less desirable properties. Instead, the C_3A reacts with the gypsum to form calcium sulfoaluminate hydrate (ettringite), as shown in Equation 6.8:

$$C_3A \quad + \quad 3\,C\bar{S}H_2 \quad + \quad 26H \quad \rightarrow \quad C_6A\bar{S}_3H_{32} \qquad (6.8)$$

Calcium + Gypsum + Water → Ettringite
Aluminate

The ettringite forms a diffusion barrier around the calcium aluminate and slows the hydration reaction. Once the gypsum has been consumed, the ettringite reacts with the tricalcium aluminate to form monosulfoaluminate ($C_4A\bar{S}H_{12}$), as shown in Equation 6.9:

$$2C_3A + C_6A\bar{S}_3H_{32} + 4H \quad \rightarrow \quad C_4A\bar{S}H_{12} \qquad (6.9)$$

The monosulfoaluminate is stable in cement but makes the cement vulnerable to attack from sulfate ions. Monosulfoaluminate will react to form additional ettringite in the presence of sulfate ions. The new ettringite causes an expansion within the cement that can result in cracking.

The tetracalcium aluminoferrite (C_4AF) has the same reaction pathway as the C_3A, but because it is far less reactive, very little of the gypsum reacts with it. Instead, most of the C_4AF reacts with the calcium hydroxide formed in C_2S and C_3S hydration.

By the time the cement paste has hardened, it consists of mostly C-S-H, C-H, and $C_4A\bar{S}H_{12}$ with about 5% unhydrated silicates, as summarized in Table 6-8. The C-S-H accounts for between 50% and 70% of the total volume and provides much of the strength. As the C-S-H coatings grow on the C_3S and C_2S grains, they begin to radiate outward as spikes. Continued hydration causes the spikes from adjacent particles to interlock, thereby physically connecting the cement grains together, as shown in Figure 6-17. The micropores present in the C-S-H structure reduce its strength, but as the water is lost from these pores during drying, the cement shrinks and strengthens.

The calcium hydroxide (CH) forms thick crystalline plates embedded in the larger C-S-H matrix, as shown in Figure 6-17. These plates grow in pores, thereby reducing the overall porosity of the cement and increasing its strength. The CH also buffers the solution, allowing it to maintain an alkaline pH. The solubility of CH in water causes it to slowly leach out of the cement and can result in increased porosity and reduced durability. The monosulfoaluminate

TABLE 6-8   Components Present in Hardened Cement Paste

Component	Volume Fraction	Density (kg/m^3)	Microstructure
C-S-H	0.50–0.70	2000	Intermingled radiating spikes of porous solids
CH	0.20–0.25	2250	Thick crystalline plates
$C_4A\overline{S}H_{12}$	0.10–0.15	1950	Thin, irregular, clustered crystalline plates
Unhydrated silicates	>0.05	3150	Still possess original grain structure

| Capillary Pores |
Open spaces between grains.

| Gel Pores |
Spaces within the C-S-H material during the hydration of cement.

plays little role in strengthening the concrete but does fill some pore spaces. The $C_4A\overline{S}H_{12}$ forms thin irregular plates.

Porosity also plays a significant role in the strength of cements. *Capillary pores* are the open spaces between grains, while *gel pores* are spaces within the C-S-H material. Most of the porosity of the cement occurs as gel pores.

Much of the current research in cement production involves using industrial by-products, including fly ash and blast furnace slag. In addition to the ecological benefit of reusing these waste products, fly ash and slag often improve the durability of the cement and reduce the heat of hydration because the spherical shape of the particles reduces internal friction. The cement flows with less water and less segregation occurs.

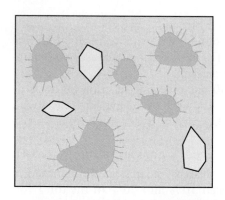

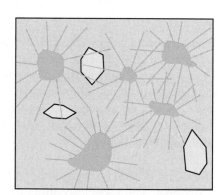

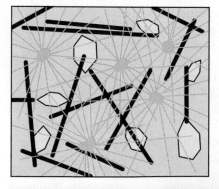

 FIGURE 6-17   Microstructure of Hardened Cement Pastes

*R*efractories are capable of withstanding high temperatures without melting, degrading, or reacting with other materials. This combination of properties makes refractory ceramics ideal for the high-temperature furnaces needed to melt glass, metal, and other materials. The iron and steel industry alone uses well over half of all refractory ceramics produced in the United States. Most commonly, refractories are sold as bricks, but they also can be found as boards, blankets, or in specially made shapes.

Refractory ceramics come in two basic types: *clay* (containing at least 12% SiO_2) and nonclay. The most common clay refractory is *fireclay*, which is made from *kaolinite*, primarily a mixture of alumina (Al_2O_3) and silica (SiO_2) that is discussed in greater detail in Section 6.7. The alumina–silica phase diagram is shown in Figure 6-18. Fireclay may contain anywhere from 50% to 70% silica, with 25% to 45% alumina. Minor constituents including CaO, Fe_2O_3, MgO, and TiO_3 combine to account for less than 5% of the total material. As seen on the phase diagram in Figure 6-18, fireclay in this range can handle temperatures of as high as 1587°C without melting. When alumina fractions above 50% are desired, bauxite serves as the base material. The presence of silica helps make the refractory resistant to attack by acids.

When extremely high (greater than 88%) alumina concentrations are desired, *mullite* bricks are used. These bricks are the most thermally stable but are considered nonclay refractories. Mullite bricks do not begin to melt until temperatures near 1890°C. Other nonclay refractories include silica bricks, which contain silica with 3% to 3.5% CaO, and *periclase*, which contains at least 90% magnesium oxide (MgO). These periclase refractories are especially resistant to attack by alkali materials and find significant use in the steel industry.

| *Refractories* |
Materials capable of withstanding high temperatures without melting, degrading, or reacting with other materials.

| *Clay* |
Refractory material containing at least 12% silicon dioxide (SiO_2).

| *Fireclay* |
Clay material containing 50% to 70% silica and 25% to 45% alumina.

| *Kaolinite* |
Clay mineral named for the Gaolin region of China where it was discovered.

| *Mullite* |
Clay material formed by high temperature aluminosilicates.

| *Periclase* |
Refractory material containing at least 90% magnesium oxide (MgO).

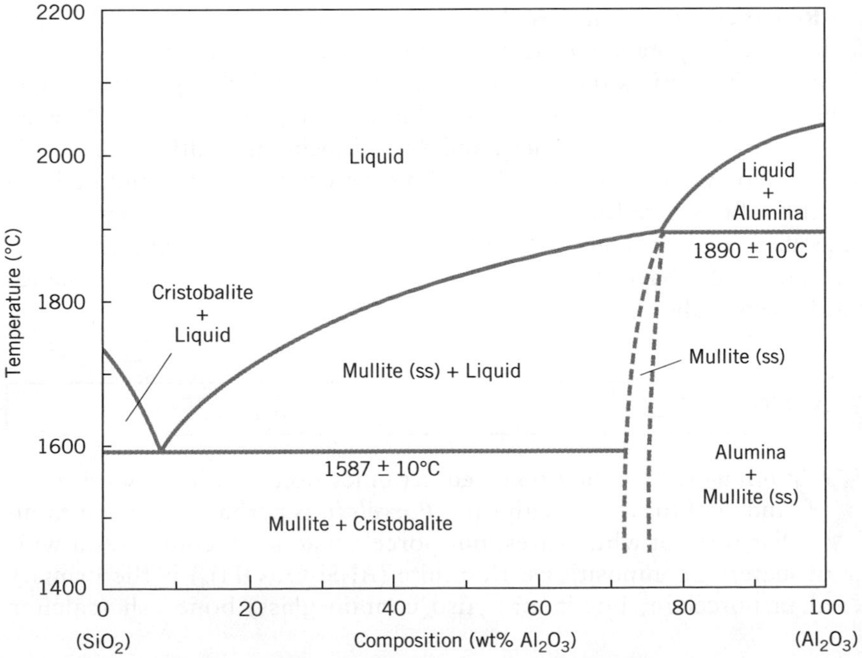

FIGURE 6-18 Alumina–Silica Phase Diagram

From F. J. Klug, S. Prochazka, and R.H. Doremus, "Alumina–Silica Phase Diagram in the Mullite Region," Journal of the American Ceramic Society, Vol. 70, No. 10 (1987): 758. *Reprinted with permission of the American Ceramic Society.*

FIGURE 6-19 The Terra-Cotta Army

Yoshio Tomii/SUPERSTOCK

6.6 STRUCTURAL CLAY PRODUCTS

| *Structural Clay Products* |
Any ceramic materials used in building constructions, including brick and terra cotta.

Structural clay products include any ceramic materials used in building constructions and most commonly involve brick and terra cotta, which is easily recognizable by its reddish orange color. *Terra cotta*, which literally means burned earth, was used in earthenware as far back as 3000 BC and in sculpture as far back as the third century BC with the terra-cotta army in China, as shown in Figure 6-19. In the United States, terra cotta is made from a Redart clay that contains about 7% iron oxide along with 64% silica, 16% alumina, 4% potassium oxide, and a mix of other metallic oxides.

| *Terra Cotta* |
Ceramic material made with an iron-oxide rich clay that is recognized by its reddish-orange color.

$ Structural integrity is the critical factor in structural clay products, especially brick. In the United States, standard bricks range from 2.75 to 4 inches in height, 4 to 6 inches in thickness, and 8 to 12 inches in width. Nearly 75% of the $2 billion worth of bricks sold each year are used in home and building construction. Bricks are formed by heating dried clay in a kiln. They are further classified into common brick and face brick, which are chemically identical, but face bricks have a uniform appearance that makes them suitable for outward-facing walls.

6.7 WHITEWARES

| *Whitewares* |
Fine-textured ceramics used in dinnerware, floor and wall tiles, and sculptures.

Whitewares are fine-textured ceramics used in dinnerware, floor and wall tiles, and sculpture. *Porcelain* is perhaps the most familiar form of whitewares, but porcelain actually comprises a wide range of material compositions. Kaolinite ($Al_2Si_2O_5(OH)_4$) is the primary material in porcelain, but it may also contain glass, bone ash (calcium

| *Porcelain* |
Whiteware with a translucence caused by the formation of glass and mullite during the firing process.

phosphate), and steatite (a glassy form of talc). Chinese porcelain also contains a material known as *China stone*, which is a blend of quartz and sericite (a small-grain mica).

Kaolinite easily is the most common clay; nearly all clay deposits will contain some translucent kaolinite particles. The name *kaolinite* comes from the town of Gaolin, China, where a large deposit exists. Kaolinite forms only microscopic crystals with a Moh hardness of between 1.5 and 2.

Typically, whiteware manufacture involves shaping kaolinite clay into the desired shape and allowing it to dry. The material is then fired in a kiln at temperatures above 800°C to convert the clay into a glasslike material called *bisque* through vitrification. Figure 6-20 shows an example of bisque. During heating, water bound to the oxides is lost. Kaolonite often loses in excess of 10% of its weight from water loss. Pores in the clay provide natural channels to vent the water vapor. The structural change to the clay in this "water smoking" process is irrevocable. The rest of the heating process involves burnout of any organics, quartz inversion analagous to that in glass manufacture, and sintering similar to that in cement production.

Upon removal from the kiln, whiteware generally is coated with a *glaze* and refired in a kiln at temperatures between 950°C and 1430°C to convert the clay to a hard glasslike substance throughout and to set the glaze. The glazes themselves are SiO_2 glassy materials along with a blend of other metallic oxides. The color of the glaze varies with firing conditions and the presence of other metal oxides. For many years whiteware glazes often contained lead oxide, which resulted in contaminated wastewater streams. Much of the ongoing research in the field focuses on eliminating the use of lead in glazes and improving methods of recycling glaze waste.

Glazes often serve a purely decorative function, providing color to the plain white fired kaolinite product, but they also can be used to provide an impermeable barrier. Glazes are applied as water-based suspensions. For most commercial products, the application of the glazing is automated, but many artisans still develop and apply their own glazes by hand.

| *China Stone* |
Blend of quartz and mica used to make Chinese porcelain.

| *Bisque* |
Glasslike material that results from the firing of kaolinite clay.

| *Glaze* |
Blend of SiO_2 and metal oxides that are used to coat bisque material and provide color when refired.

Courtesy James Newell

| Advanced Ceramics |
Engineered ceramic
materials used primarily
for high-end applications.

| Bioceramics |
Ceramics used in
biomedical applications.

Advanced ceramics are engineered materials used primarily for high-end applications. These materials are constantly evolving, and this section endeavors to provide a blend of current applications and views to the future.

Ceramic armor, ceramic nanoparticles, solid oxide fuel cells, and *bioceramics* comprise primary examples of advanced ceramics. Ceramic armor is made primarily from carbides or boron nitride, largely because of the exceptional hardness of these materials. Table 6-9 summarizes their key physical properties.

Boron nitride (BN) occurs in two distinct forms. Hexagonal boron nitride is similar to graphite and is often used as a lubricant but is not suited for armor. Cubic boron nitride forms a structure similar to diamond and is second only to diamond in hardness. Although a process to manufacture boron nitride was first discovered in 1957, there was no significant commercial production until the late 1980s. Because of the difficulty in manufacturing cubic boron nitride, it is used more as powders or insulating coatings than for armor.

Boron carbide (B_4C) is the third hardest material behind diamond and cubic boron nitride. The combination of low density, high hardness, and high Young's modulus make boron carbide ideal for use as body armor. Boron carbide is manufactured by adding carbon to B_2O_3 in an arc furnace, but it is difficult to sinter. Often other materials must be added as sintering aids.

Titanium carbide (TiC) has found most of its applications as tips on cutting tools and drill bits (often blended with tungsten carbide), but new advances have increased its use as armor. Because of its relatively high density, TiC is poorly suited for body armor but is finding use on armored fighting vehicles.

Recently more commercial uses have been found for ceramic nanoparticles. Ceramic nanoparticles can be produced through a variety of techniques, including sol-gels and *flame aerosol processes*. In the flame aerosol process, a liquid metallic organic is vaporized and brought to a flame, where it rapidly oxides and forms a nucleus for a metal oxide nanoparticle. These nuclei collide and combine to form the final particle. Light-sensitive silica-based nanoparticles are being used to target tumors in cancer patients. $BaTiO_3$ nanoparticles in the 20 nm to 100 nm size range are being used in multilayer ceramic capacitors, while nanoparticles of silicon carbide (SiC) are being tested for use in improved armor.

In general, ceramic nanoparticles show great potential in improving toughness and ductility for ceramics, improving wear and scratch resistance

| Flame Aerosol Process |
Method of producing ceramic
nanoparticles in which a liquid
metallic organic is brought to
a flame to form a nucleus for
particle growth.

TABLE 6-9 Physical Properties of Advanced Ceramics Used in Armor

Material	T_m (°C)	Young's Modulus (GPa)	Density (g/cm³)	Moh Hardness
Boron nitride	3000	20–100	2.2	9.5–10
Boron carbide	2445	450–470	2.52	9.5
Titanium carbide	3160	440–455	4.93	9–9.5

of ceramic parts, and allowing for blending of materials that are not usually miscible.

With the ongoing demand to reduce dependency on fossil fuels, the industrial interest in solid oxide fuel cells (SOFCs) has grown dramatically. An SOFC uses electrochemistry to convert chemical energy directly to electrical energy in a process similar to the galvanic cells discussed in Chapter 5. In the cell, an anodic material and a cathodic material are separated by an insulator that also conducts oxygen atoms from the cathode to the anode, where they react with the fuel source. Many fuel cells require a clean stream of hydrogen as the fuel, but SOFCs can also use hydrocarbons.

Ceramic materials including lanthanum chromite doped with strontium are used as the interconnect material between the individual cells. Unlike many fuel cells, SOFCs operate at a very high efficiency rate; because they also require high temperatures, the use of advanced ceramics is essential. Nickel oxides also are commonly used as the anode material.

Bioceramics are discussed more thoroughly in Chapter 9 (biomaterials), but primary functions of bioceramics include lubricating surfaces in prosthetics, providing clot-resistant coatings on heart valves, serving as transfer agents and collectors of radioactive species for cancer treatments, and providing frameworks to stimulate bone growth.

What Happens to Ceramic Materials at the End of Their Useful Lives?

6.9 RECYCLING OF CERAMIC MATERIALS

Because ceramic materials do not corrode, often their usable life spans are far greater than those of other materials. The great pyramids of Egypt have stood for thousands of years. However, the same physical properties that make ceramics so durable also make them extremely difficult to recycle, with the exception of glass. Most municipalities routinely collect glass recyclables as part of their regular service. These recycled bottles, glassware, lightbulbs, jars, and other items are sorted by color and crushed into a fine powder called *cullet* that can be remelted and re-formed into new glass products. Each ton of cullet used instead of fresh silica saves more than 600 pounds of carbon dioxide emissions.

| *Cullet* |
Finely ground glass powder used in ceramic recycling.

Most other ceramic materials have been disposed of by burying them in landfills. However, recent initiatives are challenging those practices. Abrasives, once thought to be impossible to recycle, are now collected, recycled, and exchanged via the Internet between different end users. Some companies have begun pulverizing unwanted Portland-based concrete from construction demolition projects and using it as aggragate for future projects. Whitewares are also commonly landfilled after their use is complete, but a pair of companies in New Zealand, Electrolux and Fisher & Paykel Appliances, have initiated project life stewardship programs to reduce waste in manufacture of whitewares and to provide end-of-life recycling.

6.10 GRAPHITE

| *Graphite* |
Allotropic carbon material consisting of six member aromatic carbon rings bonded together in flat planes, allowing for the easy occurrence of slip between planes.

G raphite makes an ideal transition from the discussion of ceramics to the discussion of carbon materials. Graphite is an allotropic form of carbon with a layered structure in which six-member aromatic rings exist in graphene planes, with each layer separated by 3.35 angstroms from the next layer, as shown in Figure 6-21. The resonance-stabilized covalent bonds between carbon atoms in the planes are incredibly strong, but only weak Van der Waals interactions exist between the planes. This difference between in-plane and between-planes bond strengths makes graphite a highly anisotropic material in which material properties are a strong function of direction. Great stresses are required to break the in-plane bonds, but relatively little stress will cause the layer planes to slide across each other. Despite having the same chemical composition as diamond, graphite is extremely soft with a Moh hardness between 1 and 2. As a result, graphite makes an excellent lubricant and functions as pencil "lead." Graphite is thermally and electrically conductive along the layer planes but insulates in the direction perpendicular to the plane. Graphite finds significant use in nuclear reactors, both as cladding for the reactor itself and as a moderator that slows neutrons down enough to induce fission in U-235.

In many ways, graphite is a difficult material to classify. It is clearly an allotropic form of carbon, but it also fits the technical definition of a polymer. It is a long chain of repeated units (the six-member rings) covalently bonded to each other. Graphite is clearly not a ceramic material because it does not possess both metallic and nonmetallic atoms, yet it possesses many of the physical properties associated with ceramics. Graphite is brittle, strong, insulating (at least in one direction), and has the highest Young's modulus of any material (1080 GPa).

Although the properties of graphite are unique, other carbon materials offer distinct blends of properties that make them ideal for specific applications. The next three sections examine diamond, carbon fibers, carbon nanotubes, and fullerenes.

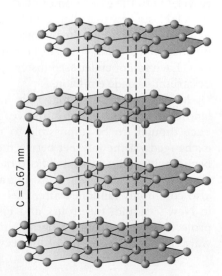

FIGURE 6-21 Structure of Graphite

6.11 DIAMOND

Diamond is an allotropic form of carbon that forms a clear FCC structure with carbon atoms also occupying four of the eight tetrahedral interstitial sites, as shown in Figure 6-22. Most of the exceptional properties of diamond result from this crystal structure. Diamond is the hardest naturally occurring material (Moh hardness = 10) and can be scratched only by other diamonds. The very name *diamond* comes from the Greek word for invincible (*adamas*).

Most diamonds are colorless, but they may develop a color due to the presence of impurities. The yellow tint associated with lower-quality diamonds results from the presence of nitrogen, the most common impurity. The nitrogen impurity occupies a lattice site normally filled by a carbon atom. Other impurities can result in a variety of colors, including white, metallic, blue, red, green, and pink.

The commercial use of diamonds falls into two broad categories: gemstones and industrial processes. More than 25,000 kilograms of diamonds are mined each year, but most lack the color, size, and clarity needed for use as gemstones. Instead, nearly 80% are classified as *bort* and used for industrial purposes, including saw blades, drill bit tips, low-friction hinges on space shuttles, and as lapping agents.

Natural diamonds form when carbon deposits are exposed to a blend of extreme temperature and pressure for long enough periods of time to allow for crystal growth. Most diamonds occur in volcanic pipes that form from deep underground eruptions of magma. These pipes provide both the temperature and pressure needed to form diamonds. Such volcanic pipes are rare; most diamonds are mined from central and southern Africa.

| *Diamond* |
Allotropic, highly crystalline form of carbon that is the hardest known material.

| *Bort* |
Diamonds lacking gem-quality that are used for industrial purposes.

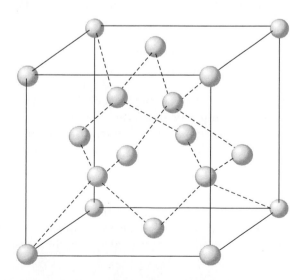

FIGURE 6-22 Crystal Structure of Diamond

The high cost of natural diamonds limits their industrial use, but synthetic diamonds offer many possibilities. The famous science fiction author H. G. Wells speculated about the manufacture of synthetic diamonds in his 1914 book *The Diamond Maker*, but the first synthetic diamonds were not produced until 1953. The largest obstacle to the manufacture of synthetic diamonds has been the need to achieve and maintain pressures as high as 55,000 atmospheres. The most common industrial method, the *HPHT method* (high-pressure, high-temperature), uses giant presses to produce and maintain pressures as high as 5 GPa at a temperature of 1500°C. This method is slow and the equipment is expensive and weighs several tons, but improved materials for containment now allow faster production of larger and higher-quality synthetic diamonds.

| *HPHT Method* |
Process for producing synthetic diamonds using elevated temperatures and high pressures.

Chemical vapor deposition (CVD) of synthetic diamonds first appeared in the 1980s and offers a promising alternative. A carbon-containing plasma is formed on a substrate molecule by microwave radiation or arc discharge, and the diamond is essentially "assembled" one atom at a time from the gas. Although promising, the CVD processes still face numerous technical barriers before large-scale manufacture is practical.

Diamonds have great potential as heat sinks in microprocessors or even as semiconductors themselves. Diamond semiconductors operate at temperatures above which silicon would melt. Many researchers expect the next generation of supercomputers to include diamond microchips.

6.12 CARBON FIBERS

| *Carbon Fibers* |
Forms of carbon made by converting a precursor fiber into an all-aromatic fiber with exceptional mechanical properties.

| *Carbonization* |
The controlled pyrolysis of a fiber precursor in an inert atmosphere.

| *Stabilization* |
The conversion of a carbon fiber precursor to a thermally stable form that will not melt during carbonization.

| *Graphene Layer Planes* |
Parallel planes consisting of conjugated six-member aromatic carbon rings.

| *Turbostratic* |
Structure in which irregularities in otherwise parallel planes cause distortion.

Carbon fibers are defined by the International Union of Pure and Applied Chemistry (IUPAC) as "fibers (filaments, tows, yarns, rovings) consisting of at least 92% (mass fraction) carbon, usually in a non-graphitic state." The ultimate properties of the carbon fiber directly depend on the selection and processing of the precursor materials, the formation of the fiber, and subsequent processing of the fiber. Depending on how they are processed, carbon fibers can obtain a wide range of physical, chemical, electrical, and thermal properties. This flexibility makes carbon fibers the main reinforcement materials in advanced composites with diverse applications, including military aircraft and missiles, automobile body panels, sports equipment, batteries and capacitors, and activated carbons.

Most commercial carbon fibers are produced through *carbonization*, although some speciality fibers are produced through growth from gaseous hydrocarbons. Extremely high temperatures are required to drive off everything but the aromatic carbon rings, but most fibers would melt before carbonizing. Therefore, the precursor fibers must be converted to a thermally stable form that cannot melt through a process called *stabilization*. During the carbonization process, inorganics and aliphatic carbons are driven off, leaving a fiber consisting of *graphene layer planes*. An idealized carbon fiber is shown in Figure 6-23.

Real carbon fibers never achieve the idealized structure shown in Figure 6-23. Instead they form *turbostratic* layers in which the distance between the layer planes varies, but the average distance is always larger than the optimal 3.35 angstroms. Voids and misalignments are common and tend to significantly reduce the strength of the fiber.

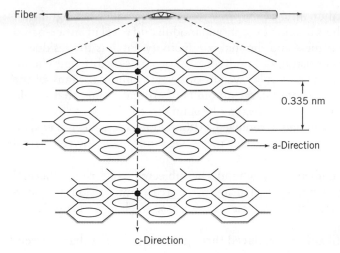

0.335 nm

a-Direction

c-Direction

FIGURE 6-23 Idealized Carbon Fiber

The majority of commercial carbon fibers are produced from polyacrylonitrile (PAN) fibers. PAN-based carbon fibers are several times stronger than steel on a per-weight basis. As the strongest commercially available carbon fibers, they dominate the market for structural applications.

Commercial polyacrlyonitrile is actually a copolymer of acrylonitrile and another monomer (vinyl acetate, methyl acrylate, or acrylic acid) that is added to lower the glass transition temperature of the material and control its oxidation resistance. Before the acrylic fibers produced in the wet-spinning process can be subjected to the elevated temperatures of carbonization, they are stabilized in air at temperatures between 200°C and 300°C and under tension to prevent fiber shrinkage.

During the stabilization process, the fiber is converted to a ladder polymer through a combination of oxidation, dehydrogenation, and cyclization reactions shown in Figure 6-24. The stabilized fiber is carbonized in an inert atmosphere at temperatures ranging from 1000°C to 3000°C, depending on the desired properties. During the pyrolysis, nearly all inorganics and nonaromatic carbons are driven off.

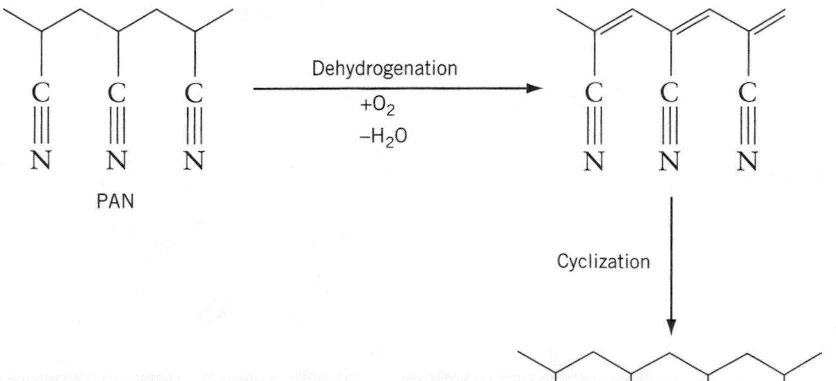

Dehydrogenation
$+O_2$
$-H_2O$

PAN

Cyclization

FIGURE 6-24 General Mechanism for the Stabilization of PAN

| Mesophase Pitch |
By-product of coal or
petroleum distillation formed
into regions with liquid
crystalline order through
heat treatment.

Carbon fibers based on *mesophase pitch* represent a smaller but significant market niche. These fibers develop exceptional moduli and excel in lattice-based properties, including stiffness and thermal conductivity. Pitch is the residue of petroleum or coal tar consisting of hundreds of thousands of different chemical species with an average molecular weight of several hundred. Many of the molecules in these pitches contain large amounts of aromatic rings that would make ideal carbon fibers. Some pitches can be spun directly into isotropic pitch fibers that can be stabilized and carbonized. Although inexpensive and easy to make, these isotropic carbon fibers tend to be weak and have lower thermal conductivities.

To get better carbon fibers, the pitch must be subjected to a series of thermal treatments and solvent extractions. During these treatments, a liquid crystalline phase (or mesophase) forms. A typical mesophase pitch molecule is shown in Figure 6-25.

Mesophase pitch fibers are produced through melt spinning that is essentially the same process used to spin commercial polymers. An extruder melts pitch particles and pumps the molten pitch through a multihole spinnerette. The fibers emerging from the spinnerette are drawn by a wind-up spool. Table 6-10 summarizes properties of mesophase pitch-based and PAN-based carbon fibers.

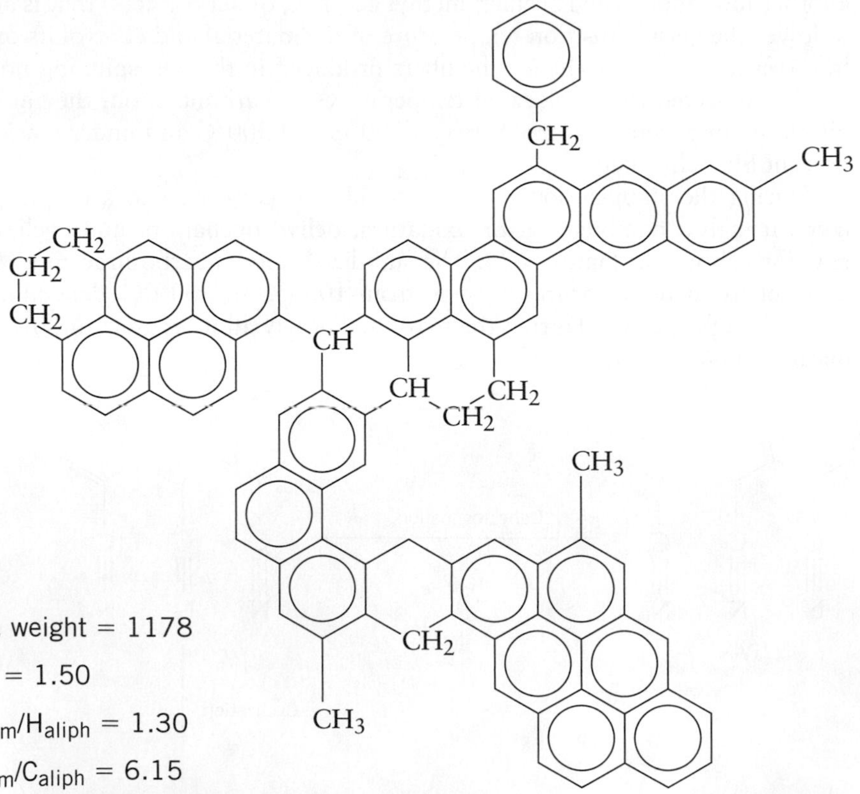

Mol. weight = 1178

C/H = 1.50

$H_{arom}/H_{aliph} = 1.30$

$C_{arom}/C_{aliph} = 6.15$

FIGURE 6-25 Representative Meosphase Pitch Molecule

TABLE 6-10	Mechanical Properties of Commercial Carbon Fibers		
Fiber Type	Precursor Material	Tensile Strength (GPa)	Young's Modulus (GPa)
T-300	PAN	3.66	231
T-650/35	PAN	4.28	241
P-100S	Pitch	2.41	759
P-120S	Pitch	2.41	828
K-1100	Pitch	3.10	931

6.13 FULLERENES (BUCKYBALLS) AND CARBON NANOTUBES

Until 1985 graphite and carbon were the only known stable allotropes of carbon. By 1990 scientists were capable of producing a new carbon form that closely resembled a soccer ball. This new allotrope, shown in Figure 6-26, contains 60 carbon atoms (C_{60}) shaped into 20 hexagons and 12 pentagons. The molecules are officially named *buckminster fullerenes* after the famous architect Buckminster Fuller, who designed a similar-looking geodesic dome, but are more commonly called *fullerenes* or *buckyballs*. Larger fullerenes with 70, 76, and 78 carbon atoms have been synthesized in the remains from plasma streams, and still more stable structures have been theorized.

| *Buckminster Fullerenes* |
Allotropes of carbon with at least 60 carbon atoms shaped like a soccer ball or geodesic dome.

The empty cagelike shape of the buckyballs has intrigued researchers who look to entrain specific molecules within the open structure. Some researchers are examining the possibility of using buckyballs to deliver specific antibiotics to bacteria that resist traditional oral or intravenous antibiotics. The structure also appears to show promise in superconduction and in inoculating steel

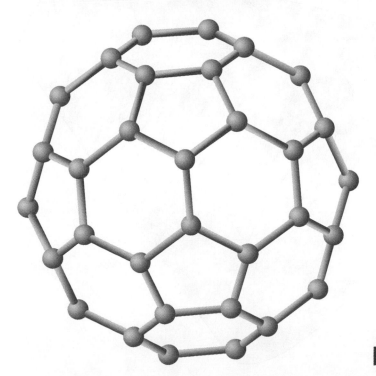

FIGURE 6-26 Structure of a Buckyball

with buckyballs that have trapped metal atoms. Some studies have indicated that buckyballs may be toxic to living organisms, but many of these studies are too preliminary to gauge the potential health effects of the molecules.

Carbon nanotubes represent the other most promising new carbon material. Single-walled carbon nanotubes (SWNTs) are essentially a single graphene plane rolled into a cylindrical tube, as shown in Figure 6-27. Carbon nanotubes are an order of magnitude stronger than steel on a per-weight basis and six times stronger than the best carbon fibers, and exhibit exceptional electrical properties. Many experts believe that SWNTs will revolutionize the microelectronic industry much in the same way the silicon wafer has. SWNTs also can be aligned to form simple structures, such as ropes and films.

Multiwalled carbon nanotubes (MWNTs) involve multiple graphene layer planes rolled into the same type of cylindrical tubes as in SWNTs. A *Russian doll model* is used to describe these tubes in which an outer graphene layer surrounds an inner graphene layer plane that surrounds another, much like nesting dolls fit one inside of the other. The inner layers of the MWNT can slide past each other with essentially no friction, which creates the first perfect rotational bearing.

MWNTs are classified by a pair of integers (n.m) that represent a pair of unit vectors that define direction along the graphene plane, as shown in Figure 6-28, much like the indices used to represent directions in a Bravais lattice. Nanotubes with m ≠ 0 are classified as *zigzag*. Those with m = n are called *armchair*, while those with m ≠ n are called *chiral*.

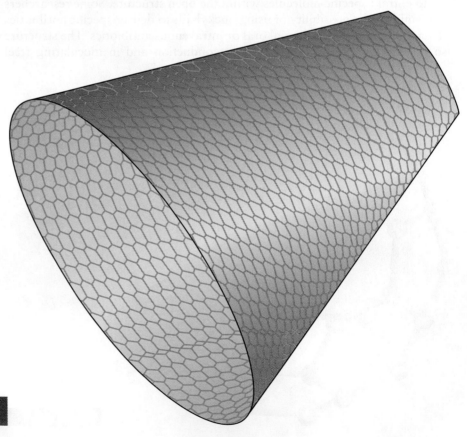

FIGURE 6-27 Single-Walled Carbon Nanotube

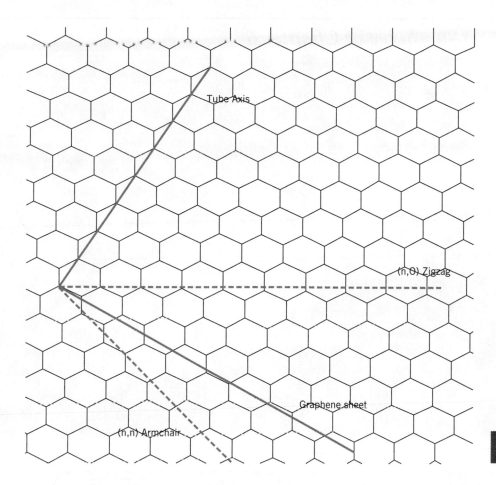

Tube Axis

(n,O) Zigzag

Graphene sheet

(n,n) Armchair

Figure 6-28 Vectors on a Graphene Plane

The structure of the tube greatly affects the electrical properties of MWNTs. MWNTs with indices that satisfy Equation 6.10,

$$2m + n = 3q, \qquad (6.10)$$

where q is any integer, conduct as efficiently as metals but are able to handle current densities several orders of magnitude greater than the most conductive metals. Those that do not satisfy the equation display semiconductor properties.

The high cost of producing carbon nanotubes limits their widespread commercial use. Small quantities of some carbon nanotubes occur naturally in soot and ash, but these nanotubes are too irregular for commercial use. Many techniques can be used to produce carbon nanotubes, including laser ablation and arc discharge, but the most promising seems to a chemical vapor deposition (CVD) process. In this process, acetylene or another carbonaceous gas is passed across a metal catalyst at high temperatures in tightly controlled conditions. Significant research is focused on reducing the processing costs and improving the quality of the product.

Summary of Chapter 6

In this chapter, we:

- Learned the crystal structures unique to ceramics and how to calculate coordination numbers

- Examined the relevant properties of abrasives with a more detailed examination of sandpaper

- Explored the chemistry of Portland cement

- Examined the process of glass formation, including the role of additives

- Discussed the relevant properties of refractories

- Reviewed the uses of structural clay products

- Examined the formation of whitewares and the use of glazes

- Considered the unique structure and properties of graphite

- Learned about the properties of diamond and explored the processes to manufacture synthetic diamonds

- Compared PAN-based and pitch-based carbon fibers

- Looked at the structures of fullerenes and carbon nanotubes while considering their potential for future applications

Key Terms

Homework Problems

1. Calculate the coordination number for ZrO_2.

2. Explain why gypsum is added to Portland cement.

3. How do network modifiers affect the properties of glasses?

4. Compare and contrast the glass transition temperature of glasses with those of polymers.

5. Describe the operating principle of a solid oxide fuel cell.

6. Find at least three commercial products made with carbon fibers. Determine whether pitch-based or PAN-based fibers are used.

7. What is mesophase pitch?

8. Calculate the coordination number of CaO.

9. Why is graphite sometimes considered a polymer and sometimes treated like a ceramic?

10. Explain why diamond is extremely hard and graphite is not, even though both materials are made of pure carbon.

11. Consider a blend of 40wt% MgO and 60wt% Al_2O_3. What phases and compositions are present at 1600°C? At what temperature does the first liquid appear?

12. Explain why ceramics tend to be strong but brittle.

13. Show that the minimum cation–anion ratio is .155 for a system with a coordination number of 3.

14. Explain the role of C-S-H in cement hydration.

15. Explain why terra cotta often is used for sewer pipes.

16. Explain the difference between gel pores and capillary pores.

17. Describe the differences among armchair, chiral, and zigzag nanotubes.

18. Why is it so easy to separate layers of graphite?

19. Explain the environmental and economic impacts of the widescale use of SOFCs.

20. What is the purpose of a glaze in whitewares?

21. Explain why molten SiO_2 forms glass instead of cooling back into a crystalline form.

22. Explain why hexagonal boron nitride is not suited for use in body armor.

23. Define a sol-gel.

24. Draw a Spinel structure and label the tetrahedral and octahedral positions.

25. Explain why diamond dust is so ideal for lapping.

26. Compare and contrast vitrification and sintering.

27. Compare and contrast the properties of silica glass and Portland cement.

28. Why is it potentially beneficial to include fly ash into Portland cement?

29. Find two additional potential uses for ceramic nanoparticles not discussed in the chapter.

30. Determine the mass fractions in the liquid phase for a system with 20% Al_2O_3 and 80% MgO at 2400°C.

31. What properties are most important in selecting a refractory?

32. Explain how isotactic powder pressing differs from standard powder pressing.

33. Explain how the need to balance charge affects the mechanical properties of ceramics.

34. Cite advantages and disadvantages for using carbon fibers instead of high-performance polymers for the following applications:
    a. Aerospace equipment
    b. Racing car frames
    c. Golf club shafts

35. Why are ceramics essentially immune to corrosion?

36. Describe the five stages of cement hardening.

7 Composites

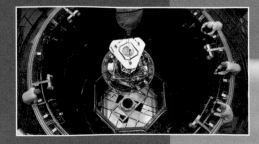

CONTENTS

What Are Composite Materials and How Are They Made?

What Happens to Obsolete Composites?

Learning Objectives

By the end of this chapter, a student should be able to:

- Identify the relevant features between fiber-reinforced, particulate, and laminar composites.

- Distinguish between a composite and an alloy.

- Compare and contrast the roles of the fiber and matrix in fiber-reinforced composites.

- Explain the significance of the quality of the bonding between the fiber and the matrix.

- Estimate the density, electrical conductivity, thermal conductivity, tensile strength, and elastic modulus of a composite if the volume fractions of the component materials (including pores) are known.

- Estimate the fraction of applied load handled by the fibers.

- Distinguish among uniaxial fibers, chopped fibers, and woven mats, and explain how the fiber orientation will impact the properties of the composite.

- Discuss the economic and mechanical factors that influence the decision of how much fiber to include in a composite.

- Describe the fundamentals of composite production techniques, including pultrusion, wet-filament winding, prepregging, and resin transfer molding.

- Examine the major types of reinforcing fibers, and consider factors that influence their selection.

- Contrast the different functions of a variety of matrix materials.

- Explain the role of aggregate in particulate composites.

- Discuss the implications of the Portland cement matrix in the properties of concrete.

- Explain the role of admixtures in concrete.

- Describe the influence of water–cement ratio, aggregate size, and aggregate shape on the mechanical properties of concrete.

- Explain why concrete is much weaker in tension than in compression.

- Measure and interpret the maximum 28-day compressive strength of a concrete column.

- Define and measure the modulus of rupture.

- Define and measure the effective secant modulus of elasticity.

- Define rebar and explain its use in reinforced concrete.

- Explain the role of asphalt in paving.

- Distinguish among hot-mix (HMAC), warm-mix (WAM), and cold-mix asphalt processes and the influence on properties.

- Describe the components and uses of plywood.

- Explain the factors that influence the selection of epoxy for laminar composites.

- Examine the fate of most obsolete composites, and consider the state of recycling.

What Are Composite Materials and How Are They Made?

| *Composites* |
Material formed by blending two materials in distinct phases causing a new material with different properties from either parent.

| *Fiber-Reinforced Composites* |
Composites in which the one material forms the outer matrix and transfers any loads applied to the stronger, more brittle fibers.

7.1 CLASSES OF COMPOSITES

Much like metal alloys, *composites* blend two or more materials together to form a material with properties that are different from either parent material. However, composites differ from alloys in that each parent material continues to exist in a distinct phase. Composite materials are classified into one of three categories: fiber-reinforced, particulate, and laminar. *Fiber-reinforced composites*, shown in Table 7-1, surround strong fibers with a typically amorphous *matrix* material that protects and orients the fibers. *Particulate composites* involve large particles dispersed in a

Composite Category	Definition	Diagram	Example
Fiber-reinforced	Composite in which the one material forms the outer matrix and transfers any loads applied to the stronger, more brittle fibers.		Kevlar-epoxy composites
Particulate	Composite that contains large numbers of coarse particles, to reinforce the matrix.		Concrete
Laminar	Composite that is made by alternating the layering of different materials affixed together with an adhesive.		Plywood
Hybrid	Composite composed of other composite materials.		Rebar-reinforced concrete

TABLE 7-1 Classes of Composites

matrix, while *laminar composites* involve alternating layers of material bonded to each other.

In some cases, *hybrid composites* are produced that involve composites of composites. For example, a steel-belted radial tire is a hybrid composite. The "tire rubber" is a particulate composite with a polymer matrix surrounding carbon black particles. The "tire rubber" encases and orients steel threads to form a fiber-reinforced composite, while multiple layers of these fiber-reinforced composites are bound together to form a laminar composite.

7.2 FIBER-REINFORCED COMPOSITES

Fiber-reinforced composites consist of two phases: the fiber and the *matrix*. In most cases, strong and stiff but brittle fibers are set in a tough but more ductile matrix, resulting in a material with excellent strength-to-weight ratio, stiffness, and fatigue resistance. The role of the fiber is to withstand significant tensile loads in the longitudinal direction. Common fibers used for reinforcement include carbon, glass, high-performance polymers, polyester, steel, titanium, and tungsten.

The matrix material surrounds the fibers, orients them to optimize their collective performance, protects them from environmental attack, and transfers the load to them. Polyester is the most common matrix material because of its relatively low cost. Epoxy resins are used when shrinkage is an issue and cost is less of a concern.

| *Matrix* |
Material in a composite that protects, orients, and transfers load to the reinforcing material.

| *Particulate Composites* |
Composites that contain large numbers of coarse particles, such as the cement and gravel found in concrete.

| *Laminar Composites* |
Composites that are made by alternating the layering of different materials bonded to each other.

| *Hybrid Composites* |
Composite materials produced with at least one phase that is a composite material itself.

The use of fiber-reinforced composites dates back to antiquity, when bricks were formed from mixtures of clay (the matrix) and straw (the fiber). Carbon-fiber reinforced composites find use in military and aerospace applications as well as in modern sailboats, race cars, performance bicycles, and golf and tennis equipment.

7.2.1 // Properties of Fiber-Reinforced Composites

Because the fibers serve as the load-bearing material in the composite, strong fibers are preferentially selected, but the relationship between the strength of the fiber and the strength of the composite is not simple. The matrix must be able to transfer the mechanical load to the fiber through the covalent bonding between the fiber and the matrix. Many factors—including the size and orientation of the fiber, the surface chemistry of the fiber, the amount of voids present, and the degree of curing—influence these bonds. However, the degree and quality of bonding between the fiber and the matrix is the most significant factor in the strength of the composite.

| *Longitudinal Direction* |
Direction of fiber alignment.

| *Transverse Direction* |
Direction perpendicular to the fibers in a composite.

Fiber-reinforced composites are anisotropic with very different properties in the direction of fiber alignment (the *longitudinal direction*) from the direction perpendicular to the fibers (the *transverse direction*). When the fibers are aligned, they all contribute to handling a longitudinal load but provide almost no reinforcement to a transverse load.

Several important factors influence the performance of the fibers, including length and diameter, fiber fraction, and orientation. Fibers can be any length from a few millimeters long in the case of chopped fibers (in which long fibers are cut into tiny pieces and randomly aligned) to several miles long in the case of continuous monofilaments. Most reinforcing fibers range from 7 microns to 150 microns in diameter. As a point of reference, a typical human hair is about 80 microns thick. In general, thinner fibers are stronger because their reduced surface area makes them less susceptible to surface imperfections, and longer fibers support load more efficiently than shorter fibers because there are fewer ends.

| *Aspect Ratio* |
Ratio of the length to the diameter of the fiber used in a fiber-reinforced composite.

The ratio of fiber length to diameter is called the *aspect ratio* (l/d). Clearly, large aspect ratios result in stronger composites, but larger fibers are more difficult to process. They are more difficult to orient and often are limited by the size of the composite material itself. Many times composite designers define a critical length (l_c) below which the fiber provides limited reinforcement but above which the fiber acts as if it were almost infinitely long. The defining equation,

$$l_c = \frac{\sigma_f d}{2\tau_i},$$ (7.1)

expresses the critical length of fiber as a function of the tensile strength of the fiber (σ_f), the fiber diameter (d), and an empirical constant (τ_i) that relates to the quality of the bonding between the fiber and the matrix, called *wet out*. However, the bonding quality is difficult to characterize, and more often the critical length is determined by trial and error rather than by theoretical analysis. If the shear yield strength of the matrix is significantly smaller than τ_i, that measure often replaces τ_i in the critical length equation.

| *Wet Out* |
Quality of bonding between the fiber and the matrix in a composite material.

Some mechanical properties can be predicted more accurately for the composite material. A simple rule of mixture applies well to densities, electrical

Example 7-1

For a Kevlar-29/epoxy resin composite, the critical length of 13 μm fiber was found to be 44 mm. When small quantities of glass fiber are added to the composite to improve wet out, the critical length reduced to 33 mm. Estimate the wet out constant for the composite both without the glass additive and with it.

SOLUTION

The wet out constant (τ_i) can be calculated from Equation 7.1. For the composite without the glass fiber,

$$\tau_i = \frac{\sigma_f d}{2 l_c} = \frac{(2.8 \text{ GPa})(.013 \text{ mm})}{2(44 \text{ mm})} = 4.14 \times 10^{-4} \text{ GPa}.$$

For the composite with the glass,

$$\tau_i = \frac{\sigma_f d}{2 l_c} = \frac{(2.8 \text{ GPa})(.013 \text{ mm})}{2(33 \text{ mm})} = 5.52 \times 10^{-4} \text{ GPa}.$$

conductivities, and thermal conductivities. If the composite is considered to consist of three materials—matrix, fiber, and pores (empty space)—then the volume fractions (f) of each of these materials must sum to 1, as shown in Equation 7.2:

$$f_f + f_m + f_v = 1 \qquad (7.2)$$

The density, thermal conductivity, and electrical conductivity of a void are all essentially zero and will not appear in subsequent equations, but care must be taken to make sure the presence of voids is accounted for in the calculation of volume fractions. Equations 7.3 through 7.5 provide the equations for density (ρ), thermal conductivity (k), and electrical conductivity (σ) for composites.

$$\rho_c = \rho_m f_m + \rho_f f_f \qquad (7.3)$$

$$k_c = k_m f_m + k_f f_f \qquad (7.4)$$

$$\sigma_c = \sigma_m f_m + \sigma_f f_f \qquad (7.5)$$

The relationships for mechanical properties are more complex. When a tensile load is applied to the composite in the direction of fiber reinforcement (longitudinally), both the fiber and the matrix start to deform. If the quality of bonding between the fiber and matrix is sufficient, they elongate at the same rate and experience the same strain. This is called the *isostrain condition*. While the applied load remains small, both the fiber and the matrix stretch elastically. When the yield strength of the matrix $(\sigma_{y,m})$ is exceeded, the matrix begins plastic deformation, but the stronger fibers remain in the elastic stretching region. At this point, more of the load is passed to the fibers, and the composite does not

| *Isostrain Condition* |
Condition where the quality of bonding between the fiber and matrix is sufficient that both elongate at the same rate and experience the same strain.

fracture even at loads that would destroy an unreinforced matrix. Because the stress on the composite (σ_c) must be borne by either the fiber or the matrix,

$$\sigma_c = \sigma_m f_m + \sigma_f f_f. \tag{7.6}$$

Because the strain (ϵ) on the fibers and matrix are equal, Equation 7.6 can be rewritten as

$$\frac{\sigma_c}{\epsilon} = \frac{\sigma_m f_m}{\epsilon} + \frac{\sigma_f f_f}{\epsilon}, \tag{7.7}$$

but the ratio of σ_c/ϵ is just the elastic modulus (E). Thus, the elastic modulus of the composite (E_c) may be estimated from

$$E_c = E_m f_m + E_f f_f. \tag{7.8}$$

The relative contribution of fiber and matrix to handling the applied load can be estimated from Equation 7.9,

$$\frac{F_f}{F_m} \approx \frac{E_f f_f}{E_m f_m}, \tag{7.9}$$

where F_f and F_m are the applied loads on the fibers and matrix, respectively.

Because the reinforcement fibers are brittle, they begin to fracture when their tensile strength ($\sigma_{s,f}$) is surpassed. Because of natural random variation in the tensile strengths of individual fibers, however, these fractures will result over a wide range of applied loads. Even when the fibers fracture, the composite itself can survive. The matrix continues to deform plastically, while the broken fiber pieces remain bonded to matrix material. Thus even the broken filaments continue to serve some reinforcement function.

All of the analysis just performed relies on the assumption of high-quality bonding between the fiber and the matrix. If the bonding is less strong, the bonds between the fiber and the matrix break, resulting in *fiber pull-out*. With no bonding between the fiber and the matrix, the load cannot be transferred to the fibers and the matrix behaves as if it were not reinforced at all.

When the load is applied in the transverse direction, the fibers provide essentially no reinforcing benefit to the matrix. As such, the stresses experienced by the fibers and the matrix are the same,

$$\sigma_c = \sigma_f = \sigma_m, \tag{7.10}$$

while the elastic modulus of the composite can be estimated from Equation 7.11,

$$E_c = \frac{E_f E_m}{f_m E_f + f_f E_m}. \tag{7.11}$$

This is called an *isostress condition*.

7.2.2 // Impact of Fiber Amount and Orientation

The amount of fiber added to a composite impacts both its cost and performance. Because the fibers are responsible for handling the applied load, the use of more fiber results in stronger composites. However, when the fiber fraction exceeds about 80%, there is not enough matrix material to completely

| *Fiber Pull-Out* |
Premature failure in a composite caused by inadequate bonding between the fiber and matrix.

| *Isostress Condition* |
Condition in which the fibers in a matrix offer essentially no reinforcing benefit to the matrix when a load is applied in the transverse direction, causing both to experience essentially the same strain.

Example 7-2

A cylindrical fiber-reinforced composite with a cross-sectional area of 100 mm^2 is comprised of 60 volume percent of a polymeric matrix (density = 1.2 kg/m^3, elastic modulus = 3 GPa, tensile strength = 300 MPa), 35% E-glass reinforcing fibers (properties found in Table 7-2 on p. 236), and 5% net voids. 20 MPa of stress is applied to the composite in the longitudinal direction.

a. Estimate the density of the composite.
b. Estimate the modulus of elasticity of the composite in the longitudinal direction.
c. Estimate the modulus of elasticity of the composite is the transverse direction.
d. Estimate the fraction of the load supported by the fibers.
e. Estimate the strain on either the fiber or the matrix, and explain why it is necessary to calculate only one of these.

SOLUTION

a. The density can be estimated from the mixing rule given in Equation 7.3:

$$\rho_c = \rho_m f_m + \rho_f f_f = \left(1.2\ \frac{kg}{m^3}\right)(0.60) + \left(2.58\ \frac{kg}{m^3}\right)(0.35) = 1.62\ \frac{kg}{m^3}.$$

Recall that the density of a pore is zero but that it still accounts for 5% of the total volume.

b. The modulus of elasticity in the longitudinal direction is calculated by Equation 7.8:

$$E_c = E_m f_m + E_f f_f = (3\text{ GPa})(.60) + (22\text{ GPa})(.35) = 9.5\text{ GPa}.$$

c. The modulus of elasticity in the transverse direction can be calculated from Equation 7.11:

$$E_c = \frac{E_f E_m}{f_m E_f + f_f E_m} = \frac{(22)(3)}{(.60)(22) + (.35)(3)} = 4.63\text{ GPa}.$$

d. Equation 7.9 is used to find the ratio of the load borne by the fibers relative to that borne by the matrix:

$$\frac{F_f}{F_m} \approx \frac{E_f f_f}{E_m f_m} = \frac{(22\text{ GPa})(.35)}{(3\text{ GPa})(.60)} = 4.3$$

This ratio indicates that the fibers bear 4.3 times more load than the matrix.

e. When it is assumed that the quality of bonding between the fiber and the matrix is high, a load applied in the longitudinal direction results in an isostrain case in which $\epsilon_f = \epsilon_m = \epsilon$. Once the strain on the fibers is known, so is the strain on the matrix. To find the strain on the fibers, it is first necessary to known the stress on the fibers. Stress is defined as the load divided by the cross-sectional area:

$$\sigma_f = \frac{F_f}{A_f}$$

To determine the load on the fibers, the total load on the composite must be calculated:

$$F_c = \sigma_c A_c = (20 \text{ MPa})(100 \text{ mm}^2) = 2000 \text{ N}.$$

The total load on the composite is the sum of the loads on the fiber and matrix,

$$F_c = F_f + F_m,$$

and part (d) of this problem concluded that the fiber took on 4.3 times more load than the matrix, so

$$F_c = 4.3 F_m + F_m$$
$$2000 \text{ N} = 5.3 F_m$$
$$F_m = 377 \text{ N}$$
$$F_f = 2000 \text{ N} - 377.4 \text{ N} = 1623 \text{ N}.$$

The cross-sectional area of the fibers must be determined from the volume fraction:

$$A_f = A_c f_f = (100 \text{ mm}^2)(.35) = 35 \text{ mm}^2.$$

So the stress on the fibers is given by

$$\sigma_f = \frac{F_f}{A_f} = \frac{1623 \text{ N}}{35 \text{ mm}^2} = 46.4 \text{ MPa}.$$

Once the strain on the fiber is known, the stress is determined from the elastic modulus:

$$\epsilon_f = \frac{\sigma_f}{E_f} = \frac{46.4 \text{ MPa}}{22,000 \text{ MPa}} = 2.11 \times 10^{-3}.$$

The identical number would be achieved using matrix fractions for the same calculation.

surround and bond with the fiber and transfer the load effectively. In most cases the reinforcing fibers are far more expensive than the surrounding matrix material, making it desirable to reduce the fraction of fiber in the composite. Although the exact fiber fraction varies with material type and application, most fiber-reinforced composites contain 35% to 50% fiber by volume.

The orientation of the fibers also plays a significant role in fiber properties. As shown in Figure 7-1, composites can be manufactured with uniaxial

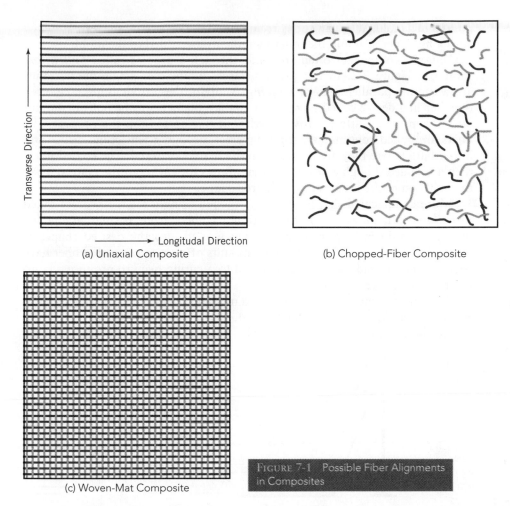

(a) Uniaxial Composite

(b) Chopped-Fiber Composite

(c) Woven-Mat Composite

FIGURE 7-1 Possible Fiber Alignments in Composites

fibers, randomly oriented chopped fibers, or complex two-dimensional or three-dimensional woven mats. Uniaxial fibers result in the composite having significantly more reinforcing capabilities in the longitudinal direction than in the transverse because of the near-perfect alignment of the fibers. Randomly oriented chopped fibers are isotropic, providing fundamentally the same properties in all directions. Because the fibers are smaller and only a small fraction is aligned to any direction of applied load, the maximum reinforcing capabilities are less than uniaxial composites. However, chopped-fiber composites are produced more easily and far more inexpensively. When both higher strength and the ability to withstand loads in multiple directions are needed, two- and three-dimensional weaves are used. With weaves, fibers can be aligned into multiple directions, and applied loads always will be perpendicular to some fraction of the fibers. However, the more complicated the weave pattern, the more complicated and expensive the manufacturing process.

7.2.3 // Manufacture of Fiber-Reinforced Composites

Composites are manufactured through a variety of processes. Simple chopped-fiber composites often are made by *resin formulation*, in which bits of chopped fiber are mixed or blown into the matrix material, along with any curing agents, accelerators, diluents, fillers, and pigments. If a polymeric matrix is used, the matrix material is melted prior to the addition of the other ingredients, then

| *Resin Formulation* |
Process in which bits of chopped-up fibers are mixed or blown into the matrix material, along with any curing agents, accelerators, diluents, fillers, or pigments, in order to form a simple chopped-fiber composite.

poured in a mold. When an epoxy resin is used as the matrix, the resin is mixed with a hardener as it and the other ingredients are added to the mold.

Uniaxial composites often are made through a process called *pultrusion*, shown in Figure 7-2. In this process, large numbers of single fiber strands are wound in parallel to form a *roving*. Many of these rovings are connected together in a device called a *creel* that enables filaments to be pulled continuously from many different rovings without having to stop the process. The fibers are pulled continuously from the creel through a tensioning device and into a bath where they are coated with the matrix material. The coated fibers then are pulled though a heated die that allows for curing of the matrix. Then the final composite is cut into the desired shape.

More complicated shapes can be developed using a related process called *wet-filament winding* in which continuous fibers from rovings are pulled through a resin impregnation bath, then wound into the desired shape, as shown in Figure 7-3. When sufficient amounts of resin-impregnated fiber have been wound around the part, it is moved to a curing oven to produce a composite of the desired shape.

Weaves and mats often are converted into composites using a *resin transfer molding* technique, shown in Figure 7-4. The woven fiber mat is placed into a space between a bottom and top mold. Resin is injected into a cavity in the

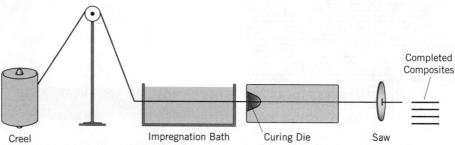

FIGURE 7-2   Pultrusion Process

Creel   Impregnation Bath   Curing Die   Saw   Completed Composites

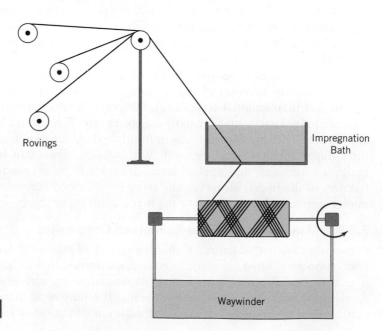

Rovings

Impregnation Bath

Waywinder

FIGURE 7-3   Wet-Filament Winding

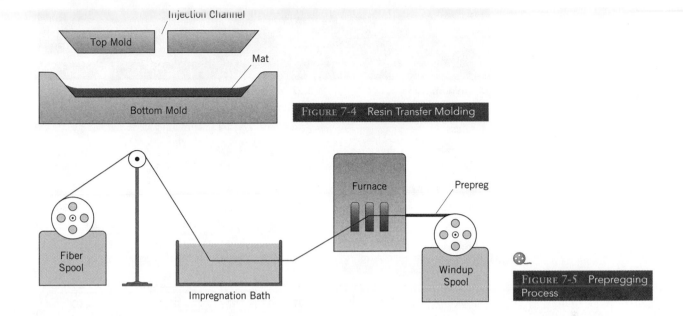

Injection Channel

Top Mold

Mat

Bottom Mold

FIGURE 7-4   Resin Transfer Molding

Furnace

Prepreg

Fiber
Spool

Impregnation Bath

Windup
Spool

FIGURE 7-5   Prepregging
Process

top mold with enough pressure to ensure that it penetrates and surrounds the woven mat. The molds are then cured using a combination of heat and pressure, creating a composite part in the shape of the molds.

All of the production techniques discussed so far involve coating the fibers with matrix materials immediately prior to manufacturing the composite part. Often, however, it is advantageous to have a fiber bundle that has already been impregnated with the matrix material and can be converted to the composite without any additional processing. These precoated fibers are referred to as *prepreg*, and the process of making them is called *prepregging*. During this process, the fibers are dipped in a resin solution or coated with small quantities of molten polymer or pitch, as shown in Figure 7-5. The coated fibers then are heated slightly in an oven to ensure that the coating sticks to the fibers. The resultant prepreg is stored in refrigeration until it is ready to be made into a composite.

Prepreg offers a significant advantage in that the matrix material is already dispersed on the fibers. Thus high-pressure injection of the resin into a mold is unnecessary. Prepreg generally is woven into the desired mat, trimmed, and placed into a mold. Often several layers are required to obtain the desired thickness. The mold is placed in a vacuum bag, then cured in an autoclave where the part is subjected to pressure and heat.

| *Prepreg* |
Fiber bundle already impregnated with matrix material, which can be converted into a composite without any additional processing.

| *Prepregging* |
Process of creating prepregs by dipping fibers into a resin bath and heating them slightly to ensure that the coating sticks.

7.2.4 // Selection of Reinforcing Fibers

All reinforcing fibers share the need for high tensile strength, but other considerations, including cost and density, exert strong influences on the selection. The specific strength (σ_{sp}) of a fiber is defined as the tensile strength of the fiber divided by its density, as shown in Equation 7.12:

$$\sigma_{sp} = \frac{\sigma_s}{\rho}.$$ (7.12)

The use of specific strength allows the strength-to-weight ratio to be factored into decisions.

Table 7-2 Properties of Common Reinforcing Fibers				
Fiber Type	Density (kg/m³)	Tensile Strength (GPa)	Specific Strength (GPa)	Elastic Modulus (GPa)
Aluminum oxide	3.97	10.0	2.5	360
Carbon fiber (PAN-based)	1.75	3.5	2.0	230
E-Glass	2.58	3.4	2.6	22
Kevlar-29	1.44	2.8	1.9	122
Kevlar-49	1.44	4.0	2.8	131
Molybdenum	10.2	2.2	0.2	327
Polyester	1.4	0.2	0.14	4.3
Silicon carbide	3.0	20.0	6.6	150
Steel (high tensile)	7.9	2.3	0.3	210
Titanium	4.5	8.3	1.8	116
Tungsten	19.3	2.9	0.15	21.1
UHMWPE	0.97	2.6	2.7	210

Reinforcing fibers tend to be made from ceramics, high-performance polymers, metals, or carbon fibers. Table 7-2 summarizes key properties for several common reinforcing fibers. Ceramic fibers such as silicon carbide and aluminum oxide tend to be stiff and strong but rather dense. Often glass fibers (E-glass) are selected because of a blend of high strength, chemical resistance, and low cost, although great care must be taken to avoid damaging the fibers during handling. Tungsten and molybdenum fibers find significant uses in space and welding applications because of their high melting points. Steel fibers tend to be heavy but add strength and thermal conductivity to the composite. In addition to possessing an exceptional specific strength for a metal, titanium fibers are inert in the body and capable of *osseointegration*, or forming a direct connection with living bone, making them ideal for composite dental and joint-replacement applications.

High-performance polymer fibers, including Kevlar and UHWMPE, generally are used with polyester or epoxy matrices and find significant use in bullet-resistant materials, sports gear, break linings, and tires. When performance needs outweigh cost concerns, carbon fibers are often the reinforcing material of choice. Carbon fibers have high specific strength and maintain their properties at elevated temperatures. Carbon fiber composites find widespread use in racecar frames, the space shuttle, and some construction applications, including the Westgate Bridge above the Yarra River in Melbourne, Australia.

7.2.5 // Selection of Matrix Materials

Most fiber-reinforced composites use polymeric materials as the matrix phase, although some applications benefit from the use of metals or ceramics. When the mechanical properties of the matrix are not crucial to the application, polyester resins provide the most economical choice. The majority of fiber-reinforced composites use an orthophthalic *polyester resin* that blends polyester monomers with styrene to reduce viscosity. An isophthalic polyester resin provides greater water resistance and is chosen when the composite will be exposed to aquatic environments, such as on the hull of a boat. When the resin is to be used, a catalyst is

| *Osseointegration* |
Process in which hydroxyapatite becomes part of the growing bone matrix.

| *Polyester Resin* |
Most economical choice for a matrix material in composites in situations where the mechanic properties of the matrix are not crucial to the application.

added to the pale, viscous liquid to initiate polymerization. An irreversible cross-linking, called *curing*, takes place to solidify the polyester resin.

$ *Epoxy resins* are much more expensive but provide improved mechanical properties and exceptional environmental resistance. Most composites used in the aircraft industry are made with epoxy resins because of their superior properties. Epoxy resins have a characteristic amber color and usually can cure at room temperature with the addition of a *hardener*, although the process can be accelerated by heating. Hardeners differ from catalysts in that hardeners becomes incorporated into the resultant polymer through an addition polymerization. Most hardeners contain amine groups.

    Vinyl ester resins represent a compromise between the economic advantages of polyester resins and the exceptional properties of epoxy resins. Vinyl esters are tougher and more resilient than polyesters and find commercial use in pipelines and storage tanks. Vinyl esters generally require elevated temperatures to cure fully.

    Although the three systems just discussed comprise the most common polymeric resins, other polymeric materials find use in special cases. *Phenolic resins* produce composites with many voids and poor mechanical properties but offer a level of fire resistance. Polyurethane resins provide a level of chemical resistance and offer significant toughness but are weak in compression. *Polyimide resins*, which are extremely expensive and find use only in high-end applications, such as missiles and military aircraft, can maintain their properties at temperature above 250°C.

    Metal matrix composites offer an alternative to the more common polymer matrices. Although once limited to military and aerospace applications because of high costs, metal matrix composites are making inroads into the sporting goods industry, the automotive industry, and electronic materials. The most common metal matrix is aluminum because of its high specific strength and relatively low cost. Compared to polymeric matrices, metal matrices provide high strength, improved environmental resistance (including the fact that they do not burn), much greater thermal conductivity, improved abrasion resistance, and the ability to operate at elevated temperatures.

    Ceramic matrix composites (CMCs) serve a different purpose from the other materials discussed in this section. When ceramic fibers are added to a matrix of a different ceramic material, the fracture toughness of the composite increases significantly while it maintains the ability to withstand high temperatures and corrosive environments. For this reason, CMCs generally replace standard ceramics for applications in which fracture toughness is a major concern. Many experts anticipate that CMCs will become standard features in advanced engines, which could make cooling fluids unnecessary and increase efficiency dramatically. Light CMCs may also replace superalloys, allowing for significant weight reduction.

7.3   PARTICULATE COMPOSITES

Particulate composites generally cannot provide the same strength as fiber-reinforced composites but are much easier to manufacture and much less expensive. Particulate composites contain a large number of randomly oriented particles called *aggregate* that help the composite withstand compressive loads. The final properties of particulate composites are easier to predict because

---

| *Curing* |
Hardening or toughening of a polymer material through a cross-linking of polymer chains.

| *Epoxy Resins* |
Resins used as matrixes in composite materials that are more expensive than polyester resin but provide improved mechanical properties and exceptional environmental resistance.

| *Hardener* |
Substance added to epoxy resin to cause it to cross-link; the hardener becomes incorporated in the resulting polymer.

| *Vinyl Ester Resins* |
Polymeric matrix material that combines the economic advantages of polyester resins and the exceptional properties of epoxy resins.

| *Phenolic Resins* |
Matrix materials that have many voids and poor mechanical properties but do offer a level of fire resistance.

| *Polyimide Resins* |
Polymeric matrix materials that are extremely expensive and used only in high-end applications, due to their ability to maintain their properties at temperatures above 250°C.

| *Metal Matrix Composites* |
Composites that use a metal as the matrix material in place of more common polymer matrices.

| *Ceramic Matrix Composites (CMCs)* |
Addition of ceramic fibers to a matrix of a different ceramic material to significantly increase the fracture toughness of the composite.

| *Aggregate* |
Hard, randomly oriented particles in a particulate composite that help the composite withstand compressive loads.

they are free from the orientation issues experienced by fiber-reinforced composites. Particulate composites tend to be isotropic, possessing the same properties in all directions.

In general, aggregate materials are much stronger than the surrounding matrix phase, but adjacent aggregate molecules cannot bind with each other. The matrix phase binds the harder aggregate particles together but also limits the strength of the composite. A composite made with strong aggregate but a weak matrix will fail under relatively low tensile loads. The aggregate particles tend to increase the modulus of the composite while reducing the ductility and permeability of the matrix material.

The aggregate molecules often reduce time-dependent deformations, including creep, and usually are far less expensive than the matrix material. Aggregate particles of less then 0.25 inches in diameter are classified as *fine aggregates*, while larger particles are classified as *coarse aggregates*.

7.3.1 // Concrete

The most important commercial particulate composite is *concrete*, a blend of gravel or crushed stone (the aggregate) and Portland cement (the matrix). Technically, the term *concrete* refers to any particulate composite that blends mineral aggregates with a binding matrix, but now the term generally applies to Portland cement concrete. The first use of Portland cement in concrete dates to 1756, when British engineer John Smeaton blended the cement with crushed brick and pebbles. Concrete is now ubiquitous in construction and paving; in the United States alone, there are more than 45,000 miles of concrete interstate highways plus countless minor roads and driveways. More than 6 billion tons of concrete are produced every year with nearly 40% being used in China.

In addition to Portland cement and aggregate, concrete uses water (to initiate the hydration reactions in the cement as discussed in detail in Chapter 6), *admixtures* (additives to the concrete designed to alter properties), and fillers (also discussed in Chapter 6). The aggregate accounts for about 75% of the total volume of the concrete, but the properties of the composite are dominated by the weaker Portland cement in the matrix. Admixtures typically account for less than 5% of the total volume of the concrete and serve four fundamental purposes:

1. *Hydration catalysts* alter the rate of hydration in the Portland cement. Those that speed up the hydration are called *accelerators*, and those that slow the rate are called *retarders*.
2. *Pigments* provide color to the concrete for aesthetic value.
3. *Air-entrainers* cause tiny air bubbles to form and distribute throughout the concrete, which enables the concrete to withstand the freeze-thaw expansion cycles without failing.
4. *Plasticizers* reduce the viscosity of the cement paste, making it easier to flow the concrete mixtures into its final form.

Concrete develops several desirable mechanical properties, including excellent compressive strength, moisture and chemical resistance, and volume stability, but has little tensile strength. The mechanical properties of concrete relate to the water–cement ratio; the size, shape, and composition of aggregate used; and the blend of cement and fillers.

| *Fine Aggregates* |
Aggregate particles with a diameter less than 0.25 inches.

| *Coarse Aggregates* |
Aggregate particles with a diameter greater than 0.25 inches.

| *Concrete* |
The most important commercial particulate composite, which consists of a blend of gravel or crushed stone and Portland cement.

| *Admixtures* |
Molecules added to a composite to enhance or alter specific properties.

| *Hydration Catalysts* |
Catalysts that alter the rate of hydration in Portland cement.

| *Accelerators* |
Hydration catalysts that speed up the rate of hydration in Portland cement.

| *Retarders* |
Hydration catalysts that decrease the rate of hydration in Portland cement.

| *Pigments* |
Coloring agents that do not dissolve into the polymer.

| *Air-Entrainers* |
Additives that cause air bubbles to form and be distributed throughout concrete to enable it to withstand freeze-thaw expansion cycles without failing.

| *Plasticizers* |
Additives that cause swelling, which allows the polymer chains to slide past one another more easily, making the polymer softer and more pliable. Also used to decrease the viscosity of cement paste to make it easier to flow the concrete into its final form.

The water–cement ratio exerts the greatest influence on the strength and durability of the concrete. When the water-to-cement ratio is high, a wet gel is produced and results in concrete that is weak and very susceptible to weathering. Lowering the water–cement ratio increases strength but presents two difficulties of its own. The *American Concrete Institute (ACI)* requires that sufficient water be present to allow the concrete to cure for at least seven days to enable any unreacted cement to hydrate. Additionally, sufficient water must be added to enable the concrete to be worked into all parts of its mold or frame. To alleviate these problems, small quantities of admixtures are added to the concrete. Air-entrainers trap small air bubbles throughout the concrete, which improve its durability and serve as a lubricant, reducing the need for water. Additionally, plasticizers are added that significantly both improve the flow characteristics of the concrete and reduce the need for water.

Aggregate sizing affects nearly every mechanical property of the concrete, including fatigue resistance, stiffness, moisture resistance, workability, and stiffness. Smaller aggregates tend to produce stronger concrete but are harder to work with and negatively impact modulus of elasticity and creep resistance. Typically, a distribution of aggregate particle sizes is used with the goal of having smaller particles fill some of the void spacing between larger particles, as shown in Figure 7-6, thereby improving the interaction between aggregate particles.

The particle size distributions are obtained by passing the aggregate through sieve trays in a process called *gradation*. In 1907 William Fuller and Sanford Thompson showed that the maximum packing density could be achieved when Equation 7.13 was satisfied:

$$p = \left[\frac{d}{D_{100}}\right]^{0.5},\qquad(7.13)$$

where P is the percentage of particles that pass through a given sieve size, d is the diameter of the particle, and D_{100} is the diameter of the largest aggregate particle. The Department of Transportation typically uses a *Fuller-Thompson parameter* of 0.45 instead of 0.5 in the exponent of Equation 7.13.

The shape of the aggregate particles also impacts properties. Smooth pebbles have a much smaller surface-area-to-volume ratio than irregular broken pieces of rock. Therefore, significantly more cement is required to achieve the same degree of bonding between the smooth-pebble aggregate and the matrix. Similarly, the surface chemistry of the aggregate impacts how effectively it can bond with the cement matrix. For example, limestone ($CaCO_3$) bonds much more easily with Portland cement than gravel or many other aggregate types.

| *American Concrete Institute (ACI)* |
Technical society dedicated to the improvement of the design, construction, and maintenance of concrete structures.

| *Gradation* |
Process of passing aggregate through sieve trays to acquire the particle size distributions.

| *Fuller-Thompson Parameter* |
Parameter used in the equation used to determine the maximum packing density in concrete.

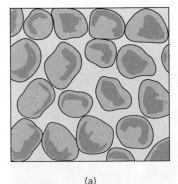
(a)

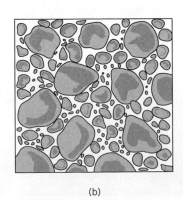
(b)

FIGURE 7-6   Alterations in Packing between (a) Large Relatively Uniform Particles and (b) Graded Particles

The compressive strength of concrete is measured by testing samples that have been hardened at constant temperature and 100% humidity for 28 days. A cylindrical sample of 6 inches in diameter and 12 inches in length is tested to failure in uniaxial compression, as described in Chapter 3. The *maximum 28-day compressive strength* (f_c) usually occurs near a strain of 0.002, but the stress-strain curve for concrete is highly nonlinear. As a result, the modulus of elasticity varies with stress level. The ACI specifies an *effective secant modulus of elasticity* (E_c) of

$$E_c = 0.043 w_c^{1.5} \sqrt{f_c'}. \tag{7.14}$$

where E_c is in MPa and w_c is the density of the concrete (typically 2320 kg/m^3).

The tensile strength of concrete varies greatly but tends to range between $0.08 f_c'$ and $0.15 f_c'$. The tensile strength is measured by loading a concrete beam to failure by concentrating a load at midspan. The maximum tensile stress at the bottom surface of the beam, called the *modulus of rupture* (f_r), can be calculated from standard beam stress equations. However, because the tensile strength of concrete is so much smaller than its compressive strength, concrete always fails in tension, even when compressive loads are applied. As such, the modulus of rupture can be correlated with the 28-day maximum compressive strength. The ACI code provides

$$f_r = 0.7 \sqrt{f_c'}. \tag{7.15}$$

The presence of many fine cracks throughout the concrete significantly impairs the tensile strength while having little effect on the compressive strength. When the concrete is placed under tension, stress concentrations develop at the crack tips, resulting in local stresses significant enough to lengthen the cracks. As the crack grows, the undamaged area shrinks, stresses rise, and failure results.

Because almost all structural applications require the ability to withstand some significant level of tensile forces, reinforcing materials usually are added to the concrete to handle the tensile loads. Although glass, polymeric, and even carbon fibers have been added to concrete, the most common reinforcing materials are steel-deformed reinforcing bars (*rebar*), like the one shown in Figure 7-7. Rebar is manufactured in diameters ranging from 0.375 inches to 2.25 inches and is categorized by *bar size*, in which each number represents an additional 0.125 inches. Thus a bar size of 3 would refer to a 0.375-inch-diameter rebar, while a bar size of 10 would represent a 1.25-inch sample. Table 7-3 summarizes the mechanical properties of rebar.

The addition of rebar creates a new composite with the rebar essentially serving the role of the fiber-reinforcing material and the concrete acting as the orienting and load-transferring matrix. Essentially all of the issues discussed in Section 7.2 with fiber-reinforced composites apply to rebar-reinforced concrete.

| *Maximum 28-Day Compressive Strength* |
Compressive strength of concrete from a tested sample that was hardened at constant temperature and 100% humidity for 28 days.

| *Effective Secant Modulus of Elasticity* |
American Concrete Institute–specified modulus of elasticity for concrete that accounts for the tendency of the modulus of elasticity for concrete varying with the stress level.

| *Modulus of Rupture* |
Maximum tension strength at the bottom surface of a concrete beam.

| *Rebar* |
Steel-deformed reinforcing bars used to reinforce the ability of concrete to handle tensile loads.

| *Bar Size* |
Categorization of the diameter of rebar, with each number representing an additional 0.125 inches.

FIGURE 7-7 Photo of Rebar Sample

Courtesy James Newell

TABLE 7-3 Properties of Rebar			
Bar Number	Diameter (in)	Cross-Sectional Area (in²)	Nominal Weight (lb_m/ft)
3	0.375	0.11	0.376
4	0.500	0.20	0.681
5	0.625	0.31	1.043
6	0.750	0.44	1.502
7	0.875	0.60	2.044
8	1.000	0.79	2.670
9	1.128	1.00	3.400
10	1.270	1.27	4.303
11	1.410	1.56	5.313

7.3.2 // Asphalt (Asphalt Concrete)

The other major commercial particulate composite is asphalt concrete (also known as asphalt cement) or, more commonly called, *asphalt*. By any name, asphalt is the familiar blacktop material used in roadways and parking lots. Like traditional concrete, asphalt concrete is a blend of mineral aggregate and a binder phase of asphalt, a high molecular weight fraction produced from petroleum distillation. Unlike traditional concrete, which cures into a permanent form through a complex series of hydration reactions, asphalt concrete can be remelted and reshaped many times. In fact, asphalt is the most recycled material in the world, although the recycled material does not maintain the same level of moisture resistance as new asphalt.

The use of asphalt in place of concrete in roadways has some disadvantages. The initial cost of asphalt is less, but a typical concrete roadway will last twice as long. Asphalt deforms more under the heavy loads of trucks and other large vehicles, making it less appropriate for major roadways used by freight haulers. Also, asphalt is less slip resistant and tends to absorb heat, often making it too hot to walk on during summer months.

Most of the asphalt used on major highways is produced through a *hot-mix asphalt concrete (HMAC) process*. The asphalt is softened by heating it to above 160°C before mixing in the aggregate, then the pavement is laid and compacted at 140°C. The process releases significant amounts of carbon dioxide and organic vapors and generates a characteristic aroma. The amount of vapor emissions can be reduced by adding zeolites (porous materials), such as sodium aluminum silicate or waxes, to the mixture. Zeolites reduce the softening temperature of the asphalt by as much as 25°C, leading to a *warm-mix asphalt concrete (WAM) process*. Although the technology reduces emissions, lowers costs, and makes a more pleasant working environment, the construction industry has been slow to embrace the change. During the 1970s the industry attempted a warm-mix process using moisture from the aggregate, but inconsistent moisture levels in the aggregate led to inconsistent properties and a near abandonment of the process for a quarter of a century.

Cold-mix asphalt concrete results from adding water and surfactant molecules to the asphalt before mixing it with the aggregate. When the water evaporates,

| *Asphalt* |
Blend of high-molecular-weight hydrocarbons left over from petroleum distillation. Also the common name used for asphalt concrete.

| *Hot-Mix Asphalt Concrete (HMAC) Process* |
Process used to produce most asphalt on major highways, in which asphalt is heated to 160°C before mixing in the aggregate and is laid and compacted at 140°C.

| *Warm-Mix Asphalt Concrete (WAM) Process* |
Process of creating asphalt concrete, by adding zeolites to lower the softening temperature by up to 25°C. This reduces the emission released and the cost, and creates more pleasant working conditions.

| *Cold-Mix Asphalt Concrete* |
Asphalt concrete created by adding water and surfactant molecules to asphalt, resulting in the creation of asphalt concrete similar to that formed through the hot-mix asphalt concrete (HMAC) process. This serves as the primary mechanism for recycling petroleum-contaminated soil.

the resultant asphalt concrete takes on properties similar to that produced through the HMAC process. This technique generally is applied only to patches on existing pavement or on low-volume roadways. Cold mix asphalt also serves as the primary recycling mechanism for petroleum-contaminated soils.

7.4 LAMINAR COMPOSITES

Laminar composites consist of alternating layers of two-dimensional materials that have an anisotropic orientation that connects together by layers of matrix materials. The most common laminar composite is *plywood*, which consists of thin layers of wood veneer bonded together by adhesives. The selection of both the veneer and the adhesive depends on the desired application. Plywood is more resistant to shrinking and warping than regular wood because of *cross-banding*, so that the grain of the wood in each sheet of veneer is offset 90 degrees from its neighbors.

When plywood is designed for decorative use, the surface veneer may come from birch, maple, mahogany, or other hardwoods, while the inner layers generally are made from less expensive soft pines. Most plywood contains three, five, or seven layers of veneers, each 0.125 inches thick. Plywood designed for indoor use typically uses a phenol formaldehyde resin that is inexpensive but can dissolve in water; outdoor plywood uses a more expensive but more robust phenol-resorcinol resin.

Other commercial laminar composites include windshield safety glass (discussed in Chapter 1), snow skis (originally made from layers of fiberglass and wood but now often manufactured from more complicated blends that include layers of sintered polyethylene, steel, rubber, carbon, fiberglass and wood), and the rubbery material in automobile tires, which contain 28% inexpensive carbon black particles in a polyisobutylene matrix.

In aerospace applications, where strength is required but weight is a significant factor, a more complex stacking pattern is used for the layers, leading to the formation of a *sandwich composite*. Often strong *face sheets* are used on the outer ends of the composite. These face sheets usually are made of very strong materials, such as titanium alloys, aluminum alloys, or fiber-reinforced composite mats, because they are responsible for handling most of the applied loads and stresses. In between the face sheets, a low-density material often is shaped in the *honeycomb structure* shown in Figure 7-8, and used to add stiffness and resist perpendicular stresses. The properties of the honeycomb materials depend on the cell size and the thickness and strength of the web material. In simple applications, cardboard may be used as the honeycomb material, but aluminum or high-performance polymers are needed for aerospace and other high-end applications.

| *Plywood* |
Most common type of laminar composite, consisting of thin layers of wood veneer bonded together with adhesives.

| *Cross-Banding* |
In plywood, the grain of the wood veneer is offset by 90 degrees from its neighbor, causing plywood to be more resistant to warping and shrinking than normal wood.

| *Sandwich Composite* |
Composite used in situations where strength is required but weight is a significant factor. Usually made up of strong face sheets on the outer ends of the composite with a low-density material sandwiched inside, often in a honeycomb structure, which adds stiffness and resistance to perpendicular stress.

| *Face Sheets* |
Typically very strong materials used as the outer ends of a sandwich composite.

| *Honeycomb Structure* |
Common shaping used for the low-density material in sandwich composites to add stiffness and resistance to perpendicular stresses.

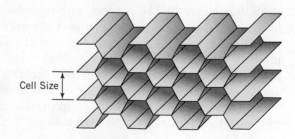

Cell Size

FIGURE 7-8 Honeycomb Structure Used in a Sandwich Composite

7.5 RECYCLING OF COMPOSITE MATERIALS

More than 98% of all obsolete composite materials are incinerated or sent to landfills; recycling has been little more than an afterthought. Historically, the most common technique for the limited recycling of composite materials has been to grind, shear, or chip the obsolete composite into small-enough pieces to be used as filler in new composites. This technique is reasonably effective when the composite materials contain significant quan_ tities of low-cost filler materials, but higher-end composite materials create both economic and environmental problems.

Acid digestion has been used to eat away the matrix and allow for the reharvesting of expensive carbon fibers, but this technique is replete with problems. The resultant acidic mixture is highly toxic, and the required processing offsets most of the economic benefits of the recycling. Major aircraft manufacturers have been performing detailed life cycle analyses to determine environmentally benign ways to reuse and recycle composite materials.

As is often the case, Europe is leading the way in mandating the recycling of composite materials. Most states in the European Union banned the landfilling of composite materials by 2004, with additional limits placed on the total amount that can be incinerated. As a result, many composite manufacturers have been forced to assume the responsibility for recycling the final products.

While the supply of obsolete composite material far exceeds demand, Europe has developed a "European Composite Recycling Concept" to fund research into developing and validating economically viable procedures for large-scale commercial recycling of composite materials.

Summary of Chapter 7

In this chapter, we:

- Learned how to classify a composite as fiber-reinforced, particulate, laminar, or hybrid
- Compared and contrasted the role of the fiber and matrix in fiber-reinforced composites
- Developed mixing rules that enable the prediction of mechanical properties of composites
- Discussed different methods of manufacturing fiber-reinforced composites
- Summarized the key properties involved in the selection of reinforcing fibers and matrix materials
- Analyzed the factors that influence the mechanical properties of particulate composites with special emphasis on concrete and asphalt
- Considered the production of laminar composites with special emphasis placed on plywood
- Discussed the slow progress in developing viable commercial strategies to recycle composite materials

Key Terms

Homework Problems

1. A composite with negligible voids and 35% fiber fraction is to be manufactured with a polyester resin ($\rho = 1.4$ kg/m^3, E = 3.5 GPa) as its matrix. Compare the density of the composite and the fraction of load taken on by the fibers if titanium fibers are used or if Kevlar-49 fibers are used. What other factors might influence the final fiber selection?

2. Explain why the tensile strength of concrete is so much lower than its compressive strength.

3. Calculate the strain on the matrix in a 40 mm^2 cylindrical composite that is subjected to a load of 30 GPa if the fiber fraction is 40%, voids are negligible, and the elastic modulus of the fiber is 35 times that of the matrix.

4. Why is it important that the compressive strength of concrete always be measured after the same amount of curing time (28 days)?

5. Estimate the modulus of elasticity for concrete with a density of 120 lb/ft^3 and a 28-day maximum compressive strength of 3500 psi.

6. Two important goals in manufacturing composites are improving the quality of bonding between the fiber and the matrix and reducing costs. Explain how enhancing the two major goals tends to result in composites that are much more difficult to recycle.

7. If the mechanical properties of the composite are governed by the properties of the reinforcing fiber, why would an engineer pay more for an epoxy resin instead of using a less expensive polyester resin?

8. Explain the advantages and disadvantage of using smaller aggregate particles in particulate composites.

9. Many bridges are made from rebar-reinforced concrete composites. Classify the composite system (fiber-reinforced, particulate, laminar, or hybrid), and explain why it is such a ubiquitous choice for construction.

10. Why is asphalt concrete often used instead of Portland cement concrete in parking lots and driveways?

11. Explain why sandwich composites with aluminum face sheets and high-performance polymeric honeycombs are used in aircraft wings.

12. Explain why fiber-reinforced composites can withstand much higher loads in the longitudinal direction than in the transverse. How does this impact the use of woven mats?

13. Describe the impact of voids on composites in terms of tensile strength, compressive strength, density, cost, and elastic modulus.

14. Explain why Equations 7.2 through 7.11 do not apply to composites with low wet out.

15. A scientist has developed a new polymeric fiber that she feels would make an excellent reinforcing fiber in composite materials. What properties of the fiber would you need to know (and why) before you could judge its potential as a reinforcement fiber?

16. Why would cold-mix asphalts be an attractive material for patching roadways but not for constructing them in the first place?

17. Explain the function of steel belts, carbon black, and polyisobutylene polymer in a commercial automobile tire.

18. What are the primary differences between plywood designed for indoor and outdoor uses?

19. What distinguishes an admixture from a filler?

20. What is the difference between a composite and an alloy?

21. What fiber fraction would be required for a composite with titanium fibers in a polycarbonate matrix ($E = 2.5$ GPa) to have the fibers handle 85% of a longitudinal tensile load? Estimate the density and transverse modulus of the composite.

22. If the effective secant modulus for a concrete sample is 450 MPa, estimate its 28-day maximum compressive strength.

23. Why would a contractor bother to prepare prepreg instead of making a composite immediately?

24. What are the advantages and disadvantages of using randomly oriented, chopped fibers in composites?

25. Compare the elastic modulus and density of two composites, one made with 40% E-glass fiber in a polyester matrix ($\rho = 1.35, E = 45$ MPa), the other made with 50% E glass in the same polymer. Why would it not be beneficial to use 85% E-glass fiber?

26. Compare the density, elastic modulus in the longitudinal direction, elastic modulus in the transverse direction, and fraction of load handled by the fibers for a composite made with 35% Kevlar-49 fibers in an epoxy resin ($\rho = 1.1$ kg/m^3, $E = 2.5$ GPa) with that made from a polyester matrix ($\rho = 1.35, E = 45$ MPa).

27. Discuss possible alternatives besides incineration or landfilling for
   a. Concrete
   b. Automobile tires
   c. Fiberglass
   d. Sandwich composites from airplane wings

28. Explain what happens to the mechanical properties of a composite (both in the transverse and longitudinal direction) if there is significant misalignment in the reinforcing fibers.

29. Discuss the primary factors that influence the selection of the optimal fiber and fiber fraction for a composite application.

30. Why are polymeric fibers generally a poor choice for high-temperature applications?

8

Electronic and Optical Materials

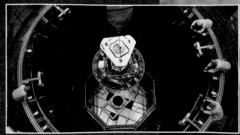

CONTENTS

Learning Objectives

By the end of this chapter, a student should be able to:

- Explain the physical and chemical basis for conduction in metals.

- Calculate drift mobility, current density, and conductivity.

- Discuss the factors that influence resistivity.

- Determine the resistivity of an alloy.

- Explain the nature and significance of the energy gap between the valence band and conduction band.

- Describe the nature of intrinsic semiconduction and the migration of both electrons and holes in response to an applied electric field.

- Explain the role of dopants on both p-type and n-type extrinsic semiconductors and the operation of diodes based on p–n junctions.

- Discuss the operational principles of BJT transistors and MOSFETs.

- Describe the manufacture of integrated circuits.

- Explain the role of dielectric materials in capacitors.

- Distinguish between ferroelectric and piezoelectric materials, and describe the unique properties associated with each.

- Explain the distinctions among transmitted, absorbed, reflected, and refracted light.

- Calculate reflectance and transmission coefficients given indices of refraction.

- Apply Snell's law to determine the angle of refraction and the critical angle for total internal reflection.

- Explain the operating principles and commercial uses of optical fibers and lasers.

Most of the analysis in this text has focused on the mechanical properties of materials. However, some materials have commercial importance because of how they conduct electricity (electronic materials) or reflect, absorb, or transmit light (optical materials). Although electronic and optical materials consist of the standard classes already discussed (polymers, metals, ceramics, and composites), their unique behaviors warrant separate consideration.

How Do Electrons Flow through Metals?

8.1 CONDUCTIVITY IN METALS

Because of the dissociated nature of the metallic bond, metals possess free electrons that are able to move when the material is subjected to an electric field at absolute zero. Any material that does not have free electrons at absolute zero is considered to be a nonmetal or *insulator*. However, at temperatures greater than absolute zero, many insulators develop the ability to conduct electrons. These materials are called *semiconductors*, and their development has ushered in a revolution in communication and other technologies that have fundamentally altered the life of most of the developed world.

The ability to conduct electrons relates directly to the configuration of electron shells around the atom. Electrons remain in orbit around the nucleus because of the Coulombic attraction between the electrons and the protons in the nucleus. Complex quantum mechanics proves that an electron must exist in a discrete energy level, or shell, that satisfies the equation

| *Insulator* |
Material that has no free electrons at absolute zero.

| *Semiconductors* |
Materials having a conductivity range between that of conductors and insulators.

$$E = \frac{n^2 h^2}{8m\,L^2} \tag{8.1}$$

TABLE 8-1 Subshells and Electron Available for Each Quantum Number

Quantum Number	Maximum Electrons	Subshell Types
1	2	1s
2	8	1s, 3p
3	16	1s, 3p, 5d
4	32	1s, 3p, 5d, 7f

where E is the energy of the electron, n is the *principal quantum number* that represents each shell, h is Planck's constant, m is the mass of the electron, and L is a characteristic length associated with wave motion. The energy of all electrons within the same shell is equal. Within each shell, distinct sublevels (s, p, d, and f) account for differences in angular momentum and spin. Ultimately, there are n^2 quantum states for each shell, with each state capable of holding one electron of positive spin and one electron of negative spin, as required by the Pauli exclusion principle. Table 8-1 summarizes the different energy levels and quantum states.

As long as atoms remain isolated from other atoms, the number of allowable states is defined rigidly. The matter becomes more complicated when large numbers of atoms come into close proximity, as in a solid Bravais lattice. As electron clouds are brought together, free electrons from one atom can move to the orbitals of another atom. This "sharing" of electrons is the basis for the metallic bond, but even this sharing has limitations. The Pauli exclusion principle prevents atoms with the same spin from existing on the same sublevel. As a result, the energy levels split slightly to form *energy bands*, each with slightly different energy levels.

Consider a Group I element that has a single unpaired electron in its outer shell. If a specific number of these Group I atoms (n) come in contact with each other, there would be n s-levels with n free electrons. The lowest energy states always must fill first and each s-state is capable of holding two electrons. As a result, the n/2 bands with the lowest energy states (or ground states) will be filled with two electrons each, while the n/2 bands with the highest energy will be vacant. Figure 8-1 shows the formation and filling of these energy bands for four atoms coming into contact. The two lowest energy bands would each contain two electrons of opposite spin, while the two highest energy bands would be vacant. The energy level of the highest occupied band is called the *Fermi energy* (E_F).

The type of band splitting shown in Figure 8-1 is not limited to Group I atoms. Because the electronic properties of materials are governed by the interactions of the outermost electron bands and because these bands can

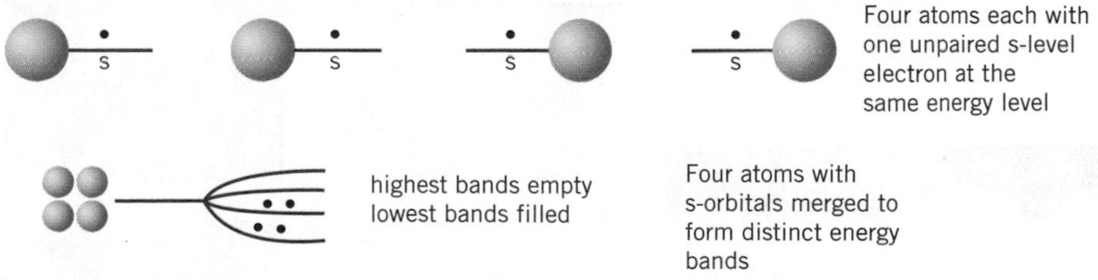

| *Principal Quantum Number* |
| Major shell in which an electron is located. |

| *Energy Bands* |
| Split of energy levels with slight variance. |

| *Fermi Energy* |
| Energy level of the highest-occupied energy band. |

Four atoms each with one unpaired s-level electron at the same energy level

highest bands empty
lowest bands filled

Four atoms with s-orbitals merged to form distinct energy bands

FIGURE 8-1 Electron Bands for Four Group I Atoms in Close Proximity

contain only electrons from the same corresponding levels, all metals with a single unpaired electron in an s-shell will form bands like those shown in Figure 8-1. Copper has one electron in its 4s shell that can interact with the 4s electron from another copper atom. The 4s shell can interact only with another 4s shell.

The amount of splitting between the electron bands varies with interatomic spacing, and the number of states depends on the type of orbitals present in each atom. If n atoms are present, each s band will have n states, each p band will have 3n states, while each d band would have 8n states.

When s and p orbitals are both present, the energy bands often overlap. For example, in aluminum some of the 3p bands have a lower energy state than some of the 3s bands, as shown in Figure 8-2. This combined spb allows aluminum to have free electrons, even though an isolated atom would have no unpaired electrons to share.

The rate of electron movement through a unit area is given by

$$J = \frac{\Delta q}{A \Delta t},$$
(8.2)

| *Current Density* |
Density of electrical current.

where J is the *current density*, Δq is the total charge passing through the area, A is the area, and t is time. Because the electrons are free to move randomly in any direction through the shared orbitals, an electric field (E) must be applied to induce a net flow of electrons in a specific direction. The unpaired electrons dissociate from the nucleus and become part of the electron cloud surrounding the metal and are free to respond to the applied field to generate a current density.

The stronger the applied field, the faster the electrons will flow. The average velocity of electrons is defined as the *drift velocity* (v_d),

| *Drift Velocity* |
Average velocity of electrons due to an applied electrical field.

$$v_d = \mu_d E,$$
(8.3)

| *Drift Mobility* |
Proportionality constant relating drift velocity with an applied electrical field.

where the proportionality constant μ_d is called the *drift mobility*.

Although there is a net electron flow in the direction opposite of the electric field, the electrons do not flow in an orderly fashion. The atoms within

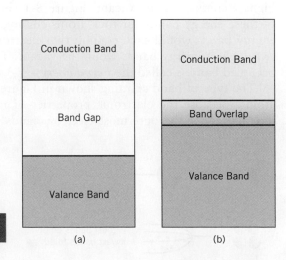

FIGURE 8-2   Energy Bands for (a) Copper and (b) Aluminum

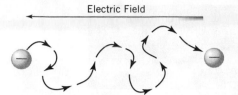

Electric Field

FIGURE 8-3 Electron Scattering in Metals

the metal continually vibrate, and the lattice itself contains many dislocations, substitutions, and other defects, so the electrons inevitably will collide with the vibrating lattice atoms. As a result, electron scattering results in which an electron is deflected between atoms in the lattice, much like a pinball bouncing off of bumpers, as illustrated in Figure 8-3. Each individual bounce may move the electron forward or backward, but the electric field will cause a net positive migration.

The amount of time between scattering collisions is accounted for by the drift mobility, which can be defined as

$$\mu_d = \frac{e\tau}{m_e},$$ (8.4)

where e is the absolute charge of an electron (1.6×10^{-16} C), τ is the mean time between scattering collisions, and m_e is the mass of an electron (9.1×10^{-31} kg).

The current density (J) is directly proportional to the drift velocity and can be expressed as

$$J = env_d,$$ (8.5)

where e again is the absolute charge of an electron, and n is the number of electrons per unit volume. Combining Equations 8.3 and 8.5 provides

$$J = en\mu_d E.$$ (8.6)

The *electrical conductivity* (σ) of a metal then can be defined as

$$\sigma = \frac{J}{E},$$ (8.7)

which is really just the microscopic form of Ohm's law. Ohm's law is more familiar in its macroscopic form,

$$\frac{1}{R} = \frac{I}{v}.$$ (8.8)

Combining Equations 8.6 and 8.7 yields

$$\sigma = en\mu_d.$$ (8.9)

Table 8-2 summarizes electrical conductivity values for several common metals and metal alloys.

| *Electrical Conductivity* |
Ability of a material to conduct an electrical current.

TABLE 8-2 Electrical Conductivity Values for Common Metals and Metal Alloys near Room Temperature

Material	Electrical Conductivity $(\Omega cm)^{-1}$
Aluminum	3.8×10^5
Carbon Steel	0.6×10^5
Copper	6.0×10^5
Gold	4.3×10^5
Iron	1.0×10^5
Silver	6.8×10^5
Stainless Steel	0.2×10^5

Example 8-1

Calculate the drift mobility of silver near room temperature.

SOLUTION

Equation 8.8 provides

$$\sigma = en\mu_d \text{ or } \mu_d = \frac{\sigma}{en}.$$

The electrical conductivity (σ) is provided in Table 8-2, and the charge of the electron (e) is a constant, so the only unknown is the number of free electrons per unit volume (n). Because each silver atom has one unpaired electron to donate to the electron cloud, n must be the same as the number of silver atoms per unit volume. Therefore, n must be given by

$$n = \frac{\rho_{Ag}N_A}{M_{Ag}},$$

where ρ_{Ag} is the density of silver, N_A is Avogadro's number, and M_{Ag} is the molecular weight of silver. Thus,

$$n = \frac{(10.49 \text{ g/cm}^3)(6.02 \times 10^{23} \text{ atoms/mol})}{(107.87 \text{ g/mol})}$$

$$= 5.85 \times 10^{22} \text{ electrons/cm}^3,$$

$$\mu_d = \frac{\sigma}{en} = \frac{6.8 \times 10^5 (\Omega cm)^{-1}}{(1.6 \times 10^{-19} \text{ C})(5.85 \times 10^{22} \text{ electrons/cm}^3)}$$

$$= 76.2 \frac{cm^2}{Vs}.$$

I f the conducting electrons were free to move unencumbered with no collisions, all metallic lattices would be free of resistance. Instead, collisions within the lattice create a barrier to conductivity. The magnitude of this barrier is called the *electrical resistivity* (ρ) and is the inverse of conductivity,

$$\rho = \frac{1}{\sigma}. \tag{8.10}$$

Anything that results in more collisions (and thus more scattering) raises the electrical resistivity and lowers conductivity. Three main factors combine to impact the resistivity of a metal: temperature, impurities, and plastic deformation. *Matthiessen's rule* states that each of these factors acts independently of the other and that the total resistivity of the material can be defined as

$$\rho = \rho_t + \rho_i + \rho_d, \tag{8.11}$$

where t, i, and d correspond to thermal, impurity, and deformation, respectively.

When temperature increases, the number of vacancies in the lattice increases, as does the rate of atomic vibration. As a result, the amount of scattering and the *resistivity* increase. From kinetic theory, the explicit relationship for the thermal portion of resistivity is defined as

$$\rho_t = \frac{m_e T}{e^2 n a}, \tag{8.12}$$

where T is temperature and a is a material-specific proportionality constant. Because the mass of an electron (m_e), the charge of an electron (e), and the number of electrons per unit volume (n) are all independent of temperature, Equation 8.12 is often applied as a simple proportionality relative to a specific reference temperature,

$$\rho_t = \rho_0 + \frac{T}{a}, \tag{8.13}$$

where ρ_0 is the resistivity at the reference temperature.

Impurity atoms are either larger or smaller than the main lattice atoms, and they distort the lattice in the region around the impurity. This irregularity in structure increases the amount of scatter and accordingly increases resistivity. In the case of metal alloys that form single-phase solid solutions, one atom functions as an impurity in the lattice of the other. As the amount of impurity increases, the term can become so large that it dominates the total resistivity, and the role of temperature becomes less important. For example, when 20% chromium is alloyed with 80% nickel, the resultant alloy (called nichrome) has a resistivity 16 times higher than pure nickel. Increased electrical resistivity is not always a bad thing; nichrome has extensive commercial applications as heating wires in furnaces, toasters, and other consumer appliances.

The resistivity of a binary alloy can be estimated with *Nordheim's rule*,

$$\rho_i = CX(1 - X), \tag{8.14}$$

where C is the *Nordheim coefficient*, a proportionality constant that represents how effective the impurity is at increasing resistivity, and X is the concentration of the impurity. The Nordheim equation works well only

| *Electrical Resistivity* |
| Barrier to conduction of electrons caused by collisions within the lattice. |

| *Matthiessen's Rule* |
| Rule that states that temperature, impurities, and plastic deformation act independently of one another in impacting the resistivity of a metal. |

| *Nordheim's Rule* |
| Method for estimating the resistivity of a binary alloy. |

| *Nordheim Coefficient* |
| Proportionality constant representing the effectiveness of an impurity in increasing resistivity. |

TABLE 8-3 Nordheim Coefficients for Copper Lattices		TABLE 8-4 Nordheim Coefficients for Gold Lattices	
Metallic Impurity	C (nΩ m)	Metallic Impurity	C (nΩ m)
Gold	5500	Copper	450
Nickel	1200	Nickel	790
Tin	2900	Tin	3360
Zinc	300	Zinc	950

with dilute solutions. Tables 8-3 and 8-4 summarize Nordheim coefficients for alloys of metallic impurities in copper or gold lattices.

Plastic deformation also increases the electrical resistivity of the system by distorting the lattice and increasing the amount of scattering. However, the impact of plastic deformation tends to be far less than either that of increasing temperature or the presence of impurities.

What Happens When There Are No Free Electrons?

8.3 INSULATORS

| Hole |
Positively charged vacant site left by an electron moving to a higher energy state.

| Conduction Band |
Band containing conductive electrons in a higher energy state.

| Valence Band |
Band containing covalently bonded electrons.

| Energy Gap |
Gap between the conduction and valence bands.

In most covalently bonded materials, the s and p states are completely filled, leaving no free electrons. With no empty states available, there is nowhere for the electron to move when exposed to an electric field. The covalent bond must be broken for the electron to be able to move. When such a breakage occurs, the electron moves to a much higher energy state and leaves behind a positively charged vacant site, called a *hole*.

The conductive electrons in the higher energy state are said to be in a *conduction band*, while the covalently bonded electrons are said to be in a *valence band*. The difference between electrical insulators and semiconductors is the size of the *energy gap* (E_g) between the bands. As shown in Figure 8-4, *insulators* have a large energy gap while semiconductors have a much smaller gap. Traditional semiconductor materials, including silicon and germanium, have small enough energy gaps that they become conductive at elevated temperatures. Materials with extensive covalent bonding tend to be exceptional insulators. Long-chain polymer molecules such as polystyrene are as much as 20 orders of magnitude less conductive than many metals.

8.4 INTRINSIC SEMICONDUCTION

Semiconductors have electrical conductivities between those of insulators and conductors and possess small enough energy gaps that electrons can be promoted from the valence band to the conduction band. An *intrinsic semiconductor* is a perfect crystal with no defects, dislocations, substitutions, or grain boundaries that has a completely filled valence band and a completely empty conduction band separated by a band gap of less than 2 eV. Elemental silicon is the most common intrinsic semiconductor. When four silicon atoms are in close proximity, their orbitals overlap and form hybrids, each containing two electrons, as shown in Figure 8-5. The orbitals form a symmetric tetrahedron to allow them to be as far apart in space as possible. Adjacent silicon atoms form covalent bonds with the valence band completely filled and the conduction band completely empty.

When an electric field is applied to a crystal of silicon, electrons in the valence band can gain energy and move to the unoccupied conduction band. When enough energy is applied to the electron to overcome the energy gap, the electron becomes free and leaves a hole behind. The energy provided also can be in the form of a photon of light or result from thermal vibration. The free electron can migrate through the crystal in response to the field. The positively charged hole also can migrate effectively because adjacent electrons can tunnel into the hole in the valence band, filling that hole but leaving a new one behind. As a result, there are two distinct charge carriers, negatively charged electrons and positively charged holes. The electrons migrate in the opposite direction of the field while holes migrate in the direction of the field, as shown in Figure 8-6.

The current in the semiconductor conducts through both the movement of holes and electrons, so the current density (J) can be defined by Equation 8.15,

$$J = enV_{d,e} + enV_{d,h}, \tag{8.15}$$

where e is the charge of the electron, n is the number of electrons in the conduction band, and $V_{d,e}$ and $V_{d,h}$ are the drift velocities of the electrons and holes, respectively.

As discussed in Section 8.2, the quantity of electrons in the conduction band is a strong function of temperature. Near absolute zero, all of the electrons would be in the valence band with the conduction band empty. As temperature increases, more electrons are able to overcome the energy gap and enter the conduction band, leaving behind a hole in the valence band.

| **Intrinsic Semiconductor** |
Pure material having a conductivity ranging between that of insulators and conductors.

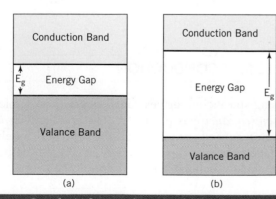

FIGURE 8-4 Energy Gaps for (a) Semiconductors and (b) Insulators

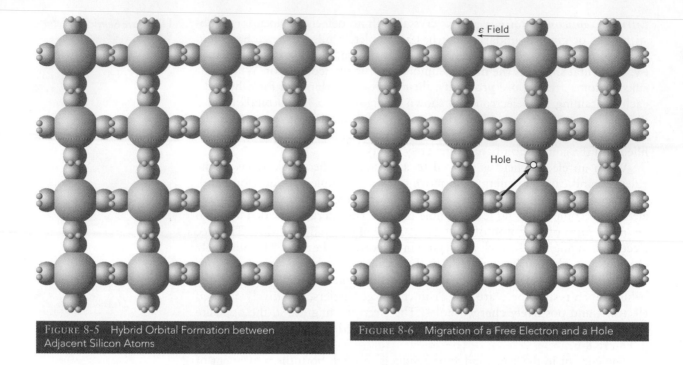

The temperature dependence of this phenomenon is represented by a simple Arrhenius relationship,

$$n = n_0 \exp\left(\frac{-E_A}{RT}\right), \tag{8.16}$$

where the pre-exponential constant n_0 is a function of Boltzmann's constant, Planck's constant, the mass of an electron, and the effective mass of a hole.

The essential exponential temperature dependence of the conductivity of intrinsic semiconductors limits their applicability in microelectronics. Small changes in temperature can dramatically affect the conductivity of the material and, accordingly, the performance of the device. Instead, commercial applications require semiconductors whose conductivity is not primarily a function of temperature.

8.5   EXTRINSIC SEMICONDUCTION

| *Dopants* |
Impurities deliberately added to a material to enhance the conductivity of the material.

| *Extrinsic Semiconductor* |
Created by introducing impurities called dopants into a semiconductor.

By introducing specific impurities called *dopants* to semiconductors, an *extrinsic semiconductor* is produced. The conductivity of an extrinsic semiconductor depends primarily on the type and number of dopants and is nearly independent of temperature within limited temperature ranges. The nature of the dopant material greatly impacts the properties of the semiconductor.

In an intrinsic semiconductor, the number of electrons (n) promoted to the conduct band is always equal to the number of holes (p) left behind in the

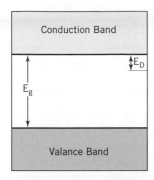

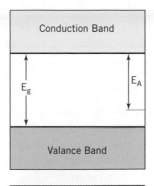

FIGURE 8-7 Difference in Energy Levels between Donor Electrons and Regular Valence Electrons

FIGURE 8-8 Energy Barrier (EA) for a p-Type Semiconductor

valence band, but this is not true for extrinsic semiconductors. If the dopant atom donates electrons to the conduction band, the number of holes in the conduction band will be less than the number of electrons in the conduction band, and the semiconductor is classified as an *n-type semiconductor*. Similarly, it is possible to add a dopant that removes electrons from the valence band, resulting in more holes in the valence band than electrons in the conduction band. This dopant creates a *p-type semiconductor*.

The dopant molecule added to an n-type semiconductor functions as an *electron donor* and typically involves a Group VA atom (phosphorous, arsenic, or antimony) with a valence of 5. Four of the electrons in such an atom are active participants in covalent bonding, but the fifth electron has nothing to bond with and hovers at an energy level just below the conduction band. The energy required to promote that electron to the conduction band (E_D) is substantially smaller than the energy required to promote a typical valence band electron (E_g), as shown in Figure 8-7.

This behavior has significant impact on the properties of the semiconductor. When temperatures are too low to provide enough energy to overcome even the reduced energy barrier (E_D) of the donor electron, there is almost no conductivity. However, for the entire range of temperatures that provide enough energy to overcome E_D but not enough to overcome E_g, the entire conductivity on the semiconductor is controlled by the number of dopant atoms (and therefore donor electrons) present. This makes extrinsic semiconductors ideal for microelectronic applications because the material properties are governed by a controllable feature (number of dopant atoms) rather than an external variable like temperature.

The dopant molecule added to a p-type semiconductor functions as an *electron acceptor* and typically involves a Group IIIA atom (boron, aluminum, gallium, or indium) with a valence of 3. With only three electrons available, these dopants cannot fully participate in the covalent bonding process without drawing an electron from elsewhere in the valence band, which leaves a hole behind. These hole sites have an energy level slightly higher than the rest of the valence band, so a smaller energy barrier (E_A) must be overcome to create a hole in the valence band, as shown in Figure 8-8. This hole then can migrate and carry charge.

| *n-Type Semiconductor* |
Semiconductor in which a dopant donates electrons to the conduction band, causing the number of holes to be less than the number of electrons in the conduction band.

| *p-Type Semiconductor* |
Semiconductor in which a dopant removes electrons from the valence band, causing there to be more holes than electrons in the valence band.

| *Electron Donor* |
Molecule that donates electrons to another substance.

| *Electron Acceptor* |
Molecule that accepts electrons from another substance.

8.6 DIODES

| **Diode** |
Electronic switch that allows electrons to flow in one direction only.

| **Recombination** |
Process in which holes in a p-type semiconductor and electrons in an n-type semiconductor cancel each other out.

| **p–n Junction** |
Area in which the p-type and n-type areas meet.

| **Depletion Zone** |
Nonconductive area between the p–n junction, in which recombination takes place.

| **Forward Bias** |
Connecting a battery with the positive terminal corresponding with a p-type site and the negative terminal corresponding with an n-type site.

| **Reverse Bias** |
Connecting a battery with the positive terminal corresponding with an n-type site and the negative terminal corresponding with a p-type site.

| **Zener Breakdown** |
Rapid carrier acceleration caused by the reverse bias becoming too large, which excites other carriers in the region to cause a sudden large current in the opposite direction.

Both p-type and n-type semiconductors are capable of conducting charge, but when they are put close together, they can be used to form a *diode*, an electronic switch that allows current to flow in only one direction. Consider a single semiconductor that is doped to be p-type on one side and n-type on the other, as shown in Figure 8-9. The material on the p-type side conducts charge through the movement of holes while the material on the n-type side conducts charge by the movement of electrons. At the boundary between the two sides, the electrons and the holes are immediately attracted to each other. The electron drops in energy and reoccupies the empty electronic state (the hole). The hole disappears and the electron leaves the conduction band and returns to the valence band. This process is called *recombination*. The area where the p-type and n-type regions meet is known as a *p–n junction*, and the nonconducting layer between them in which recombination occurs is called the *depletion zone*.

The behavior of the p–n junction can be influenced by the connection of a battery. When the positive terminal is connected to the p-type end and the negative terminal is connected to the n-type end, as shown in Figure 8-10, the holes in the p-type region and the electrons in the n-type region migrate toward the junction, and the depletion zone becomes thinner. As the distance between the electrons and the holes decreases, the electrons are capable of passing through the depletion zone and create a significant current. Connecting the battery in this fashion is called a *forward bias*.

When a *reverse bias* is used (the positive terminal of the battery is connected to the n-type side and the negative terminal to the p-type side), the behavior is quite different. This time the electrons and holes are drawn away from the depletion zone, as shown in Figure 8-11, and almost no recombination occurs. As a result, the junction becomes a powerful insulator. If the reverse bias becomes too large, any carrier that manages to break through the depletion zone becomes rapidly accelerated, which excites the other carriers in the region and causes a sudden large current in the opposite direction. This phenomenon is known as *Zener breakdown*. Some diodes are designed specifically to break down at specific voltages to protect electronic circuits.

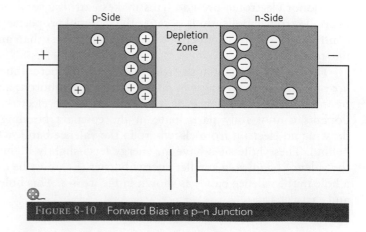
FIGURE 8-10   Forward Bias in a p–n Junction

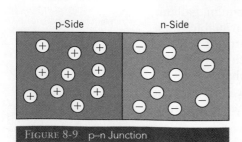
FIGURE 8-9   p–n Junction

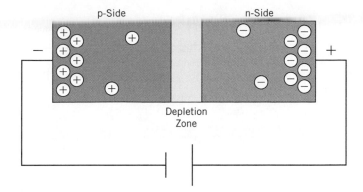

8.7 TRANSISTORS

Transistors serve as amplifying or switching devices in microelectronics. The *bipolar junction transistor (BJT)* revolutionized the microelectronics industry when it was developed at Bell Labs in 1948. The BJT consists of three regions: emitter, *base*, and *collector*, and comes in two distinct forms: p–n–p and n–p–n. In either case, the transistor is essentially a sandwich of three doped regions. The base is located between the collector and transmitter and consists of a lightly doped and high-resistivity material. The collector region is much larger than the emitter and completely surrounds it, as shown in Figure 8-12, so that essentially all electrons (or holes in the case of a p–n–p transistor) given off by the emitter are collected. As a result, a small voltage increase between the emitter and base results in a large increase in current in the collector.

Junction transistors have been almost entirely phased out in favor of more modern *MOSFETs* (metal oxide semiconductor field effect transistors). The original MOSFETs used metal oxides as the semiconducting material, but now almost all use either silicon or blends of silicon and germanium. Transistors without metal oxides are more appropriately called *IGFETs* (insulated gate field effect transistors), but the terms IGFET and MOSFET are used interchangeably.

MOSFETs can be produced as p-type or n-type, but the operating principles are similar for each. In a p-type MOSFET, two small regions of p-type semiconducting material are deposited on a large region of n-type material. These small regions are connected by a narrow p-type channel, as shown in Figure 8-13. Metal connections are affixed to each region, then a thin layer of silicon dioxide is deposited on top of the p-type material to serve as an insulator. One region of p-type material will serve as a *source* while the other becomes a *drain*. An additional metal connector, called a *gate*, is attached to the insulating layer between the source and drain.

| *Transistors* |
Devices that serve as amplifying or switching devices in microelectronics.

| *Bipolar Junction Transistor (BJT)* |
Prototypical electronic device developed in 1948 used to amplify signals.

| *Base* |
Central piece of a BJT transistor that is made from a lightly doped and high-resistivity material.

| *Collector* |
Largest region in a BJT transistor that surrounds the emitter and prevents the escape of all electrons or holes.

| *MOSFETs* |
Transistor originally made with metal oxides, but the term is now interchangeable with IGFET.

| *IGFETs* |
Transistor made without the use of metal oxides, but the term is now interchangeable with MOSFET.

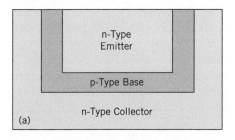

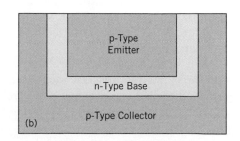

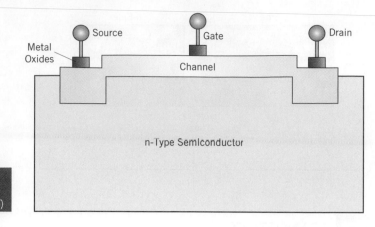

Source   Gate   Drain

Metal
Oxides

Channel

n-Type Semiconductor

Figure 8-13
Schematic of a
MOSFET (or IGFET)

| *Integrated Circuits* |
Many transistors integrated
on a single microchip.

| *Analog Circuits* |
Type of integrated circuit that
can perform functions such as
amplification, demodulation,
and filtering.

| *Digital Circuits* |
Type of integrated circuit
capable of performing
functions such as flip-flops,
logic gates, and other more
complex operations.

| *Mixed Signal* |
Type of integrated circuit
containing both analog and
digital on the same chip.

| *Photoresist* |
Coating that becomes
soluble when exposed to
ultraviolet light.

| *DNQ–Novolac* |
Photoresist-light-sensitive
polymer blend.

| *Softbaking* |
Process of removing residual
solvents from the photoresist.

| *Mask* |
Transparent glass plate used
in the photoresist process.

| *Developer* |
Alkaline solution that removes
exposed material when
applied to a microchip.

| *Projection Lithography* |
Process of projecting
ultraviolet light onto a
microchip in a manner similar
to a slide projector.

When an electric field is applied to the gate, the conductivity in the channel is impacted directly. Small alterations in applied field result in substantial difference in current. The operation of MOSFETs is quite similar to BJTs, but much smaller gate currents are required, and MOSFETs are not subject to breakdown.

8.8   INTEGRATED CIRCUITS

By the middle of the twentieth century, companies were able to integrate many transistors into a single microchip. By 2006, a single chip could be imprinted with 1 million transistors per square millimeter. *Integrated circuits* can be classified roughly into three categories: *analog circuits*, *digital circuits*, and *mixed signal*, which contain both analog and digital on the same chip. Analog circuits perform functions including amplification, demodulation, and filtering; digital integrated circuits can include flip-flops, logic gates, and other more complex operations.

Integrated circuits are manufactured from large single crystals of extremely high purity silicon that are cut into thin wafers. After the wafer has been thoroughly cleaned, a thin layer of SiO_2 is deposited on the surface followed by a *photoresist* coating that becomes soluble when exposed to ultraviolet light. The photoresist is a light-sensitive polymer or polymer blend, such as *DNQ–Novolac* (diazonaphthoquinone and novolac resin). Residual solvents are removed from the photoresist through a process called *softbaking*, then a transparent glass plate, or *mask*, is placed over the wafer, and an emulsion of metal film forms a pattern on one side of the glass. When the ultraviolet light is shined through the mask, the photoresist coating is dissolved away, leaving the underlying wafer exposed. Then the entire wafer is exposed to an alkaline solution called the *developer*, which removes the exposed material, as shown in Figure 8-14. Through this method, complex circuit patterns can be built one layer at a time.

When this method was developed, the mask was applied directly on the silicon wafer, but this process tended to leave microscopic damage that became more significant as the circuitry became more compact. Today *projection lithography*, in which the ultraviolet light is projected onto the wafer much like a slide projector, has become far more common.

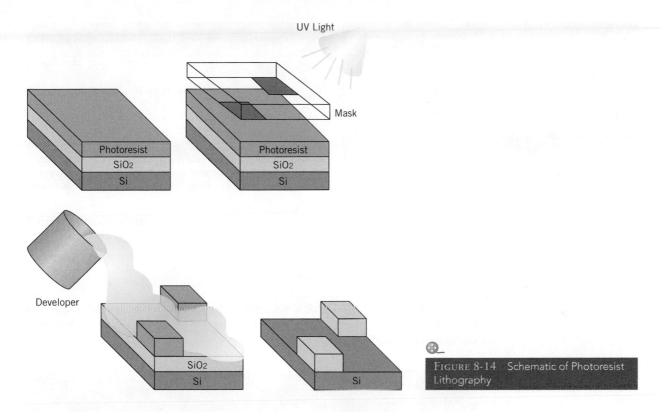

FIGURE 8-14 Schematic of Photoresist Lithography

The density of transistors on integrated circuits expands rapidly, making electronic devices smaller, faster, and less expensive. Gordon Moore, a cofounder of the chip-making giant Intel, observed as early as 1965 that the density of transistors doubles every 18 to 24 months. This empirical observation has become known as *Moore's law* and has become a target for the entire microelectronics industry.

| ***Moore's Law*** |
Empirical observation that the density of transistors doubles every 18 to 24 months.

8.9 | DIELECTRIC BEHAVIOR AND CAPACITORS

When two parallel plates are connected to a battery, the plates become charged, and an electric field develops running from the positive plate to the negative plate, as shown in Figure 8-15. When these charged plates are used as part of an electric circuit, they are called *capacitors* and primarily serve as a means of storing energy. The capacitance (C) of the parallel plates is defined as

$$C = \frac{Q}{A},$$
(8.17)

where Q is the charge on the plates (typically in Coulombs) and A is the cross-sectional area of one plate.

If only air separated the parallel plates, then a strong electric field would result in sparking between the plates. To prevent this from happening, a *dielectric* material is placed between the two plates to reduce the strength of the electric field without lowering the voltage. The dielectric is a polarized material that opposes the electric field between the plates, as shown in Figure 8-16. The ability of the dielectric material to oppose the field is given by a dimensionless *dielectric constant* (K) that is a function of composition, microstructure, temperature, and

| ***Capacitors*** |
Two charged plates separated by a dielectric material.

| ***Dielectric*** |
Material placed between the plates of a capacitor to reduce the strength of the electric field without reducing the voltage.

| ***Dielectric Constant*** |
Dimensionless value rerpresenting the ability of a dielectric material to oppose an electric field.

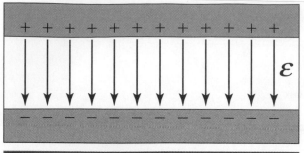

FIGURE 8-15   Parallel Plate Capacitor

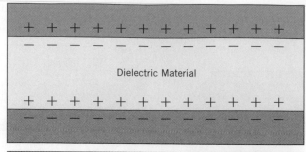

FIGURE 8-16   Role of a Dielectric Material in a Capacitor

electrical frequency. Values of K for common dielectric materials are summarized in Table 8-5. The capacitance for capacitors containing dielectrics is given by

$$C = \frac{K\epsilon_0 A}{d},$$ (8.18)

where ϵ_0 is the *permittivity* of a vacuum (8.85×10^{-12} F/m) and d is the distance between the plates. The parallel plates are generally either nickel or a silver-palladium alloy.

TABLE 8-5   Dielectric Constants of Common Materials

Material	K (at 60 Hz)	K (at 1,000,000 Hz)
Alumina	9.0	6.5
Barium Titanate (BaTiO$_3$)	n/a	2000–5000
Nylon	4.0	3.6
Polyethlyne	2.3	2.3
Polystyrene	2.5	2.5
Polyvinyl chloride	3.5	3.2
Rubber	4.0	3.2
Soda-lime glass	7.0	7.0

What Other Electrical Behaviors Do Some Materials Display?

8.10   FERROELECTRIC AND PIEZOELECTRIC MATERIALS

| Ferroelectric Materials |
Materials with permanent dipoles that polarize spontaneously without the application of an electric field.

Most of the electronic materials discussed in this chapter require the application of an electric field for their electronic properties to become evident. *Ferroelectric materials*, however, have permanent dipoles that cause them to polarize spontaneously even without an applied electric field. The polarization can be reversed if an electric field is applied to the material and will stop if temperatures are raised above a critical limit called the *Curie temperature* (T_c). Ferroelectric materials find uses in dynamic random

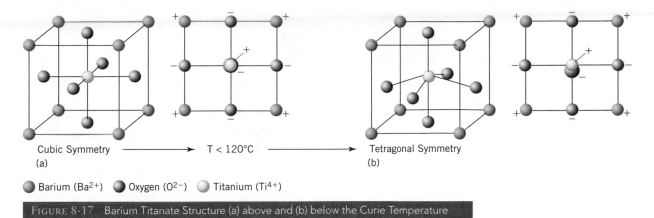

Cubic Symmetry ⟶ T < 120°C ⟶ Tetragonal Symmetry
(a)   (b)

● Barium (Ba^{2+}) ● Oxygen (O^{2-}) ○ Titanium (Ti^{4+})

FIGURE 8-17   Barium Titanate Structure (a) above and (b) below the Curie Temperature

access memory systems (DRAMs) and infrared sensors. Common ferroelectric materials include barium titanate (BaTiO$_3$) and lead zirconium titanate (PZT).

The ferroelectric nature of these materials arises from the asymmetric structure of the unit cells. Above its *Curie temperature* (120°C), barium titanate exist as an FCC (face-centered cubic) system with a titanium ion (Ti^{4+}) sitting in the center, with oxygen (O^{2-}) on the faces and barium (Ba^{2+}) at the corners, as shown in Figure 8-17. This structure is symmetric and incapable of being ferroelectric. When the temperature of the material drops below 120°C, the unit cell shifts to simple cubic, resulting in a net shift of the relative position of the ions. Positive charges tend to accumulate near the top of the unit cell, with negative charges gathering near the bottom.

Piezoelectric materials convert mechanical energy to electrical energy or electrical energy to mechanical energy and are widely used in *transducers* that convert sound waves to electric fields. The production of an electrical field in response to a mechanical force is called the *piezoelectric effect*. The change in thickness of a material in response to an applied electric field is called the *converse piezoelectric effect*.

Like ferromagnetic materials, piezoelectrics are polarized, but their positive and negative charges are distributed symmetrically. When a mechanical stress is applied, the crystal structure deforms slightly and causes a potential difference to form across the material. The converse effect, in which the crystal structure deforms in response to an applied electric field, was postulated from thermodynamics and proven by Marie Curie, the early twentieth-century French scientist who remains the only person to win the Nobel prizes in two different branches of science: physics in 1903 and chemistry in 1911.

| *Curie Temperature* |
Temperature above which a material no longer displays ferromagnetic properties.

| *Piezoelectric Materials* |
Materials that convert mechanical energy to electrical energy or vice versa.

| *Transducers* |
Devices that convert sound waves to electric fields.

| *Piezoelectric Effect* |
Production of an electric field in response to a mechanical force.

| *Converse Piezoelectric Effect* |
Change in thickness of a material in response to an applied electric field.

What Are Optical Properties and Why Do They Matter?

8.11   OPTICAL PROPERTIES

Optical materials are special because of the ways they reflect, transmit, or refract light. Visible light is only a tiny section of the overall electromagnetic spectrum, shown in Figure 8-18. Gamma rays, X-rays

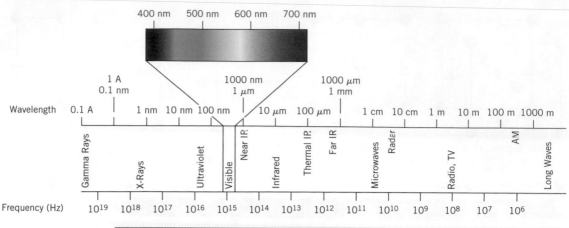

FIGURE 8-18   Electromagnetic Spectrum

(discussed in Chapter 2), microwaves, radio waves, infrared, and ultraviolet radiation vary in frequency and wavelength but travel at the same velocity (3×10^8 m/s in a vacuum). Electromagnetic radiation displays properties of both particles and waves, and the velocity of the radiation is defined as the product of the wavelength times the frequency, as shown in Equation 8.19:

$$c = \lambda v. \tag{8.19}$$

| *Photons* |
Discretes units of light.

All electromagnetic radiation travels in discrete units called *photons*, which are governed by the fundamental equation

$$E = h v \ = \frac{hc}{\lambda}, \tag{8.20}$$

where E is the energy of the photon and h is Planck's constant (6.62×10^{-34} J · s). As a consequence, the photon may be considered simultaneously a particle with energy (E) or a wave with a set frequency and wavelength.

| *Fresnel Equation* |
Mathematical relationship describing the quantity of light reflected at the interface of two distinct media.

When a photon of electromagnetic energy interacts with any material, four possible things result, as shown in Table 8-6.

Some fraction of the intensity of an incident light beam is reflected when it reaches a surface while the balance passes through. French physicist Augustin-Jean Fresnel developed an equation to quantify the fraction of light reflected. The *Fresnel equation* states that for light with an angle of incidence close to zero,

| *Reflection Coefficient* |
Fraction of light reflected at the interface of two media.

$$R = \left(\frac{n_1 - n_2}{n_1 + n_2}\right)^2, \tag{8.21}$$

| *Transmission Coefficient* |
Fraction of light not reflected at the interface of two media and instead enters the second media.

where R is the *reflection coefficient* and n is the index of refraction of the two media. The *transmission coefficient* (T) is just $1 - R$.

| *Index of Refraction* |
Material-specific term representing the change in the relative velocity of light as it passes through a specific medium.

The *index of refraction* (n) represents the change in the relative velocity of light as it passes through a new medium. Light traveling through a vacuum has an index of refraction of 1. Most materials have indexes of refractions that are greater than 1; most glasses and polymers have values in the range of 1.5 to 1.7. Dutch mathematician Willebrord Snellius proved that the change in

TABLE 8-6 Modes of Interaction between Electromagnetic Energy and Materials

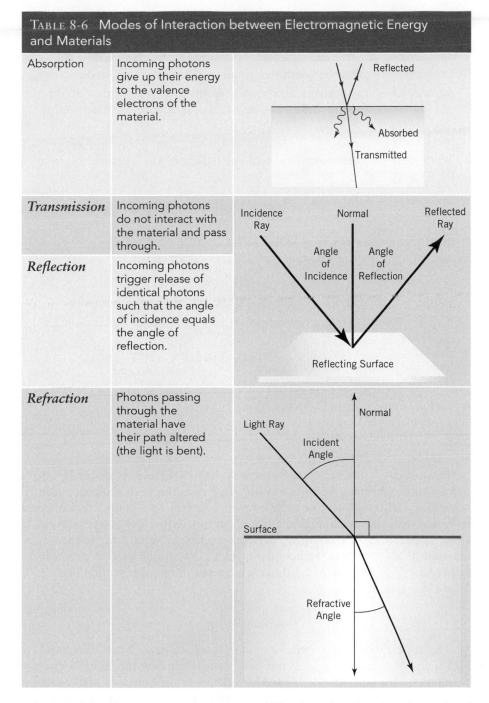

Absorption	Incoming photons give up their energy to the valence electrons of the material.	
Transmission	Incoming photons do not interact with the material and pass through.	
Reflection	Incoming photons trigger release of identical photons such that the angle of incidence equals the angle of reflection.	
Refraction	Photons passing through the material have their path altered (the light is bent).	

velocity of the electromagnetic waves could be directly related to the angle of refraction. *Snell's law* provides that

$$\frac{n_1}{n_2} = \frac{\sin \theta_2}{\sin \theta_1}.$$

(8.22)

Thus both the fraction of light reflected and the angle of the refracted light can be predicted by knowing only the angle of incidence and the refractive indexes of the two materials.

| *Snell's Law* |
Equation describing the change in velocity of electromagnetic waves passing between two media.

Example 8-2

Determine the reflection coefficient, transmission coefficient, and angle of refraction for a ray of visible light passing from air ($n = 1$) into a thin sheet of soda-lime glass ($n = 1.52$) with an angle of incidence of 10 degrees.

SOLUTION

The reflection coefficient (R) is calculated from the Fresnel equation (Equation 8.21),

$$R = \left(\frac{n_1 - n_2}{n_1 + n_2}\right)^2 = \left(\frac{1.52 - 1}{1.52 + 1}\right)^2 = 0.043,$$

so 4.3% of the light is reflected at the surface of the soda-lime glass. Accordingly, the transmission coefficient (T) is given by

$$T = 1 - R = 0.957,$$

and 95.7% of the light passes into the glass.

The angle of refraction is determined by Snell's law (Equation 8.23), such that

$$\text{Sin}\,\theta_2 = \frac{n_1 \sin \theta_1}{n_2} = \frac{(1)(\sin 10)}{1.52} = 0.114, \text{ and}$$

$$\theta_2 = \sin^{-1}(0.114) = 6.55°.$$

In most cases, the refracted beam will pass into the new material, but in certain cases when the beam is traveling in a material with a higher index of refraction and comes to one with a lower index of refraction, total internal reflection will occur. As Figure 8-19 illustrates, when the incident beam is closer to parallel to the boundary between the two materials, the refracted ray may not be able to cross the boundary. Snell's law can be used to determine a *critical angle of incidence* (θ_c) above which total internal reflection occurs,

$$\theta_c = \sin^{-1}\left(\frac{n_2}{n_1}\right). \tag{8.23}$$

| *Critical Angle of Incidence* |
Angle beyond which a ray cannot pass into an adjacent material with a different index of refraction and instead is totally reflected.

| *Specular Reflection* |
Reflection from smooth services with little variation in the angle of reflection.

| *Diffuse Reflection* |
Broad range of reflectance angles resulting from electromagnetic waves striking rough objects with a variety of surface angles.

The appearance of a material is impacted significantly by its reflections. A high index of refraction ($n > 2$) results in the opportunity for multiple internal reflections of the light. Diamonds and lead glasses sparkle because of these multiple reflections. Materials with highly polished, smooth surfaces reflect light with very little variation in the angle of reflection. This is called *specular reflection*. Rougher materials have significant local variations in surface angles, and the resulting *diffuse reflection* leads to a wider range of reflectance angles, much like the imperfect crystals in a material lead to spreading in X-ray diffraction peaks, as discussed in detail in Chapter 2.

Optical materials are classified based on their tendency to transmit light. Transparent materials allow enough light to pass through for a clear image to be seen. Opaque materials do not allow enough light to pass through for

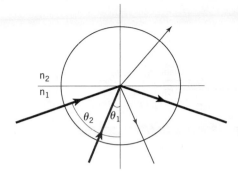

FIGURE 8-19   Total Internal Reflection

an image to be seen. Translucent materials span almost the entire spectrum between transparent and opaque, allowing only a diffuse image to be seen. Frosted glasses or milk jugs are translucent. Translucency results from the scattering of light, either because of rough surfaces or because of pores or other impurities present in the material.

The color of a material is based on the selective absorption of specific wavelengths of light. An object will have the color of the light reflecting off it. When a material reflects the entire visible spectrum equally, it appears white. When it absorbs the entire spectrum, it appears black. The addition of coloring agents to glasses (discussed in Chapter 6) is designed to change which colors are absorbed to alter the appearance of the glass.

8.12   APPLICATIONS OF OPTICAL MATERIALS

The optical properties of materials find special uses in a number of important commercial applications. Two specific ones that will be considered here are *optical fibers* and *lasers*.

Optical fibers are thin glass or polymeric fibers used to transmit light waves across distance and are especially important in the communication industry. Until the end of the 1970s, most communication signals were carried as electrical signals through copper wire. However, optical fibers offer significant advantages, including no interference between adjacent fibers, lower loss, and most important, a much higher capacity for carrying data. A single optical fiber can provide the data-carrying capacity of thousands of electrical links.

Optical fibers are photonic materials that transmit signals by photons rather than by electrons. Most commercial optical fibers involve a thin glass core surrounded by a cladding made from a material that has a lower index of refraction. The cladding is wrapped in a protective polymeric outer coating. As previously discussed, a total reflection of the beam will occur when light passes through a material with a higher index of refraction. The total reflection allows the photonic signal to propagate along long lengths of fiber. As long as the angle of incidence is greater than the critical angle, total internal reflection will cause the optical signal to propagate down the length of the core, as shown in Figure 8-20.

As the photon travels down the cable, some of the signal strength is lost because of photon absorption or scattering. This process is called *attenuation*. The optical signal must be enhanced periodically to overcome the attenuation. Early fiber optic systems required repeaters, devices that converted optical

| *Optical Fibers* |
Thin glass or polymeric fibers that are used to transmit light waves across distances.

| *Attenuation* |
Loss of strength during the transmission of an optical signal resulting from photon absorption or scattering.

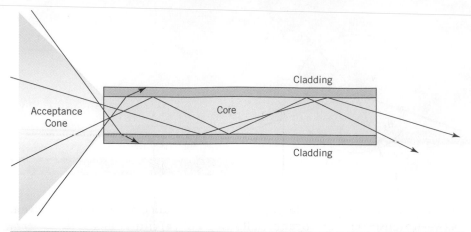

signals to electrical signals, enhanced them, then converted them back to optical signals. However, modern systems use optical amplifiers (essentially lasers) to enhance optical signals without converting them to electrical signals.

In addition to transmitting optical signals across long distances, fiber optic cables are used in everyday applications, such as connecting audio equipment. The most common connector, *TOSLINK*, was created by Toshiba and has found wide acceptance as a primary connector for compact disc players. A TOSLINK connector is shown in Figure 8-21.

Optical fibers have applications beyond the communications industry. Medical endoscopes used in imaging on internal body systems use optical fibers as a light-delivery system. When colored light signals are used, optical fibers sometimes serve purely decorative purposes, like the lamp shown in Figure 8-22.

| *TOSLINK* |
Common connector for optical cables.

FIGURE 8-21   TOSLINK Connector

Courtesy James Newell

FIGURE 8-22   Decorative Device Using Optical Fibers

Courtesy James Newell

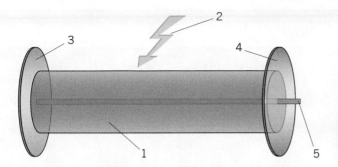

FIGURE 8-23 Schematic of a Laser: (1) Gain Medium, (2) External Power Source Used to Excite Gain Medium, (3) and (4) Optical Cavity, and (5) Laser Beam

Generally, the light transmitted through an optical fiber is produced by a laser. The word *laser* is an acronym for "light amplification by stimulated emission of radiation." Although lasers have existed since 1960, many people still picture a Hollywood ray gun when they hear the term. In practice, lasers have found far more use in bar code readers and DVD players, where their properties enable them to read the optical images on the surface of the bar code or DVD.

A laser is a surprisingly simple device that consists of a *gain medium* and an *optical cavity*. The gain medium is a substance that can pass from a higher to a lower energy state and transfers the associated energy into the laser beam. Gases such as neon, helium, and argon make excellent gain media, as do crystals doped with rare earth atoms, such as yttrium, aluminum, and garnet. An external power source (usually electrical current) stimulates the gain medium to the excited state. The optical cavity is essentially two mirrors that repeatedly reflect a beam of light through the gain medium. Each time the beam passes through the gain medium, it increases in intensity. A schematic of a laser is shown in Figure 8-23.

Conceptually, the operation of the laser is similar to an incandescent lightbulb; however, the light generated by a laser is *coherent*, meaning that it has a single wavelength and is emitted in a well-defined beam.

| *Laser* |
Device that produces light of a single wavelength in a well-defined beam. Acronym for "light amplication by stimulated emission of radiation."

| *Gain Medium* |
Substance that passes from a higher to lower energy state and transfers the associated energy into a laser beam.

| *Optical Cavity* |
Pair of mirrors that repeatedly reflect a beam of light through the gain medium of a laser.

Summary of Chapter 8

In this chapter, we:

- Discussed why electrical and optical properties warrant separate coverage
- Examined the physical and chemical basis for conduction in metals
- Calculated drift mobility, current density, and conductivity
- Explored the factors that influence resistivity and how to determine the resistivity of an alloy
- Learned about the energy gap between the valence band and the conduction band
- Discussed the nature of intrinsic semiconduction and the migration of both electrons and holes in response to an applied electric field
- Analyzed the role of dopants on both p-type and n-type extrinsic semiconductors
- Examined the operation of diodes based on p–n junctions
- Discovered the operational principles of BJT transistors and MOSFETs
- Learned about the manufacture of integrated circuits
- Determined the role of dielectric materials in capacitors
- Explored the unique behaviors of ferroelectric and piezoelectric materials
- Examined the interaction between electromagnetic waves and materials
- Applied the Fresnel equation and Snell's law to examine refraction
- Examined applications of optical fibers and lasers

Key Terms

Homework Problems

1. Why does the electrical conductivity of pure metals decrease with temperature while that of semiconductors increases?

2. Calculate the resistivity of a copper lattice with 2% gold.

3. Calculate the drift mobility of gold near room temperature.

4. How would the calculation of drift mobility differ if performed for zinc instead of gold or silver?

5. Explain the purpose of a dielectric material in a capacitor.

6. Describe the role of the emitter in a BJT transistor.

7. Discuss the advantages of using MOSFETs instead of BJT transistors.

8. Discuss the role of the energy gap in conduction and explain what physical parameters influence its magnitude.

9. Compare and contrast n-type and p-type semiconductors.

10. Compare and contrast piezoelectric materials with ferroelectric materials, and explain why barium titanate fits both definitions.

11. Explain the relationship between the current density equations derived in Equations 8.12 and 8.15.

12. How would increasing voltage effect the size of the depletion zone in a forward-bias p–n junction?

13. Explain why the addition of an electron from a dopant atom to an n-type semiconductor does not result in the creation of a hole.

14. Would phosphorous serve as an electron donor or an electron acceptor if used as a dopant molecule? Explain.

15. What properties of silicon make it suitable for use in semiconductors?

16. Why do holes migrate in the direction of the applied electric field?

17. Explain why polymeric materials make such effective insulators.

18. Calculate the drift mobility of copper near room temperature.

19. Explain how transistors can be used to amplify an electric signal.

20. What considerations should be taken into account when selecting a dopant molecule?

21. Most integrated circuits are manufactured in clean-room conditions. Explain why this level of cleanliness is necessary and how it impacts the cost of the circuits.

22. As discussed in Chapter 6, diamonds are being considered as a possible "next generation" of semiconductor materials. What properties of diamonds make them attractive candidates?

23. Discuss the advantages of projection lithography over simple photoresist lithography.

24. Explain why piezoelectric materials are useful in transducers.

25. Generate a plot of transistor densities versus time for 286, 386, Pentium, Pentium 2, Pentium 3, and Pentium 4 processors. (You will need to find these data.) How well has Moore's law applied? Extrapolate the results to estimate densities five years from now. What technological and economic barriers exist for the continuing applicability of Moore's law?

26. What physical properties are most important in selecting a gain medium for a laser?

27. Why must the cladding of an optical fiber have a lower index of refraction than the core?

28. Determine the critical angle of incidence for air with an index of refraction of 1 and water with an index of refraction of 1.333.

29. Determine the critical angle of incidence for PMMA with an index of refraction of 1.49 and polystyrene with an index of refraction of 1.60.

30. If a light beam traveling through air strikes an ice cube (n = 1.309) at an angle of 15 degrees, what will be the angle of the refracted beam?

9

Biomaterials and Biological Materials

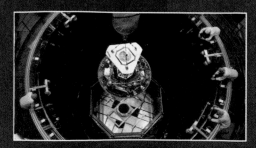

CONTENTS

Learning Objectives

By the end of this chapter, a student should be able to:

- Explain in their own words the differences among biomaterials, biological materials, bio-based materials, and biomimetic materials.

- Distinguish between structural and functional biomaterials.

- Define biocompatibility and how it impacts the design of biomaterials.

- Explain the structure of bone and the role of the four major types of bone cells.

- Describe the self-repairing of bones through remodeling.

- Explain the use of metal fasteners and bone fillers.

- Justify the selection of materials used in hip replacements, and explain their role.

- Evaluate design considerations and materials of construction for prosthetic limbs.

- Describe the role of vascular stents in balloon angioplasty.

- Compare and contrast the impact of nitinol or stainless steel as the material of choice for a vascular stent.

- Describe the use of a Foley catheter, and explain the basis for the selection of material used to make the catheter.

- Contrast the materials used in silicone and saline breast implants, and describe the controversies surrounding their use.

- Compare and contrast the financial, mechanical, and cosmetic differences between the use of traditional metal amalgams and dental composites.

- Describe the role of the membrane in kidney dialysis, and compare possible membrane materials.

- Discuss the material selection issues in artificial hearts and ventricular assist devices.

- Discuss the mechanical properties of the dermis and epidermis and how artificial skin emulates these properties.

- Explain the manufacture of synthetic bladder from biological materials.

- Describe the different mechanical heart valves that are implanted and how the choice of material impacts design.

- Calculate a performance index to compare the performance of different heart valves based on hemodynamic properties.

- Distinguish among homografts, autografts, and xenografts and their advantages and disadvantages relative to each other and mechanical heart valves.

- Discuss the two experimental approaches to developing artificial blood (oxygen therapeutics), and explain why hemoglobin cannot be used directly to increase oxygen flow in the blood.

- Describe the major mechanisms for controlled released of pharmaceuticals.

- Explain the construction and use of a transdermal patch.

What Types of Materials Interact with Biological Systems?

| Biotechnology |
Branch of engineering involving the manipulation of inorganic and organic materials to work in tandem with one another.

| Biological Materials |
Materials produced by living creatures including bone, blood, muscle, and other materials.

| Biomaterials |
Materials designed specifically for use in the biological applications, such as artificial limbs and membranes for dialysis as well as bones and muscle.

9.1 BIOMATERIALS, BIOLOGICAL MATERIALS, AND BIOCOMPATIBILITY

Biomaterials and biological materials are really polymers, metals, ceramics, and composites used in living systems, but because of the accelerating growth in *biotechnology*, the topics warrant their own chapter.

The distinction among biological materials, biomaterials, bio-based materials, and biomimetic materials serves as an important starting point for any discussion of such materials. *Biological materials* are produced by living creatures and include bone, blood, muscle, and a host of other materials that serve many different functions. *Biomaterials* are defined by the European Society for Biomaterials as "materials intended to interface with biological systems to evaluate, treat, augment, or replace any tissue, organ, or function of the body." *Bio-based materials* do not serve a function for an organism but are materials derived from living tissue such as cornstarch or polymers made from soybean oil. *Biomimetic materials* are not produced by living organism but are chemically and physically similar to materials that are. As such, they often find use as replacements for biological materials.

Material Classification	Definition
Biological materials	Materials produced by living creatures including bone, blood, muscle, and other materials.
Biomaterials	Materials designed specifically for use in biological applications, such as artificial limbs and membranes for dialysis as well as bones and muscle.
Bio-based materials	Materials that are derived from living tissue but do not serve a function for an organism.
Biomimetic materials	Materials that are not produced by a living organism but are chemically and physically similar to the ones that are.

Biomaterials can be classified roughly into two primary categories: *structural (or inert) biomaterials*, whose primary function is to provide a physical support for the body, and *functional (or active) biomaterials*, which perform a function in the body other than physical support. Artificial bone, stents in arteries, and artificial limbs would be considered structural biomaterials; artificial organs, pacemakers, and controlled release implants would fall under functional biomaterials.

Biomaterials face the same design requirements as materials used for conventional applications: They must possess a desirable blend of mechanical properties and cost efficiency; they must maintain their properties over the expected life of the material; and they must perform the function for which they were designed. However, biomaterials also must possess *biocompatibility*, the ability to function within a host organism without triggering an *immune response*. White blood cells (including macrophages and natural killer cells) are specifically designed to identify foreign materials in the body and destroy them. If the biomaterial triggers the immune response, it will be rejected by the body.

The topic of biological and biomaterials is broad and evolving continuously. This chapter categorizes the materials by their primary function, either structural or functional, then compares and contrasts the natural biological material with the biomaterials designed to replace or interact with it.

| *Bio-Based Materials* |
Materials that are derived from living tissue but do not serve a function for an organism.

| *Biomimetic Materials* |
Materials that are not produced by a living organism but are chemically and physically similar to the ones that are.

| *Structural Biomaterials* |
Materials designed to bear load and provide support for a living organism, such as bones.

| *Functional Biomaterials* |
Materials that interact or replace biological systems with a primary function other than providing structural support.

| *Biocompatibility* |
Ability of a biomaterial to function within a host without triggering an immune response.

| *Immune Response* |
White blood cells identifying a foreign material in the body and attempting to destroy it.

What Biological Materials Provide Structural Support and What Biomaterials Interact with or Replace Them?

9.2 STRUCTURAL BIOLOGICAL MATERIALS AND BIOMATERIALS

9.2.1 // Bone

The most important structural biological material is *bone*, a naturally occurring fiber-reinforced composite that comprises the skeletal systems of most animals. Bone consists of organic *collagen*, fibrils in a primarily calcium phosphate

| *Bone* |
Structural biological material that consists of a fiber-reinforced composite and comprises the skeletal system of most animals.

| *Collagen* |
High-tensile-strength structural protein found in bone and skin.

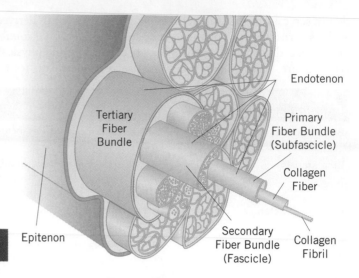

FIGURE 9-1 Collagen Fiber Schematic

Tertiary Fiber Bundle

Epitenon

Endotenon

Primary Fiber Bundle (Subfascicle)

Collagen Fiber

Secondary Fiber Bundle (Fascicle)

Collagen Fibril

| *Scleroprotein* |
High-tensile-strength structural protein.

| *Woven Bone* |
During growth and repair, bone produced in which the collagen fibrils are aligned randomly.

| *Lamellar Bone* |
Bone that replaces woven bone, in which the collagen fibrils align along the length of the bone.

| *Hydroxyapatite* |
Biomimetic material that is often used as a bone filler when osteoblasts cannot reconnect disjointed pieces of bone without aid.

| *Osteoblasts* |
Cells located near the surface of bone that produce osteoid.

| *Osteoid* |
Blend of structural proteins (containing primarily collagen) and hormones that regulate that growth of bones.

| *Bone-Lining Cells* |
Cells that serve as an ionic barrier and line the bone.

| *Osteocytes* |
Osteoblasts trapped in the bone matrix that facilitate the transfer of nutrients and waste materials.

| *Osteoclasts* |
Cells that dissolve the bone matrix using acid phosphatase and other chemicals to allow the body to reabsorb the calcium in the bone.

| *Mineralization* |
Growth of the bone matrix onto fibrils of collagen.

matrix. Collagen is a high-tensile-strength structural protein (or *scleroprotein*) that forms a triple helix substructure with cross-linking between the individual strands. Collagen accounts for 40% of the protein in most mammals and also provides the major organic component in skin and teeth. A typical collagen fiber is shown in Figure 9-1.

Bones can be classified as either woven or lamellar, depending on the orientation of the collagen fibrils. During growth or repair, the collagen fibrils align randomly, much like a chopped fiber composite. This *woven bone* has a comparatively low tensile strength. As the growth continues, the woven bone is gradually replaced with *lamellar bone*, in which the collagen fibrils align along the length of the bone.

The predominantly calcium phosphate matrix provides a high compressive strength to the bone but is relatively brittle. The calcium phosphate forms in a hexagonal lattice structure similar to *hydroxyapatite* $(Ca_{10}(PO_4)_6(OH)_2)$. Bone cells manage the lifelong production, reabsorption, and repair of these composite bones.

Four distinct types of bone cells exist:

1. *Osteoblasts* are cells located near the surface of the bone that produce *osteoid*, a blend of structural proteins containing primarily collagen and hormones that regulate the growth of bones.

2. *Bone-lining cells* serve as an ionic barrier and line the bone.

3. *Osteocytes* are the most numerous cells present in bone. Osteocytes began as osteoblasts that became trapped in the matrix and developed into a series of star-shaped cells connected by narrow channels that facilitate the transfer of nutrients and waste products.

4. *Osteoclasts* are bone-destroying cells that migrate to specific bone areas and cluster in pits on the surface. These cells release acid phosphatase and other chemicals that dissolve the bone matrix and allow the calcium to be reabsorbed by the body.

When it is time for bone to grow, osteoblasts secrete fibrils of collagen along with osteoid. During the growth of the matrix, called *mineralization*, the osteoblasts release sealed vesicles containing the enzyme alkaline phosphatase, which cleaves phosphate bonds. Phosphate and calcium begin to deposit on the vesicles, which then split apart and serve as heterogeneous nuclei to facilitate crystal growth.

Bone is a relatively brittle material, despite the flexibility provided by the collagen fibrils, and tends to suffer cracks and surface fractures. However, throughout the life of the organism, bone is continually reabsorbed and replaced through a process called *remodeling*. Deformation resulting in microcracking or other damage produces a small chemical potential because of the piezoelectric nature of the bone. When the signal is received, the osteoclasts migrate and cluster on the damaged areas and begin to reabsorb the damaged bone. Ultimately, osteoblasts produce new collagen and osteoid to replace the bone without any significant change in shape.

When specific areas are exposed to repeated stresses (such as the forearms of professional tennis players), the rate of bone growth increases, and the bone thickens. *Wolf's law* indicates that bone will adapt to repeated environmental stresses, becoming stronger when exposed to high stress levels and becoming weaker when stress is reduced. This law has potential implications for astronauts, who may be subject to bone loss when freed from the strain of gravity for extended periods of time.

Even though bone is a self-healing, continuously adapting material system, biomaterials are used to help individuals with specific problems. Small breaks will heal themselves and often require no more than a splint or cast to hold the bone in place during remodeling. When severe breaks occur, the surgical addition of titanium screws, rods, or plates may be necessary to hold the bone in place.

When large segments of bone must be removed, the osteoblasts cannot reconnect the disjointed pieces without help. Often powdered or beaded hydroxyapatite is used as a bone filler, creating a framework to support the growth of new bone. The hydroxyapatite is a biomimetic material, not produced by the body but similar enough in chemistry and mechanical properties to be accepted by the body. Although the hydroxyapatite is not strong enough itself to support mechanical loads, it is thermodynamically stable at body pH and is close enough in chemical composition that it becomes part of the growing bone matrix through a process called *osseointegration*.

In young people, the rate of bone formation exceeds the rate of reabsorption. The maximum bone density and strength occurs at around age 30. When the rate of reabsorption exceeds the production rate of new bone, a medical condition called *osteoporosis* can result, leaving weakened porous bones. Lack of calcium in the diet and lack of exercise are contributing factors, but osteoporosis affects more than 10 million Americans. The most common result of osteoporosis is bone fractures. More than 300,000 hip fractures occur in the United States each year.

Hip replacement surgery has become the norm in such cases. The hip is a ball-and-socket joint where the femoral head extends into the *acetabulum socket*, as shown in Figure 9-2. When hip replacement is performed, the femoral head is removed and replaced with a titanium or cobalt stainless steel implant that includes a ball-shape femoral head and a long stem that extends into the marrow in the center of the femur, as shown in Figure 9-3. The replacement femoral head usually is made of titanium or alumina, but recent advances in *bioceramics* have resulted in the use of yttria-stabilized tetragonal zirconia polycrystal *Y-TZP femoral heads*, which offer better wear rates and better strength and allow for heads with smaller diameters to be used.

The implant is coated with hydroxyapatite to reduce any chance of an immune response to facilitate osseointegration, because hydroxyapatite is

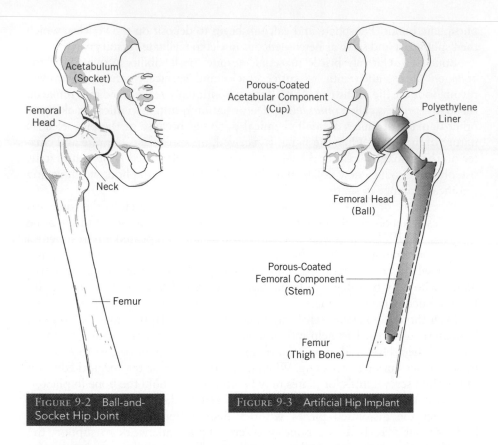

FIGURE 9-2 Ball-and-Socket Hip Joint

FIGURE 9-3 Artificial Hip Implant

already present in the body and does not trigger a response. The titanium femoral head is capped with a polyethylene liner and a polyethylene or ceramic (aluminum oxide) cup, which is fixed to the acetabular socket using bone cement (polymethylmethacrylate) lubricated with synovial fluid, the naturally occurring stringy fluid found in joints. The titanium ball can rotate within the cup, providing mobility. Recent developments include the use of cementless implants that rely on osseointegration to secure the cup in place.

The biomaterials used in hip replacements still face many challenges. Because the biocompatible metals used in the replacement are much stronger than natural bone, less load is applied to the remaining bone. Wolf's law dictates that less load on a bone will result in less bone, so the bone around the metal becomes weaker and thinner. Natural joint lubrication is often insufficient, and attempts to coat the implants with Teflon and other semipermanent lubricants have failed. The naturally occurring synovial fluid can corrode the metal, leading to potentially toxic compounds. Hip replacements are a temporary solution that will not last more than 10 to 20 years with current materials.

9.2.2 // Prosthetic Limbs

| **Prosthetic Limbs** |
Replacement artificial limbs.

Prosthetic limbs represent another significant application for structural biomaterials. For many years artificial legs consisted of wooden or metal stakes secured to the knee through a series of straps that allowed an amputee to walk (with difficulty) but little else. The first nonlocked prosthesis appeared in 1696 and used external hinges to allow pivoting around the ankle. By 1800 a pulley system that accounted for the movement of the foot was developed. This system allowed for a more normal walking stride, but the technology stagnated for 150 years.

In 1946 the American Orthotics and Prosthetics Association was chartered to improve prosthetics. Over the past 60 years, a new generation of prosthetic materials and designs have revolutionized the field and offer a richer quality of life for amputees. Even modern prosthetics cannot fully match the performance of the natural limb, but prostheses have advanced to the point that many amputees can participate in athletic events.

Transtibial protheses, artificial limbs beginning below the knee like the one shown in Figure 9-4, are by far the most common. The top portion of the prosthesis must fit securely against the residual limb. The limb is often sensitive to pressure, so a hard socket is custom molded from polypropylene and/or carbon fiber composite materials. A soft silicone liner is placed between the socket and the residual limb for cushioning.

To replace the missing tibia, all transtibial protheses contain a long bar called the *keel* that must withstand the significant compressive forces associated with walking. Most times the keel is made out of titanium or carbon fiber composites. It must be stiff but have some capability to flex. If the elastic modulus of the material is too high, walking and running would be painful, as more of the force of walking would be transmitted to the residual limb. If the elastic modulus is too low, maintaining balance during walking or running would be difficult.

Most transtibial prostheses are nonarticulated (they do not flex at the ankle); the foot is aligned at a 90 degree angle from the ankle. The most common foot is a *solid-ankle–cushion-heel (SACH) foot* that is durable and comparatively low cost. The heel contains a compressible rubber wedge in the heel with a solid keel (often wooden) at the bottom of the foot to provide stability, as shown in Figure 9-5.

| *Transtibial Prostheses* |
Artificial limbs beginning below the knee.

| *Keel* |
Long bar that replaces the missing tibia in a transtibial prosthesis limb.

| *SACH Foot* |
Solid-ankle–cushion-heel foot; the most common transtibial prosthesis, which contains a rubber wedge in the heel and a solid keel, often wooden.

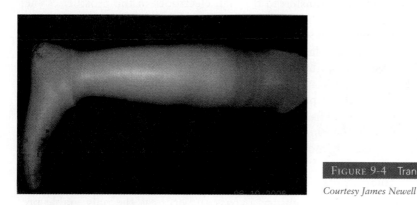

FIGURE 9-4   Transtibial Prosthesis

Courtesy James Newell

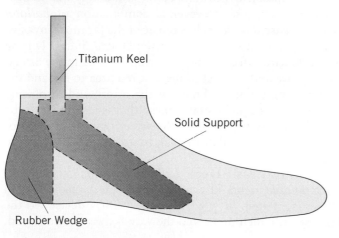

Titanium Keel

Solid Support

Rubber Wedge

FIGURE 9-5   Schematic of a SACH Foot

Ankle-foot combinations designed for athletes must have a wider range of motion and are articulated but are less stable and have greater weight.

Prosthetic arms present different challenges. Although they do not have the same load-bearing responsibilities as prosthetic legs, and the elbow joint is no more complicated than the knee, the range of motions and requirements for the hands are far more complex than for the feet. Until 1909 choices for prosthetic arms consisted of a leather socket that fit on the residual limb and a cumbersome steel frame that supported either a simple hook (that remains popular in horror movies today) or a synthetic hand that looked more realistic but offered little functionality.

In 1909 a physician named D. W. Dorrance developed the first prosthetic arm capable of prehension. His split-hook system could be opened and closed, giving the wearer some ability to grasp objects. The basic design of the system uses a harness and a cable that hooks around the opposite shoulder to provide control of the limb. Many modern transradial amputees use this *Dorrance hook* structure, but with more modern materials: socket is usually made from polypropylene rather than leather used in the original; steel in the frame is aluminum or titanium alloys or even carbon-fiber composites in some cases; frame is covered with a custom-measured polyurethane foam to emulate the shape of the other arm; and sock (usually made of Lycra) covers the foam and is colored to match the skin tone of the amputee.

The greatest innovation in *transradial* prosthetics has been the development of *myoelectric arms*, artificial arms that respond to the amputees' muscle impulses to control the function of the prosthetic. Electrical impulses from the residual upper arm are transferred through electrodes in the prosthetics shell to circuits in the fingers that trigger responses that emulate a regular hand. Although these myoelectric hands are now available commercially and are covered by some insurance plans, so far they do not have the durability of other transradial prosthetic options.

Dean Kamen, the inventor of the popular Segway and a myriad of other items, has developed the prototype for what is likely the next generation of transradial prosthetics. His "Luke Arm," named after the bionic hand used by Luke Skywalker in *The Empire Strikes Back*, offers 18 degrees of freedom and is capable of mimicking most of the subtler motions of the human hand.

9.2.3 // Vascular Stents

For much of the twentieth century, bypass surgery was the only viable procedure to address arterial blockages. However, a combination of *balloon angioplasty* and insertion of *vascular stents* has replaced the far more invasive bypass surgery in over 1 million cases per year in the United States. During the angioplasty, a thin guide wire is fed through the clogged area in the artery. A balloon is fed along the wire and inflated in the clogged area to expand the opening of the vessel, breaking the inner lining of the artery and displacing the blockage. A vascular stent is a small metallic mesh that is inserted into the blood vessel during an angioplasty to help keep the artery open after the procedure. Without the stent, the artery collapses back to its original diameter in nearly 40% of angioplasties.

Vascular stents come in two basic types: *balloon expandable* and *self-expanding*. *Balloon-expandable stents* fit over the angioplasty balloon and expand when the balloon is inflated. In 1987 Dr. Julio Palmaz developed the first balloon-expandable stent, which pioneered the way for all that followed.

| **Dorrance Hook** |
Split-hook system for transradial amputees that offers some prehensile ability.

| **Transradial** |
Below the elbow.

| **Myoelectric Arms** |
Transradial prosthetics that use muscle impulses in the residual arm to control the function of the prosthetic.

| **Balloon Angioplasty** |
Procedure used to address arterial blockages that can replace the more invasive bypass surgery procedure. A balloon is fed along a thin guide wire and inflated in the clogged area to expand the vessel opening.

| **Vascular Stents** |
Small metallic meshes inserted into blood vessels during angioplasty to keep the artery open after the procedure.

| **Balloon-Expandable Stents** |
Stents that fit over the angioplasty balloon and expand when the balloon is inflated.

Self-expanding stents are not introduced on top of the angioplasty balloon but are deployed by the use of a catheter.

The *Palmaz stent*, pictured in Figure 9-6, was made of stainless steel, which was the material of choice for many years and is still used in some stents. Stainless steel stents often trigger *restenosis*, the buildup of scar tissue around the stent that leads to restriction of blood flow. However, the unique properties of a nickel-titanium alloy called *nitinol* have revolutionized stent manufacture. Nitinol undergoes a diffusionless transformation from an FCC (face centered cubic) structure similar to austenite to a BCC (body-centered cubic) structure analogous to martensite. This transformation causes the nitinol stent to experience a *shape memory effect* in which a deformed nitinol alloy does not change shape when the load is removed but returns to its initial lattice positions when heated. When the nitinol stent is employed in the vessel, the body temperature of 38°C is sufficient to cause a phase transformation back to the austentic state.

Nitinol is not inherently biocompatible, and prolonged exposure to bodily fluids could result in the leaching of toxic nickel into the bloodstream, so a passivating layer of metal approved by the Food and Drug Administration (FDA) must coat the surface of nitinol stents. Additionally, most stents are coated with thin layers of polymers that have been infused with retinosis-retarding pharmaceuticals that slowly diffuse out and reduce the development of scar tissue in the artery. A typical nitinol stent is shown in Figure 9-7.

Catheters are tubes that are inserted into vessels or ducts in the body, generally to promote either the injection or the drainage of fluids. Although catheters can be used for a variety of purposes, including draining abscesses and introducing intravenous fluids, medication, or anesthesia, the use of a *Foley catheter* to replace a damaged urethra has significant materials selection issues. The Foley catheter, pictured in Figure 9-8, is a flexible tube inserted through the tip of the penis into the bladder to drain urine.

Foley catheters can be made from a variety of materials, including latex, polyvinyl chloride (vinyl), and silicone. Latex catheters are the most common but have disadvantages, including increasingly common allergic reactions to latex among patients and a tendency to support bacterial colonies. Urinary tract infections occur within four days after a latex Foley catheter is used while silicone catheters may last several weeks without infection. Silicone catheters offer the best overall performance but are significantly more expensive than either latex or vinyl. Vinyl

| *Self-Expanding Stents* |
Stents that are deployed by the use of a catheter.

| *Palmaz Stent* |
Balloon-expandable stent made of stainless steel.

| *Restenosis* |
Buildup of scar tissue around a stent, leading to a restriction of blood flow.

| *Nitinol* |
Nickel-titanium alloy used in stent manufacture that experiences a shape memory effect.

| *Shape Memory Effect* |
Effect in which the alloy does not change shape when the load is removed but does return to its initial lattice position when heated.

| *Catheters* |
Tubes inserted into vessels or ducts of the body, usually to promote the injection or drainage of fluids.

| *Foley Catheter* |
Tube used to replace a damaged urethra.

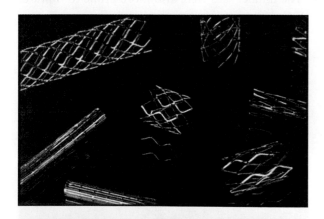

FIGURE 9-6   Palmaz Stent

Courtesy James Newell

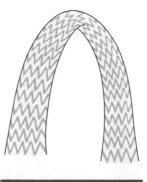

FIGURE 9-7   Nitinol Stent

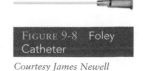

FIGURE 9-8   Foley Catheter

Courtesy James Newell

catheters offer a compromise in performance and cost but present challenges of their own. Foley catheters must be soft and flexible. PVC is extremely hard unless plasticizers are added to soften it, as discussed in Chapter 4. Concerns linger about the likelihood of these plasticizers leaching out of the catheter and into the body.

9.2.4 // Breast Implants

Breast implants are used either to augment breast size or to replace breast tissue that has been surgically removed, often due to cancer. Breast augmentation is now the third most common cosmetic surgery performed in the United States with nearly 300,000 such operations performed each year. The total cost of the surgery ranges from $4,000 to $10,000, making breast implants a billion dollar a year industry.

The first breast implants occurred shortly after World War II and involved the direct injection of silicone into the breasts. The results were disastrous. Numerous complications resulted, leading many recipients to undergo mastectomies, and at least three women died from blood vessel blockages caused by the injected silicone.

In 1962 the Dow Corning company produced the first commercial *silicone gel implant* that used a silicone rubber sac filled with a silicone gel. Figure 9-9 shows a photo of silicone gel implants. Because medical devices were not regulated in the United States until the mid-1970s, there are no detailed studies of the health effects of these early implants. However, several issues are known to have occurred. As part of the immune response, scar tissue forms around an implant, creating a barrier from the rest of the body. In some cases this barrier tightens, leading to *capsular contracture*, which compresses and potentially ruptures the implant. Capsular contracture causes significant pain and can cause the breast to become misshapen.

When a sac does leak, the silicone gel that escapes the scar tissue barrier becomes free to migrate throughout the body. This *extracapsular silicone* can result in additional health consequences, although the nature, frequency, and severity of these consequences remain a topic of great debate.

In the 1970s the silicone implants were altered to have thinner gels, thinner sacs, and a polyurethane coating designed to reduce capsular contracture. The FDA banned polyurethane-coated implants after they were found to break down into a carcinogenic chemical. The thinner sacs were also more prone to rupture.

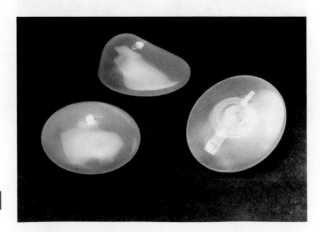

FIGURE 9-9   Silicone Gel Implants

Courtesy of the Food and Drug Administration

These ultimately were replaced with a design involving two chambers, an inner chamber filled with silicone gel and an outer chamber filled with saline to reduce the chance of rupture leading to extracapsular silicone.

The health concerns resulted in silicone implants being banned entirely in Canada and having their use restricted in the United States. From 1992 through 2006 the FDA limited the use of silicone gel implants to replacement of existing silicone gel implants that had ruptured or in controlled clinical studies for reconstruction after mastectomies. Current technology utilizes *gummy bear silicone breast implants* with a stable form. The gel is cohesive, so it is far less likely to leak into the body. These implants are used extensively outside of North America and show fewer health concerns than their predecessors but were not been approved in the United States until November 2006. The FDA requires manufacturers to perform a 10-year postapproval study following 40,000 women who receive the new implants.

Saline breast implants, pictured in Figure 9-10, offer an alternative to silicone. Saline implants first appeared in the mid-1960s but were discontinued in the 1970s because of persistent problems with rupture. When the controversies regarding silicone implants erupted in the 1990s, saline implants emerged as the leading implant choice. The new saline implants utilize thicker, room-temperature-vulcanized (RTV) shells that resist rupture. Although saline implants offer fewer complications, they cannot replicate the more natural appearance of silicone gel implants.

9.2.5 // Teeth and Dental Fillings

The living soft tissue in the center of the tooth is called *dental pulp*. It is protected by three layers of materials—*enamel*, *dentin*, and *cementum*—as shown in Figure 9-11. Enamel is the hardest substance in the body and consists primarily of hydroxyapatite, much like bone. The thickness of the enamel layer varies with location but can be as much as 2.5 mm. Although extremely hard, tooth enamel is also quite brittle.

The dentine is a yellow, porous material comprised of collagen and other structural proteins blended with a hexagonal calcium phosphate mineral called *dahlite*. The cementum is a bony material whose primary function is to provide a point of attachment for the periodontal ligaments. It is yellow, softer than either dentine or enamel, and consists of approximately 45% hydroxyapatite, 33% structural proteins, and 22% water.

| *Gummy Bear Silicone Breast Implants* |
| Silicone breast implants that potentially eliminate or at least significantly reduce the leakage of silicone gel. Named because the texture of the pouch resembles that of the famous candy. |

| *Saline Breast Implants* |
| Alternative to silicone breast implants that uses saline instead of silicone in the pouch. |

| *Dental Pulp* |
| Living soft tissue in the center of a tooth. |

| *Enamel* |
| Hardest substance in the body, which covers the teeth and consists primarily of hydroxyapatite. |

| *Dentin* |
| Yellow, porous material comprised of collagen and other structural proteins blended with dahlite. |

| *Cementum* |
| Bony material layer in teeth that primarily provides a point of attachment for the periodontal ligaments. |

| *Dahlite* |
| Hexagonal calcium phosphate mineral in dentin. |

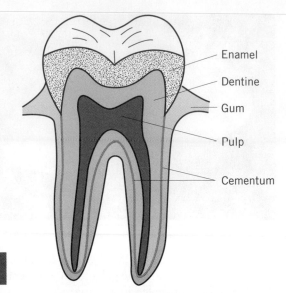

Enamel

Dentine

Gum

Pulp

Cementum

FIGURE 9-11 Schematic of a Tooth Showing Different Material Layers

The hyrdoxyapatite present in all three protective layers is subject to attack by acids. Many bacteria that reside in the mouth can interact with sugars to produce lactic acid, which lowers the pH of the mouth and begins to dissolve the enamel. As more of the hydroxyapatite degrades, the surface of the tooth softens and a cavity often results.

| Amalgams |
Mercury-based alloy used for dental fillings.

Dental fillings date back at least to the sixteenth century, when lead and cork were used to fill cavities. The first standard material for dental fillings were mercury alloys called *amalgams*. Beginning in 1895, the standard dental amalgam was the *gamma-2 phase amalgam*, a mixture containing 50% mercury and 50% of an alloy powder containing at least 65% silver, less than 29% tin, about 6% copper, and small quantities of mercury and zinc. In about 1970, the standard formulation for dental amalgams changed to a *high-copper amalgam*, primarily for economic reasons. The new amalgam contained 50% liquid mercury, but the new alloy powder was comprised of around 40% silver (the most expensive metal in the alloy), 32% tin, 30% copper, and small amounts of mercury and zinc.

| Gamma-2 Phase Amalgam |
Amalgam containing 50% mercury and 50% of an alloy powder made up of at least 65% silver, less than 29% tin, about 6% copper, and small amounts of mercury and zinc.

| High-Copper Amalgam |
Amalgam containing 50% liquid mercury and 50% of an alloy powder made of 40% silver, 32% tin, 30% copper, and small amounts of mercury and zinc.

Amalgams remain in use because of their blend of hardness, ease of manufacture, and low cost, but some concerns have been raised about the health effects of the mercury. Even low levels of mercury in the body have been linked to birth defects, nervous system difficulties, and mental disorders. Many scientists argue that the mercury is bound into the amalgam and that only negligible amounts can leach into the body. Nevertheless, the use of amalgams is declining as more dentists switch to *dental composites* that look more like natural teeth.

| Dental Composites |
Replacements for amalgams that look more like natural teeth but are more expensive and tend not to last as long.

Dental composites are more expensive than amalgams and tend not to last as long. A typical amalgam would be expected to last 10 to 15 years, while 8 is more typical for a dental composite. The most common dental composite involves a *bis-GMA (bisphenol glycidylmethacylate acrylic) resin* along with filler materials, such as powdered glass. The filler materials reduce the cost of the process but also reduce resin shrinkage during curing. The composite is manufactured one layer at a time with the resin being cured at each level by light. The formation of free radicals is essential to the curing of the resin, so photochemical catalysts such as camphoroquinone are added to the composite.

| bis-GMA Resin |
The binder material in the most common dental composite, along with filler materials such as powdered glass.

Unlike amalgam fillings, composite fillings bond with the remaining tooth enamel. As a result, composite fillings are less likely to fracture even though they have less inherent strength than amalgam fillings. Additionally, amalgam fillings expand with age and can crack the tooth itself. By contrast, composite fillings shrink and are more likely simply to fall out.

What Biomaterials Serve a Nonstructural Function in the Body?

9.3 FUNCTIONAL BIOMATERIALS

Functional biomaterials serve a purpose beyond structure in the body. They can be as involved as artificial organs that replace the all or part of the functioning of a body part or simpler items such as pacemakers that help control the electrical signals that cause the heart to beat. In many cases, functional biomaterials are implanted in the body, but sometimes materials are taken from the body, passed through the functional biomaterial, then returned.

9.3.1 // Artificial Organs

In most cases, the best replacement for a defective organ is a transplanted organ from another person. However, the limited availability of compatible transplanted organs has made the use of artificial organs a medical necessity.

The kidneys are responsible for removal of urea, other waste products, and excess fluid from the bloodstream. When the kidneys are not capable of fully managing this function, patients are treated using a membrane filtration system called *dialysis*. In the most common form of dialysis, *hemodialysis*, the blood of a patient is removed by a catheter and passed through a semipermeable membrane. *Dialysis fluid* (a sterilized, highly purified water with specific mineral salts) runs on the other side of the membrane, as shown in Figure 9-12. Differential pressure causes water to pass through the membrane, reducing the excess fluid and electrolyte concentration in the blood. This process is called *ultrafiltration*. Simultaneously, concentration gradients cause the urea and other toxins to diffuse across the membrane and out of the blood. Dialysis generally requires several hours per session and many times is required as often as three times per week.

The membrane is the cornerstone of the dialysis process. The membrane must be:

- Permeable to urea and other waste.
- Impermeable to blood cells, plasma, and other key components.
- Strong enough to withstand the differential pressures.
- Biocompatible enough to not trigger an immune response from the blood.

The most common material used in hemodialysis is cellulose acetate, also called *regenerated unmodified cellulose*. In fact, the earliest dialysis membranes were made from cellophane sausage casings. Cellulose acetate is relatively inexpensive

| *Dialysis* |
Membrane filtration system used in patients whose kidneys are not capable of fully managing the removal of urea, other waste products, and excess fluid from the bloodstream.

| *Hemodialysis* |
The most common form of dialysis, in which the blood of the patient is removed by a catheter and passed through a semipermeable membrane to remove toxins and excess water.

| *Dialysis Fluid* |
Sterilized, highly purified water with specific mineral salts used in dialysis.

| *Ultrafiltration* |
Process in which the differential pressure causes water to pass through the membrane in dialysis, reducing the excess fluid and electrolyte concentration in blood.

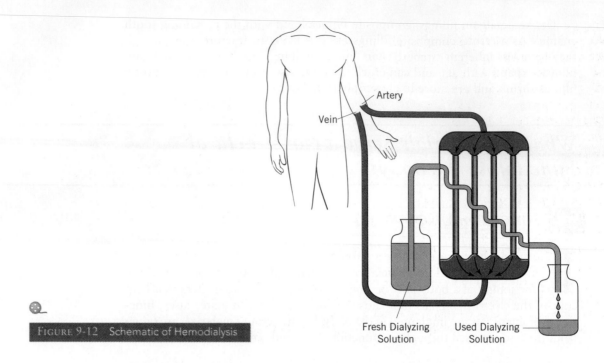

Artery

Vein

Fresh Dialyzing
Solution

Used Dialyzing
Solution

FIGURE 9-12   Schematic of Hemodialysis

and is porous to the fluid, but the pore sizes are small enough to prevent blood cells, proteins, and other key factors from passing through.

The biocompatibility of cellulose acetate membranes has been problematic; over the years significant numbers of negative reactions have been reported. Surface treatments are performed on the cellulose acetate to make it less hydrophilic and more biocompatible. Alternative membrane materials include polysulfone, PMMA, and PAN. These polymers are generally prepared as hollow tubes. The patient's blood runs through the center of the tube with the dialysis fluid running across the outside. Although these polymeric membranes are more expensive, they have fewer biocompatibility issues.

The heart is the organ responsible for pumping the blood throughout the body. When heart disease is too severe for the patient to wait for a donor organ, an artificial heart is sometimes used. As early as 1969, a patient was kept alive for 60 hours using a mechanical heart, but the pinnacle moment for artificial hearts came in 1982, when a *Jarvik-7 artificial heart* was implanted into a patient named Barney Clark, who survived for 112 days. The Jarvik-7 heart required an external power supply that the patient could wear in a backpack.

Because of the numerous issues—biocompatibility, power sources, blood cell destruction, and the like—associated with truly artificial hearts, much of the current research emphasis focuses on *ventricular assist devices (VADs)* that help a damaged heart increase its functionality and throughput. In some designs, the VAD is implanted directly into the left ventricle of a damaged heart. Inside a titanium case, a motor causes a titanium impeller to spin, thereby increasing the blood flow through the heart. In other cases, the VAD sits in the abdominal cavity, pulls blood from the left ventricle, then pumps it directly into the aorta, effectively supplementing the output from the heart.

External VADs are used during open heart surgery, but implanted models are generally viewed as bridges to transplant and are designed to remain in patients for several months while donor hearts are located. The FDA has

| *Jarvik-7 Artificial Heart* |
Artificial heart implanted into a patient named Barney Clark in 1982 that kept him alive for 112 days.

| *Ventricular Assist Devices (VADs)* |
Devices that help a damaged heart increase its functionality and throughput.

approved a VAD for permanent implantation in terminally ill patients who cannot qualify for transplant, and the American Heart Association has recommended that VADs become permanent measures for late-stage heart disease patients. This mirrors the developments in Europe, where the Jarvik 2000 VAD has been approved for implantation by the European Union.

Skin provides the barrier to protect internal organs from pathogens, keeps water in, and accounts for more than 15% of the total weight of the human body. Skin is comprised of two distinct layers: the *epidermis* and the *dermis*. The epidermis is the outer layer and contains no blood cells. Instead, it receives nutrients by diffusion from the inner layer (dermis). More than 90% of the epidermis is comprised of *keratinocytes*, cells that contain a large quantity of the hard structural protein *keratin*. The strength of skin is enhanced by the presence of keratin along with a pair of other structural proteins: collagen and *elastin*. As skin cells advance upward through the epidermis, they change shape and produce more keratin until they eventually reach the surface layer. This hard layer of keretinocytes serves as a primary moisture barrier. After about 30 days, epidermal cells dry and fall off the body to make room for the next tier through a process called *keratinization*. The lower layer, the dermis, contains a much more diverse set of materials, including blood vessels, sweat glands, nerve cells, hair follicles, oil glands, and muscles.

Skin is largely self-healing. The keratinization process automatically repairs damage that does not extend past the epidermis, but a more active healing mechanism is necessary when the damage reaches the dermis. The body responds to damage to the dermis by laying down new collagen fibers across the wound site. The fibers provide a matrix for new skin growth but also result in the presence of a scar. The regrown dermal tissue will support epidermal cells and provide a barrier to water loss and infection, but hair follicles and sweat glands will not be replaced.

When large segments of skin are damaged (e.g., in the case of burn victims), the damaged sites are too large for the body to repair itself effectively. Prior to 1986, victims who experienced third-degree burns on more than 50% of their body were almost certain to die. But since 1986 an artificial skin has offered new hope.

Instead of relying on the body to lay down collagen fibers, a synthetic matrix of fibers is applied to the wound area. The porous mesh provides a framework for the growth of new skin cells and dramatically reduces scarring. In some cases a blend of collagen fibers from cows and glycosaminoglycan fibers (derived from shark cartilage) are interwoven to provide the matrix for the growth of new dermal cells while a simpler polysiloxane mesh supports the growth of new epidermal cells. These soft scaffolds mimic natural biological materials and enable the skin to grow without the scarring associated with the natural collagen lattice manufactured by the body. Once the layer has healed sufficiently, it can be replaced by skin grafts from other parts of the body. Figure 9-13 shows a comparison of the dermal and epidermal layers in natural skin and artificial skin.

In the past, patients who experience bladder damage, either through trauma or bladder cancer, were forced to have an external urine collection bag and a small hole added to their bodies. However, the first successful artificial bladder was made in 1999. Unlike many artificial organs, the artificial bladder is a biological material rather than a biomaterial because it is made from a section of the patient's own small intestine. Surgeons now can remove approximately three feet of small intestine and shape it like a grapefruit. The reshaped intestine is placed back in the body and connected to the ureter and kidneys, thus functioning much like the original bladder.

| *Epidermis* |

Outer layer of skin, which contains no blood cells but receives its nutrients from diffusion from the dermis.

| *Dermis* |

Inner layer of skin containing blood vessels, sweat glands, nerve cells, hair follicles, oil glands, and muscles.

| *Keratinocytes* |

Cells that comprise 90% of the epidermis and contain a large quantity of keratin.

| *Keratin* |

Hard structural protein contained in keratinocytes.

| *Elastin* |

Structural protein that enhances the skin's strength and ability to stretch.

| *Keratinization* |

Process in which after about 30 days, epidermal cells dry and fall off the body to make room for the next tier of cells.

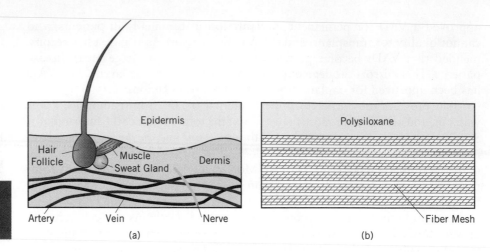

Figure 9-13
Comparison of
(a) Natural Skin and
(b) Artificial Skin

Epidermis

Hair
Follicle

Muscle
Sweat Gland

Dermis

Polysiloxane

Artery   Vein   Nerve

Fiber Mesh

(a)   (b)

Research is ongoing into improved biomaterials to replace many other organs, including eyes, muscles, and the larynx. Although many challenges remain, the field of biomaterials as replacements for damaged organs offers great promise for the future.

9.3.2 // Functional Biomaterials in Support of Functional Organs

Many biomaterials are used to enhance the performance of functional, natural organs rather than to replace them. Although countless biomaterials fit this broad classification, key applications include mechanical heart valves, artificial blood, pacemakers, and polymer implants for controlled release of pharmaceuticals.

Each of the four chambers of the heart is serviced by a check valve that prevents blood from seeping from one chamber to another, allows high flow rates without adding unnecessary resistance to the flow, and handles the high pressure of the flowing blood. As shown in Figure 9-14, the *mitral valve*

| *Mitral Valve* |
Heart valve separating the left ventricle from the left atrium.

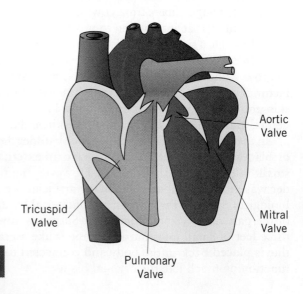

Aortic
Valve

Tricuspid
Valve

Mitral
Valve

Pulmonary
Valve

Figure 9-14   Schematic of the Heart
Showing the Four Valves

separates the left ventricle and left atrium of the heart, the *tricuspid valve* separates the right atrium and right ventricle, the *pulmonary valve* controls the flow of blood into the lungs, and the *aortic valve* regulates the entry of the oxygenated blood into the body.

Problematic heart valves tend to experience one of two problems: *regurgitation* or *stenosis*. Regurgitation occurs when the valve fails to seal properly, allowing some blood to leak back into the previous chamber. As a result, the heart must work harder to pump some of the same blood twice. When the leak is severe, the heart cannot compensate for the decreased efficiency, and congestive heart failure results. Stenosis is a hardening of the valve that prevents it from opening properly. High blood pressures build up behind the valve, and the heart must work harder to move the blood through the body, leading to an increase in the size of the heart.

The first artificial valves were implanted into 10 patients in 1952 by Dr. Charles Hufnagel. Six of the 10 patients survived the operation. Hufnagel used a *caged-ball valve*, like the one shown in Figure 9-15, made from acrylic to replace defective aortic valves. The original ball was made from Silastic silicone rubber. The patients who died generally experienced *thrombosis* (blood clotting) in part because of the body's reaction to the acrylic. Today approximately 70% of artificial valve recipients survive for at least five years after implantation.

The caged-ball design includes an orifice through which the blood flows and a series of metal struts that keep the ball in place. The only modern caged-ball design approved for use is the *Starr-Edwards valve*, which uses a silicone-rubber polymer impregnated with barium sulfate as the ball and a cobalt-chromium alloy for the struts of the cage. The ball moves to the end of the cage when sufficient pressure builds up at the orifice, which permits flow in one direction. When the pressure reduces, the ball descends from the cage and fills the orifice, preventing regurgitation.

Tilting disk valves, like the one pictured in Figure 9-16, have a single circular disc that regulates flow. The most common tilting disk valve, the *Medtronic Hall valve*, is made from titanium with four struts regulating a flat carbon-coated disk, which can tilt 75 degrees for both aortic valves and mitral valves. The orifice is made from pyrolitic carbon and is covered by a seamless polyester ring. The titanium struts support and guide the disk, which opens during flow, then seals to prevent regurgitation. The design of this valve is largely unchanged since the early 1980s.

The *St. Jude bileaflet valve* has been implanted in over 500,000 patients since its first use in 1977. This valve, pictured in Figure 9-17, features two leaflets that swing apart when the valve is opened and create three separate regions of flow. The valve housing consists of pyrolitic carbon deposited over graphite. Two pivot guards come out on the inlet side to control the opening and closing of the leaflets. The blood flow through a bileaflet valve comes much closer to the hemodynamic conditions of flow through a normal heart valve.

All mechanical heart valves lose material over time. Erosion and frictional losses impact the surface of the ball and cage, the hinges of bileaflet valves, and the contact region between the struts and disk in tilting disk valves. Those valves that use titanium struts are also susceptible to fatigue failure after prolonged use, but this is not a factor in bileaflet valves made from graphite.

| *Tricuspid Valve* |
Heart valve separating the right atrium from the right ventricle.

| *Pulmonary Valve* |
Heart valve that controls the flow of blood into the lungs.

| *Aortic Valve* |
Heart valve located between the left ventricle and the aorta that regulates the entry of oxygenated blood into the body.

| *Regurgitation* |
Blood leaking back into the previous chamber of the heart when a valve fails to seal properly.

| *Stenosis* |
Hardening of a heart valve, which prevents it from opening properly.

| *Caged-Ball Valve* |
Artificial heart valve that uses a ball to seal the valve.

| *Thrombosis* |
Blood clotting.

| *Starr-Edwards Valve* |
The only modern caged-ball artificial heart valve design approved for use by the FDA.

| *Tilting Disk Valves* |
Artificial heart valves with a circular disk that regulates the flow of blood.

| *Medtronic Hall Valve* |
Most common tilting disk valve used as an artificial heart valve.

| *St. Jude Bileaflet Valve* |
Artificial heart valve featuring two leaflets (semicircular discs) that swing apart when the valve is open to create three separate regions of flow.

| Bernoulli Equation |
A form of the mechanical energy balance relating a pressure drop with density and velocity changes within a fluid.

| Effective Orifice Area (EOA) |
Estimate that measures the efficiency of a valve.

| Performance Index (PI) |
Dimensionless term that is calculated by dividing the effective orifice area by a standard.

The reduction in area as blood passes through the valve results in a drop in pressure. Because the volumetric flow rate (Q) of blood must be the same on both sides of the orifice,

$$Q = A_1 v_1 = A_2 v_2, \tag{9.1}$$

or

$$v_2 = \frac{A_1 v_1}{A_2}, \tag{9.2}$$

where v is the blood velocity, 1 signifies before the valve entrance, 2 signifies after the valve entrance, and A is the cross-sectional area of the vessel. The *Bernoulli equation* provides

$$P_1 + \frac{\rho_1 v_1^2}{2} = P_2 + \frac{\rho_2 v_2^2}{2}, \tag{9.3}$$

where P is the blood pressure and ρ is density. Equation 9.3 can be rewritten as

$$P_2 - P_1 = \frac{\rho_1 v_1^2}{2} - \frac{\rho_2 v_2^2}{2}, \tag{9.4}$$

which allows the pressure drop across the valve to be calculated.

The performance of a mechanical heart valve often is estimated by the *effective orifice area (EOA)*, which measures the efficiency of the valve. A higher EOA means a lower energy loss by the blood passing through the valve. The defining equation for EOA is

$$EOA = \frac{Q}{51.6\sqrt{P}}. \tag{9.5}$$

Because Equation 9.5 does not account for the size of the valve, a dimensionless term called the *performance index (PI)* is calculated by dividing the effective orifice area by a standard,

$$PI = \frac{EOI}{EOI_{std}}. \tag{9.6}$$

The performance index provides a measure of effectiveness that is independent of the size of the valve.

Larger pressure drops require the heart muscle to work harder and are undesirable. The caged-ball valves cause a significantly greater pressure drop than either the tilting disk or bileaflet valves. However, caged-ball valves experience almost no regurgitation compared with tilting disk and bileaflet valves, which both experience more.

All three mechanical valves subject the blood to stresses that tend to induce the formation of blood clots (thrombosis), so patients who receive the valves must take anticoagulants. High shear stresses and the formation of regions where the blood can stagnate are major causes of thrombosis. The caged-ball design develops high stresses along the walls of the struts, while the hinge area of bileaflet valves has both high stresses and stagnation, making clot formation common. Tilting disk valves develop the most clots behind the struts.

Because of the complications associated with mechanical valves, some doctors use valves made from biological materials. These valves come in three fundamental forms:

| Homografts |
Tissues that, after being
removed from a cadaver and
frozen in liquid nitrogen,
are thawed and installed
as a direct replacement
in the body.

| Autografts |
Replacement tissues formed
by the removal of tissue from
other parts of a patient's
body that is molded around
a stainless steel stent.

| Xenografts |
Implants of tissues from
another species.

1. *Homografts.* Heart valves are removed from a cadaver and frozen in liquid nitrogen. The valves then are thawed and installed as a direct replacement in the body.
2. *Autografts.* Surgeons remove tissue from other parts of the patient's body (usually the pericardium) and mold it around stainless steel stents to form new valves.
3. *Xenografts.* Surgeons implant valves made from other species (usually pigs but cows have also been used). These may be natural heart valves or pericardial tissue molded around stents.

Homografts offer several clear advantages. Because they are natural heart valves, they create almost no clotting and do not require the use of anticoagulants. Their hemodynamic properties are far superior to mechanical valves. However, the supply of donor valves is limited. The valves must be the right size and be compatible for transplant. The recipient must take immunosuppressants to prevent rejection, and the surgical technique to implant the replacement valve is difficult and dangerous.

Autografts offer the advantage of not triggering an immune response because the patient's own tissue is used. The material used to make autografts is not specifically designed to function as a valve, however, so the efficiency is often less than with homografts or mechanical valves.

Xenografts generally involve threading a hollow polypropylene cylinder through the porcine valve and using a cobalt-chromium alloy stent inside the cylinder to maintain structural integrity. Porcine valves are more readily available than those from human donors, and the implantation surgery is not as complex as that for artificial valves. Patients must take immunosuppressants after xenografts. Xenografts offer hemodynamic properties between mechanical valves and homografts.

Blood serves many purposes within the body, but its primary functions involve the transport of oxygen to body cells and the transport of waste products away from body cells. Blood consists of a yellow fluid called *plasma* that makes up 60% of the total volume and *corpuscles* (white blood cells, red blood cells, and platelets) that comprise the other 40%. The corpuscles are produced in the bone marrow and are replaced continuously. However, when trauma or illness causes an acute deficiency of red blood cells, a blood transfusion is required.

| Plasma |
Yellow fluid that makes up
60% of the total volume
of blood.

| Corpuscles |
White blood cells, red
blood cells, and platelets
in the bloodstream.

The most common source of replacement blood comes from donors. The American Red Cross and regional blood banks routinely solicit blood donations from the public, and donated blood can be stored for as long as 42 days before transfusion. However, in acute emergencies, the demand for blood often exceeds the supply, especially because the presence of antigens on the red blood cells limits the compatibility of blood from person to person.

The greatest need for a patient who has experienced substantial blood loss is to get oxygen to the cells of the body. Normally the red blood cells perform this function because of the presence of an iron-containing protein called *hemoglobin*. Four oxygen molecules attach to each hemoglobin molecule while the blood is in the lungs, where the partial pressure of oxygen is high. As the red blood cells

| Hemoglobin |
Iron-containing protein
that supplies oxygen to
the cells of the body.

travel throughout the body, the lower partial pressure of oxygen triggers the gradual release of the oxygen molecules. An artificial blood (more properly called *oxygen therapeutics*) capable of absorbing oxygen from the lungs and releasing it gradually throughout the body would have great medical value.

The quest for artificial blood has led to two competing technologies, although currently neither is approved for medical use in the United States. The first technique involves the use of perfluorocarbons (PFCs), which are organic molecules that have fluorine replacing hydrogen in several locations. Oxygen is almost 100 times more soluble in PFCs than in regular blood plasma, so theoretically a PFC solution should be capable of sustaining the life of the patient until a normal blood transfusion is possible. A Japanese company attempted clinical trials with a PFC emulsion as artificial blood, but the results were not as effective as hoped. Newer technology using smaller PFC molecules that develop more stable emulsions show promise.

The alternative technique involves extracting the hemoglobin from blood products that have passed their 42-day shelf life and normally would be discarded. The idea of using hemoglobin in a dilute solution that would flow easily has been around for decades, but initial trials proved disastrous. The direct injection of free hemoglobin into the bloodstream caused capillaries to collapse, thereby shutting off the blood supply to the cells. In addition to binding oxygen, free hemoglobin is capable of bonding the nitric oxide molecules that help keep blood vessels open. The hemoglobin molecule must be treated through encapsulation, cross-linking, and polymerization for it to approach the oxygen transport properties that it displays as part of a red blood cell.

In late 2003 the FDA began clinical trials on an improved hemoglobin-based blood substitute called *polyheme*. This new product restores hemoglobin levels in the bloodstream, can be transfused into all patients regardless of blood type, and remains usable for at least 12 months. The Northfield Laboratories product is awaiting final FDA approval.

Artificial pacemakers are one of the most successful functional biomaterial applications. In a normal heart, the pacing is controlled by the *sinus node*, which causes the heart to beat 60 to 80 times per minute (about 100,000 times per day) and at higher rates during exercise. When the sinus node malfunctions or the electrical signals generated by the node cannot reach the heart, an irregular heartbeat called *bradycardia* results and can cause complications as severe as death.

An artificial pacemaker is a small device implanted directly in the heart. A biocompatible titanium casing houses a long-lasting battery, a microprocessor, and a header that sends out small wires that are connected directly to the right atrium, the right ventricle, or both. When the microprocessor senses an irregular heartbeat, it sends an electrical signal through the wires, causing the heart to beat normally. Over 2,250,000 pacemakers have been implanted in patients in the United States since 1958.

Controlled release agents for drug delivery are the final functional biomaterials to be discussed in this chapter. Traditionally most pharmaceutical products are ingested orally or injected into the bloodstream with a hypodermic syringe. These drug delivery methods offer several specific limitations, however. First, the dosage is difficult to control. All of the pharmaceutical is provided at once, meaning that the initial concentration is higher than optimal and gradually declines. Additionally, the pharmaceutical must be transported through the body to the desired site.

| *Oxygen Therapeutics* |
Artificial blood capable of absorbing oxygen from the lungs and releasing it throughout the body.

| *Polyheme* |
Improved hemoglobin-based blood substitute that the Food and Drug Administration began clinical trials on in late 2003.

| *Artificial Pacemakers* |
Small devices implanted in the heart that, upon sensing an irregular heartbeat, send an electrical signal to cause the heart to beat normally.

| *Sinus Node* |
Cluster of cells that generate an electric signal that controls the pacing within a normal heart.

| *Bradycardia* |
Irregular heartbeat that can cause complications as severe as death.

| *Controlled Release Agents* |
Systems implanted into the body that gradually release a pharmaceutical product at a prescribed rate.

Controlled release systems, which provide the pharmaceutical in a form that allows for the continual release of the optimal dosage level, offer significant advantages. The systems fall into three primary categories:

1. *Diffusion systems.* The pharmaceutical is suspended in a polymer matrix or encapsulated in a membrane and gradually diffuses out.
2. *Solvent-activated systems.* An initially dry implant absorbs water, causing the matrix or membrane to swell. Osmotic pressure drives the pharmaceutical into the body at a controlled rate.
3. *Polymer degradation systems.* As in diffusion systems, the pharmaceutical is housed in a polymeric matrix or membrane, but here the polymer gradually degrades, allowing the pharmaceutical to escape.

Figure 9-18 illustrates the typical time release of a pharmaceutical product through each controlled release system.

One of the first applications of polymer degradation systems dates back to 1986, when a styrene-hydroxymethacrylate copolymer was used to encapsulate insulin and vasopressin, two drugs generally injected directly into the bloodstream. The copolymer protected the drugs from digestion in the stomach and could withstand the low pH of the stomach acids, but bacterial enzymes in the intestines gradually eroded the polymer to release the pharmaceuticals in the target region of the body.

Polymeric systems offer several other advantages. Peptide linkages that can control where the polymer is adsorbed in the body can be attached to the surface

Diffusion System

Solvent-Activated System

Polymer Degradation System

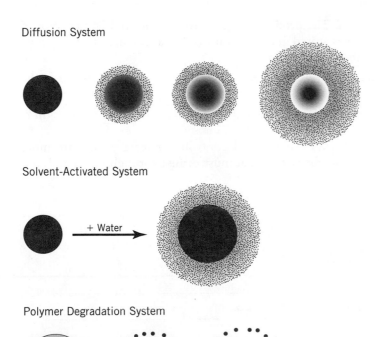

FIGURE 9-18 Controlled Release of a Pharmaceutical

of polymeric materials. Significant research has focused on using controlled release agents to provide anticancer medications directly to the site of tumors.

Another specific controlled release example involves ocular implants to release the drug pilocarpine, which reduces the pressure in the eyes of glaucoma patients. An implant consisting of pilocarpine sandwiched between two thin, transparent membranes made of ethylene-vinyl acetate copolymer is placed directly in the eye, much like a contact lens. Tears produced in the eye diffuse through the membrane and dissolve the pilocarpine, which is released at a consistent rate for seven days. After seven days, the empty membrane is removed and a new one is inserted in its place.

A variety of polymers are used in controlled release applications. Vinyl acetate often is selected because its permeability is easy to regulate. Polyurethanes provide elasticity; polyethylene does not swell and adds strength to the matrix or membrane. Methylmethacrylates are transparent and strong, and polyvinylalcohol has a strong affinity for water.

Not all controlled release agents must be implanted or ingested. One of the largest commercial applications of controlled release is the nicotine patch, which is worn on the arms of individuals who are trying to stop smoking. More than 35 transdermal patches have been approved by the FDA for a range of applications, including contraception and the prevention of motion sickness. A transdermal patch contains five main parts:

1. *Liner.* Protects the patch during shipping. It is removed and discarded prior to use.
2. *Adhesive.* Holds the patch together and attaches it to the skin. In the simplest patches, the adhesive also serves as the matrix housing the pharmaceutical.
3. *Pharmaceutical.* The product designed to be delivered at a controlled rate. It usually is stored in a reservoir inside the patch, although can be distributed through the adhesive.
4. *Membrane.* Controls the release of the pharmaceutical through the reservoir.
5. *Backing material.* Protects the patch from damage while on the skin.

The diffusion processes for transdermal controlled release agents are more complex because the the pharmaceutical must diffuse through patch and into the skin.

What Ethical Issues Are Unique to Biomaterials?

⚖️ 9.4 ETHICS AND BIOMATERIALS

The ethical issues raised in the use of biomaterials are legion. Cost, fairness, access, and safety all take on increased significance when a specific human life is at stake. How much testing is enough? If too little testing is done, a potential harmful product could be placed on the market. However, additional testing requires both time (thus delaying the introduction of the product and costing potential users access to it) and money (increasing the cost of the final product when approved).

Many of the applications for biomaterials discussed in this chapter have the potential to prolong lives and/or improve the quality of life for many people, but how do you determine who gets access to these products: need? ability to pay? age? general health? cause of the problem? value to society?

The issues have become so complex that an entire field called *bioethics* has been created to deal with these ethical dilemmas. Bioethicists are not hired because of their personal opinions but, rather, because of their understanding of the history and research in the field. Particular challenges exist for religious bioethicists, who must balance scientific advances with religious doctrines. This situation becomes particularly important in nonwestern civilizations where religion is not as clearly separated from government or scientific inquiry as it is in the West.

| *Bioethics* |
Field of study examining moral, professional, and legal repercussions of advances in biology and medicine.

Summary of Chapter 9

In this chapter, we:

- Learned to distinguish between biological materials, biomaterials, bio-based materials, and biomimetic materials

- Classified biomaterials and biological materials as structural or functional

- Learned about bone (a biological material) and hydroxyapatite (a biomimetic material)

- Dicussed several structural biomaterial applications, including artificial hips, vascular stents, catheters, breast implants, prosthetic limbs, and dental amalgams and composites

- Examined artificial organs, including hearts (ventricular assist devices), kidneys (hemodialysis), skin, and bladders

- Considered the use of other functional biomaterials, including mechanical heart valves, artificial blood (oxygen therapeutics), pacemakers, and controlled release agents for drug delivery

- Examined issues in bioethics

Key Terms

Homework Problems

1. Explain why most prosthetic limbs do not flex at the ankle but those designed for athletes do.

$ 2. Compare and contrast the advantages and disadvantages of fillings made from dental composites with those made from amalgam.

3. Describe the difference between functional and structural biomaterials.

4. Compare the design and operational mechanism of caged-ball valves, tilting disk valves, and St. Jude bileaflet valves.

5. Why is there no single test for biocompatibility?

6. Explain the three basic mechanisms for controlled release of pharmaceuticals.

7. Describe the roles of the osteoblasts and the osteoclasts in re-formation of bone.

8. Why is titanium such a commonly used biomaterial?

9. What is the purpose of coating some biomaterials with pyrolitic carbon?

10. Why is pure hemoglobin not injected into the bloodstream of patients needing transfusions?

11. Does artificial blood really perform all of the functions of blood? If not, what is its primary function, and why does it not perform others?

12. What is the advantage of using a biocompatible lattice in the case of skin injuries?

13. Compare and contrast the role of the dermis and epidermis, and discuss how these roles impact the selection of artificial skin materials.

14. Discuss the mechanical properties that govern the material selection and design of the keel in artificial limbs.

15. Explain the purpose of the four heart valves.

16. Explain the advantage of nitinol stents over those made from stainless steel.

17. Classify the listed items as biomaterials, biological materials, bio-based materials, or biomimetic materials:
    a. Hydroxyapatite
    b. Bone
    c. Seashells used as aggregate in concrete
    d. Nitinol used in a vascular stent

18. Calculate the percent pressure drop across a partially blocked artery if the healthy section of the artery is 0.5 cm diameter and the blockage reduces the area by 30%.

19. Explain the operating principle of hemodialysis.

20. Why are most biomaterials not reused or recycled?

21. Compare and contrast silicon gel, gummy bear, and saline breast implants.

22. Why are ventricular assist devices (VADs) used instead of artificial hearts?

23. Describe the function of a pacemaker, and explain how the materials of which it is constructed are selected.

24. What material properties factor into the selection for the acetabulum replacement in artificial hips?

25. Why are Y-TZP femoral heads gaining in popularity?

26. Why is cost a larger factor in the selection of materials for dialysis membranes than in the construction of artificial limbs?

27. What are the advantages and disadvantages of using mechanical heart valves instead of those made from biological materials?

28. Describe how the material concerns that impact the selection of a membrane for hemodialysis are different from those for a membrane used in a transdermal patch.

29. Why is the performance index used to measure the efficiency of heart valves?

30. Why must a dentist drill out additional enamel when preparing to fill a cavity?

31. Explain the roles of collagen in skin repair and bone formation.

32. How does Wolf's law impact the life span of artificial hips?

33. What advantages do controlled release agents offer over injection of pharmaceuticals by a syringe?

34. Why are artificial eyes so much more difficult to develop than other artificial organs?

35. Explain the keratinization process.

36. Describe the different layers of the tooth and the properties of the biomaterials that make up the outer two layers.

37. Compare and contrast autografts, homografts, and xenografts.

Appendix A
Major Producers of Metals and Polymers

Table A-1 Major Domestic Aluminum Producers

Company Name	State	Major Products	Website
Alcoa, Inc.	Pennsylvania	Ingots, alumina, packaging, building products	www.alcoa.com
Alcan, Inc.	Montreal	Bauxite and alumina, primary metal, engineered products, and packaging	www.alcan.com
Century Aluminum Company	California	Molten aluminum, ingots, billet, and sow	www.centuryca.com
Columbia Falls Aluminum Company	Montana	Aluminum and smelter grade aluminum ore	www.cfaluminum.com
Noranda Aluminum, Inc.	Missouri	Aluminum rod, billet, foundry ingot, and primary sow	www.norandaaluminum.com
Ormet Corporation	Ohio	Aluminum foil, billet, smelter grade alumina	www.ormet.com

Table A-2 Top 10 Worldwide Gold Producers

Rank	Company Name	Country	2007 Gold Production (thousands of ounces)	Website
1	Barrick Gold	Canada	2,964	www.barrick.com
2	Newmont Mining	U.S.	2,362	www.newmont.com
3	Rio Tinto	U.K.	681	www.riotinto.com
4	Kinross Gold	Canada	627	www.kinross.com
5	Anglogold Ashanti	South Africa	189	www.anglogold.com
6	Sumitomo Metal Mining	Japan	133	www.smm.co.jp
7	Quadra Mining	Canada	108	www.quadramining.com
8	Teck Cominco	Canada	104	www.teckcominco.com
9	Goldcorp	Canada	94	www.goldcorp.com
10	Golden Cycle	U.S.	93	www.goldencycle.com

Data are from the National Mining Association. www.nma.org

TABLE A-3 Top 10 Silver Producers Worldwide

Rank	Company Name	Country	2007 Tin Production (millions of ounces)	Website
1	BHP Billiton	Australia	45.7	www.bhpbilliton.com
2	Industrias Penoles	Mexico	44.5	www.penoles.com.mx
3	KGHM Polska Miedz	Poland	39.1	www.kghm.pl
4	Cia. Minera Volcan	Peru	21.1	www.volcan.com.pe
5	Khazakmys	Khazakstan	19.0	www.khazakmys.com
6	Pan American Silver	Canada	17.1	www.panamericansilver.com
7	Goldcorp	Canada	17.0	www.goldcorp.com
8	Cia. De Minas Buenaventura	Peru	16.0	www.buenaventura.com
9	Polymetal	Russia	15.9	en.polymetal.ru
10	Southern Copper Company	U.S.	15.2	www.southerncopper.com

Data are from the Silver Institute. www.silverinstitute.org

TABLE A-4 Top 10 Steel Producers Worldwide

Rank	Company Name	Country	2007 Crude Steel Production (millions of metric tons)	Website
1	Arcelor Mittal	Luxemborg	116.4	www.arcelormittal.com
2	Nippon Steel	Japan	35.7	www.nsc.co.jp
3	JFE	Japan	34.0	www.JFE-steel.co.jp
4	Posco	Korea	31.1	www.posco.com
5	Baosteel	China	28.6	www.baosteel.com
6	Tata Steel	India	26.5	www.tatasteel.com
7	Anshan-Benxi	China	23.6	www.anbensteel.net
8	Jian-Su Shagang	China	22.9	www.huaigang.com
9	Tangshan	China	22.8	www.tangsteel.com
10	U.S. Steel	U.S.	21.5	www.ussteel.com

Data are from the International Iron and Steel Institute. www.worldsteel.org

TABLE A-5 Top 10 Tin Producers Worldwide*

Rank	Company Name	Country	2007 Tin Production (thousands of metric tons)	Website
1	Yunnan Tin	China	61.1	www.ytl.com.cn
2	PT Timah	Indonesia	58.3	www.timah.com
3	Minsur	Peru	35.9	www.JFE-steel.co.jp
4	Malaysia Smelting Corporation	Malaysia	25.5	www.minsur.com.pe
5	Thaisarco	Thailand	19.8	www.thaisarco.com
6	Yunnan Chengfeng	China	18.0	en.yhtin.cn
7	Liuzhou China Tin	China	13.2	n/a
8	EM Vinto	Bolivia	9.4	Nationalized in 2007
9	Metallo Chimique	Belgium	8.4	www.metallo.com
10	Gold Bell Group	China	8.0	www.shjzong.com

*Note than no tin has been mined or smelted in the U.S. since 1983.

Data are from ITRI (formerly the Tin Research Council). www.itri.co.uk

TABLE A-6 Major Titanium Dioxide Suppliers Worldwide

Company Name	Country	Website
DuPont	U.S.	www.dupont.com
Huntsman Tioxide Europe	U.K.	www.huntsman.com
Ishihara Sangyo Kaisha	Japan	www.iskweb.co.jp
Kronos Worldwide Inc.	Belgium	www.kronostio2.com
Millenium Chemicals	U.K.	www.milleniumchem.com
Precheza	Czech Republic	www.precheza.cz
Sachtleben Chemie	Germany	www.sachtleben.de
Tayca	Japan	www.tayca.co.jp
Tronox Pigments	Netherlands	www.tronox.com

Table A-7 Major Polymers Brand Names and Producers*

Polymer	Brand Name	Company Name	Website
Polyphenylene benzobisoxazole (PBO)	Zylon	Toyobo	www.toyobo.co.jp
Polychloroprene	Neoprene	DuPont	www.dupont.com
Polyethylene (PE)	Alathon	Lyondellbasell	www.lyondell.com
	Chemplex	Chemplex	www.chemplex.com
	Dylan	Sinclair Koppers	www.koppers.com
	Marlex	Chevron Phillips	www.cpchem.com
	Paxon	ExxonMobil Chemical	www.exxonmobil.com
	Rexene	Huntsman	www.huntsman.com
	Tyvek	DuPont	www.dupont.com
	Unival	Dow Chemical	www.dow.com
Polyethylene Terephthalate (PET)	Dacron	Invista	www.invista.com
	Diolen	Diolen	www.diolen.com
	Mylar	Dupont Teijin	www.dupontteijinfilms.com
	Polyclear	Polyclear	www.polyclear.com.uk
Polymethyl methacrylate (PMMA)	Acrylex	Acrylex	n/a
	Lucite	Lucite	www.lucite.com
	Oroglass	Rohm and Haas	www.rohmhaas.com
	Plexiglas	Atoglas	www.plexiglas.com
Polypropylene (PP)	Marlex	Chevron Phillips	www.cpchem.com
	Moplen	Lyondellbasell	www.lyondell.com
	Norchem	Quantum Chemical	www.quantum.com
	Profax	Lyondellbasell	www.lyondell.com
	Tenite	Eastman	www.eastman.com
Polyphenylene terephthalate (PPTA)	Kevlar	Dupont	www.dupont.com
	Twaron	Teijin	
Polystyrene (PS)	Styron	Dow Chemical	www.dow.com
	Lustrex	Monsanto	www.monsanto.com
Polytetrafloroethylene (PTFE)	Teflon	Dupont	www.dupont.com
Polyvinylidene Chloride (PVDC)	Saran	Dow Chemical	www.dow.com
Rayon	Bemberg	Asahi-Kasei	www.asahi-kasei.jp.co
	Danufil	Kelheim Fibres	www.kelheim-fibres.com
	Tencel	Lenzing Fibers	www.lenzing.com
	Viloft	Kelheim Fibres	www.kelheim-fibres.com
Spandex	Dorlacren	Bayer	www.bayer.com
	Lycra	Invista	www.invista.com

*Note: This list is representative, but certainly not exhaustive.

Appendix B
Properties of Major Metals and Alloys

Data are from a variety of sources including:

Robert C. Weast and Melvin J. Astle, eds., *CRC Handbook of Chemistry and Physics, 63rd edition* (Boca Raton, FL: CRC Press Inc., 1982)

Matweb: The Material Property Database. www.matweb.com

Note that material properties vary with exact chemical content and processing conditions. The values presented in the appendix are representative of the material across a spectrum of temperatures and testing conditions. Data from the manufacturer should be used for design purposes.

TABLE B-1 Properties of Pure Metals				
Material	Density (g/cm^3)	Tensile Strength (MPa)	Melting Temperature (°C)	Thermal Conductivity $\dfrac{w}{(m\,K)}$
Aluminum	2.6989	45	660.37	210
Antimony	6.618	11.4	630.74	18.6
Cadmium	8.64	75	321	92
Calcium, rolled	1.54	110	841	126
Calcium, annealed	1.54	40	841	126
Cerium	6.70	100	795	109
Chromium	7.19	413	1860	69.1
Cobalt	8.80	225	1493	69.21
Copper, annealed	8.96	210	1083.2	385
Copper, Cold Drawn	8.96	344	1083.2	385
Dysprosium	8.54	246	1409	10
Erbium	9.05	136	1522	9.6
Gadolinium	7.89	190	1310	8.8
Gold	19.32	120	1064.43	301
Hafnium	13.31	485	2207	22
Holmium	8.80	259	1470	16.2
Iridium	7.31	4.50	156.61	83.7
Iron	7.87	540	1535	76.2
Lanthanum	6.166	130	915	14
Lead	11.34	18	327.5	33
Lithium	0.53	15	180.54	71.2
Lutetium	9.84	140	1651	16.4
Magnesium, Sand Cast	1.74	90	648.3	159

(continues)

Material	Density (g/cm³)	Tensile Strength (MPa)	Melting Temperature (°C)	Thermal Conductivity $\frac{w}{(m\ K)}$
Manganese	7.44	496	1244	8
Mercury	13.546	n/a	−38.87	8.50
Molybdenum, annealed	10.22	324	2617	138
Molybdenum, recrystallized	10.22	324	2617	138
Neodymium	7.01	170	1010	13
Nickel	8.88	317	1455	60.7
Niobium	8.60	585	2468	52.3
Osmium	22.5	1000	3050	91.67
Palladium	12.02	180	1552	71.2
Platinum	21.45	165	1769	69.1
Plutonium	19.0	400	640	8.4
Potassium	0.860	n/a	63.25	99.2
Praseodymium	6.77	100	927	11.7
Promethium	7.264	160	1042	17.9
Rhenium	21.03	2100	3180	39.6
Rhodium	12.4	700	1960	151
Ruthenium	12.3	540	2310	116
Samarium	7.54	120	1067	13.3
Scandium	3.0	255	1539	6.3
Selenium	4.81	n/a	220	Asymmetric: 1.31 parallel to c-axis; 4.52 perpendicular to c-axis.
Silver	10.491	140	961.93	419
Sodium	0.971	10	97.8	135
Tantalum	16.6	345	2890	59.4
Technetium	11.5	1510	2200	50.6
Tellurium	6.23	11	449.5	Asymmetric: 3.38 parallel to c-axis; 1.97 perpendicular to c-axis.
Terbium	8.27	140	1356	11.1
Thallium	11.85	7.5	304	39.4
Thorium	11.3	200	1800	37.7
Thulium	9.33	140	1530	16.9
Tin	5.765	220	231.968	63.2
Titanium	4.50	220	1650	17.0
Tungsten	19.3	980	3370	163.3
Uranium	19.07	400	1132.3	26.8
Vanadium	6.11	911	1735	31.0
Ytterbium	6.98	72	824	34.9
Yttrium	4.472	150	1515	14.6
Zinc	7.10	37	419.58	112.2
Zirconium	6.53	330	1852	16.7

Alloy Specification	Density (g/cm³)	Brinnel Hardness	Tensile Strength (MPa)	Yield Strength (MPa)	Modulus of Elasticity (GPa)	Thermal Conductivity $\frac{w}{(m\,K)}$
1190-O	2.70	12	45	10	62.0	243
1199-H18	2.70	31	115	110	62.0	240
2011-T3	2.83	95	379	296	70.3	151
2014-O	2.80	45	186	96.5	72.4	193
2017-O	2.79	45	179	68.9	72.4	193
2024-O	2.78	47	186	75.8	73.1	193
2024-T3	2.78	120	483	345	73.1	121
2024-T6	2.78	48	427	345	72.4	151
2090-O	2.59	57	210	190	76.0	88
2090-T3	2.59	86	320	210	76.0	88
2219-O	2.04	46	172	75.8	73.1	171
2219-T31	2.84	100	359	248	73.1	112
2219-T87	2.84	130	476	393	73.1	121
3003-O	2.73	28	110	41.4	68.9	193
3003-H12	2.73	35	131	124	68.9	163
3003-H18	2.73	55	200	186	68.9	154
3004-O	2.72	45	179	68.9	68.9	163
3004-H18	2.73	65	240	225	69.0	160
4043-O	2.69	39	145	70	69.0	163
5005-O	2.70	28	124	41.4	68.9	200
5005-H12	2.70	38	138	131	68.9	200
5005-H38	2.70	51	200	186	68.9	200
5050-O	2.69	36	145	55.2	68.9	193
5050-H38	2.69	63	221	200	68.9	193
5052-O	2.68	47	193	89.6	70.3	138
5052-H38	2.68	77	290	255	70.3	138
5154-O	2.66	58	241	117	70.3	125
5154-H38	2.66	80	331	269	70.3	125
5454-O	2.69	62	248	117	70.3	134
5454-H34	2.69	81	303	241	70.3	134
6053-O	2.69	26	110	55	69.0	171
6053-T6	2.69	80	255	220	69.0	163
6061-O	2.70	30	124	55.2	68.9	180
6061-T8	2.70	120	310	276	69.0	170
6463-O	2.69	25	90	50	69.0	200
6463-T6	2.69	74	241	214	68.9	200
7005-O	2.78	53	198	80	72.0	166
7005-W	2.78	93	345	205	72.0	140
7005-T53	2.78	105	390	345	72.0	148

Table B-3 Carbon Steel Properties

Alloy Specification	Density (g/cm³)	Brinell Hardness	Tensile Strength (MPa)	Yield Strength (MPa)	Modulus of Elasticity (GPa)
1006 Cold Drawn	7.872	95	330	285	205
1010 Cold Drawn	7.87	105	365	305	205
1020 As Rolled	7.87	143	450	330	200
1020 Cold Drawn	7.87	121	420	350	205
1020 Hot Rolled	7.87	111	380	205	200
1030 As Rolled	7.85	179	550	345	205
1030 Cold Drawn	7.85	149	525	440	205
1030 Hot Rolled	7.87	137	470	260	200
1040 As Rolled	7.845	201	620	415	200
1040 Cold Drawn	7.845	170	585	515	200
1040 Hot Rolled	7.845	149	525	290	200
1050 As Rolled	7.85	229	725	415	205
1050 Cold Drawn	7.85	197	690	580	205
1050 Hot Rolled	7.87	179	620	345	200
1060 As Rolled	7.85	241	814	485	205
1060 annealed at 790°C	7.85	179	625	370	205
1060 Hot Rolled	7.87	201	660	370	200
1080 As Rolled	7.85	293	965	585	205
1080 annealed at 790°C	7.87	174	615	350	200
1080 Hot Rolled	7.85	229	772	425	205
1095 As Rolled	7.85	293	965	570	205
1095 annealed at 790°C	7.85	192	665	380	205
1095 Hot Rolled	7.87	248	827	455	200
1117 As Rolled	7.85	143	490	305	205
1117 Cold Drawn	7.87	143	485	415	200
1117 Hot Rolled	7.87	116	400	220	200
1141 As Rolled	7.87	192	675	360	205
1141 Cold Drawn	7.85	212	725	605	205
1141 Hot Rolled	7.85	187	650	360	205
1211 Cold Drawn	7.87	163	515	400	200
1211 Hot Rolled	7.87	121	380	230	200
1547 Cold Drawn	7.87	207	710	605	200
1547 Hot Rolled	7.87	192	650	360	200

TABLE B-4	Stainless Steel Properties			
Alloy Specification	Density (g/cm³)	Tensile Strength (MPa)	Yield Strength (MPa)	Modulus of Elasticity (GPa)
301	8.03	515	205	212
302	7.86	640	250	193
304	8.00	505	215	200
330	8.00	586	290	197
348, annealed	8.00	620	255	195
384	8.00	510	205	193
405	7.80	469	276	200
420	7.80	2025	1360	200
440	7.80	1720	1280	200
651	7.94	838	579	200
661	8.25	824	362	200

A

Abrasives   Materials used to wear away other materials.

Accelerated Aging Studies   Tests that approximate the impact of an environmental variable on a material over time by exposing the material to a higher level of that variable for shorter times.

Accelerators   Hydration catalysts that speed up the rate of hydration in Portland cement.

Acetabulum Socket   Hip socket into which the femoral head extends.

Acrylic   One type of polymer that contains at least 85% of polyacrylonitrile (PAN).

Addition Polymerization   One of the two most common reaction schemes used to create polymers, involving three steps: initiation, propagation, and termination. Also called *chain growth polymerization* and *free-radical polymerization*.

Additives   Molecules added to a polymer to enhance or alter specific properties or molecules added to concrete for purposes other than altering a specific property.

Admixtures   Molecules added to a composite to enhance or alter specific properties.

Advanced Ceramics   Engineered ceramic materials used primarily for high-end applications.

Age Hardening   Process that utilizes the temperature dependence of the solubilities of solid solutions to foster the precipitation of fine particles of impurities. Also called *precipitation hardening*.

Aggregate   Hard, randomly oriented particles in a particulate composite that help the composite withstand compressive loads.

Air-Entrainer   Additive that causes air bubbles to form and be distributed throughout concrete to enable it to withstand freeze-thaw expansion cycles without failing.

Alloy Steels   Carbon–iron solid solutions with additional elements added to change properties.

Alloys   Blends of two or more metals.

Alternating Copolymer   Polymer comprised of two or more different monomer units that attach to the chain in an alternating pattern (A-B-A-B-A-B).

Amalgam   Mercury-based alloy used for dental fillings.

American Concrete Institute (ACI)   Technical society dedicated to the improvement of the design, construction, and maintenance of concrete structures.

Amorphous Materials   Materials whose order extends only to nearest neighbor atoms.

Analog Circuit   Type of integrated circuit that can perform functions such as amplification, demodulation, and filtering.

Annealing   Heat-treatment process that reverses the changes in the microstructure of a metal after cold working; occurs in three stages: recovery, recrystallization, and grain growth.

Anode   The site at which oxidation occurs in an electrochemical reaction.

Aortic Valve   Heart valve located between the left ventricle and the aorta that regulates the entry of oxygenated blood into the body.

Aramid   Polymer in which more than 85% of the amide groups are bonded to two aromatic rings.

Arrhenius Equation   Generalized equation used to relate the temperature dependence on various physical properties.

Artificial Pacemaker   Small device implanted in the heart that, upon sensing an irregular heartbeat, sends an electrical signal to cause the heart to beat normally.

Ashby-Style Chart   Heuristics used to provide a quick and simple means of looking at how different classes of materials tend to perform in terms of specific properties.

Aspect Ratio   Ratio of the length to the diameter of the fiber used in a fiber-reinforced composite.

Asphalt   Blend of high-molecular-weight hydrocarbons left over from petroleum distillation. Also the common name used for *asphalt concrete*.

Asphalt Concrete   Blend of mineral aggregate and a binder phase of asphalt. A major particulate composite used in roadways and parking lots, which can be remelted and reshaped. Also called *asphalt cement*.

ASTM Standards   Guidelines published by the American Society for Testing and Materials that provide detailed testing procedures to ensure that tests performed in different laboratories are directly comparable.

Atactic   Term used to describe a polymer that contains significant numbers of both syndiotactic and isotactic dyads.

Atomic Arrangement   Second level of the structure of materials, describing how the atoms are positioned in relation to one another as well as the type of bonding existing between them.

Atomic Packing Factor   Amount of the unit cell occupied by atoms as opposed to void space.

Atomic Structure   First level of the structure of materials, describing the atoms present.

Attenuation   Loss of strength during the transmission of an optical signal resulting from photon absorption or scattering.

Austenite   Phase present in steel in which the iron is present in an FCC lattice with higher carbon solubility.

Austenitic Stainless Steels Carbon–iron solid solutions with at least 12% chromium that contain at least 7% nickel.

Austenizing Process through which the iron lattice in steel reorganizes from BCC to FCC structure.

Autograft Replacement tissues formed by the removal of tissue from other parts of a patient's body that is molded around a stainless steel stent.

B

Bainite Nonequilibrium product of steel with elongated cementite particles in a ferrite matrix.

Balloon Angioplasty Procedure used to address arterial blockages that can replace the more invasive bypass surgery procedure. A balloon is fed along a thin guide wire and inflated in the clogged area to expand the vessel opening.

Balloon Expandable Stents Stents that fit over the angioplasty balloon and expand when the balloon is inflated.

Barrel Piece of the extrusion apparatus that contains a heated screw that is used to melt the polymer and force the polymer forward into the next chamber.

Bar size Categorization of the diameter of rebar, with each number representing an additional 0.125 inches.

Base Central piece of a BJT transistor that is made from a lightly doped and high-resistivity material.

Bauxite Class of minerals rich in aluminum oxides that serves as the primary ore for aluminum production.

Begin Curve Line on an isothermal transformation diagram representing the moment before the phase transformation starts.

Bend Test Method used to measure the flexural strength of a material.

Bernoulli Equation Form of the mechanical energy balance relating a pressure drop with density and velocity changes within a fluid.

Binary Eutectic Phase diagram containing six distinct regions: a single-phase liquid, two single-phase solid regions (α and β), and three multiphase regions (α and β, α and L, and β and L).

Binary Isomorphic Two-component alloy with one solid phase and one liquid phase.

Bio-Based Materials Materials that are derived from living tissue but do not serve a function for an organism.

Bioceramics Ceramics used in biomedical applications.

Biocompatibility Ability of a biomaterial to function within a host without triggering an immune response.

Bioethics Field of study examining moral, professional, and legal repercussions of advances in biology and medicine.

Biological Materials Materials produced by living creatures including bone, blood, muscle, and other materials.

Biomaterials Materials designed specifically for use in the biological applications, such as artificial limbs and membranes for dialysis as well as bones and muscle.

Biomimetic Materials Materials that are not produced by a living organism but are chemically and physically similar to ones that are.

Biotechnology Branch of engineering involving the manipulation of inorganic and organic materials to work in tandem with one another.

Bipolar Junction Transistor (BJT) Prototypical electronic device developed in 1948 to amplify signals.

bis-GMA (Bisphenol GlycidylMethacylate Acrylia) Resin Involved in the most common dental composite, along with filler materials such as powdered glass.

Bisque Glasslike material that results from the firing of kaolinite clay.

Blends Two or more polymers mechanically mixed together but without covalent bonding between them.

Blister Copper Intermediate product during copper refining from which all iron has been removed.

Block Copolymer Polymer comprised of two or more different monomers that attach to the chain in long runs of one type of monomer, followed by long runs of another monomer (AAAAAAABBBBBBBBBAAAAA).

Blown-Film Apparatus Type of polymer processing that makes products such as garbage bags and food wraps by forcing air up through molten polymer, forming a bubble that is cooled and collapsed to form a thin film.

Body-Centered Cubic (BCC) One of the Bravais lattices that contains one atom in each corner of the unit cell as well as one atom in the center of the unit cell.

Bohr Model Classical representation of atomic structure in which electrons orbit the positively charged nucleus in distinct energy levels.

Bone Structural biological material that consists of a fiber-reinforced composite and comprises the skeletal system of most animals.

Bone-Lining Cells Cells that serve as an ionic barrier and line the bone.

Bort Diamonds that lack gem-quality used for industrial purposes.

Boules Large, artificially produced monocrystals.

Bradycardia Irregular heartbeat that can cause complications as severe as death.

Braggs' Equation Formula that relates interplanar spacing in a lattice to constructive interference of diffracted X-rays. Named after the father and son (W.H. and W.L. Bragg) who proved the relationship.

Branching Formation of side chains along the polymer backbone.

Brass Alloy of copper and tin.

Bravais Lattice One of 14 crystal structures into which atoms arrange themselves in materials.

Breaking Strength Stress at which the material breaks completely during tensile testing.

Breast Implants Liquid-filled pouches surgically inserted into the body to augment breast size or to replace breast tissue that has been surgically removed.

Brinell Hardness One of the many scales used to evaluate the resistance of a material's surface to penetration by a hard object under a static force.

Brittle Materials that fail completely at the onset of plastic deformation. These materials have linear stress versus strain graphs.

Bronze Alloy of copper and zinc.

Buckminster Fullerenes Allotropic forms of carbon made up of a network of 60 carbon atoms bonded together in the shape of a soccer ball. Also known as *buckyballs* and *fullerenes* after the architect Buckminster Fuller who developed the geodesic dome.

Burgers Vector Mathematical representation of the magnitude and direction of distortions in a lattice caused by dislocations.

C

Caged-Ball Valve Artificial heart valve that uses a ball to seal the valve.

Calcination Second stage in the processing of Portland cement in which calcium carbonate is converted to calcium oxide.

Calcium Fluoride (CaF_2) Structure FCC-lattice system with cations occupying lattice sites, anions in tetrahedral sites, and octahedral sites left vacant.

Cantilever Beam Test Method used to determine fatigue by alternating compressive and tensile forces on the sample.

Capacitor Two charged plates separated by a dielectric material.

Capillary Pores Open spaces between grains.

Capsular Contracture Tightening of the scar tissue around a breast implant that can lead to its rupture.

Carbon Fibers Forms of carbon made by converting a precursor fiber into an all-aromatic fiber with exceptional mechanical properties.

Carbon Nanotubes Synthetic tubes of carbon formed by folding one graphene plane over another.

Carbon Steel Common alloy consisting of interstitial carbon atoms in an iron matrix.

Carbonization The controlled pyrolysis of a fiber precursor in an inert atmosphere.

Cast Reshaped by being melted and poured into a mold.

Catheter Tube inserted into vessels or ducts of the body, usually to promote the injection or drainage of fluids.

Cathode The site at which reduction occurs in an electrochemical reaction.

Cathodic Protection Form of corrosion resistance provided by the use of a sacrificial anode.

Cement Any material capable of binding things together.

Cementite Hard, brittle phase of iron carbide (Fe_3C) that precipitates out of the steel past the solubility limit for carbon.

Cement Paste Mixture of cement particles and water.

Cementum Bony material layer in teeth that primarily provide a point of attachment for the periodontal ligaments.

Ceramic Matrix Composite (CMC) Addition of ceramic fibers to a matrix of a different ceramic material to significantly increase the fracture toughness of the composite.

Ceramics Compounds that contain metallic atoms bonded to nonmetallic atoms such as oxygen, carbon, or nitrogen.

Chain Growth Polymerization See *addition polymerization*.

Chalcocite (Cu_2S) The most common copper ore.

Chalcopyrite ($CuFeS_2$) Iron-containing mineral comprising about 25% of copper ores.

Charpy Impact Test A single-blow test named after Georges Charpy in which a notched test sample is broken by a swinging pendulum.

China Stone Blend of quartz and mica used to make Chinese porcelain.

cis-Conformation Occurs when the substituents around a carbon chain are directly aligned, causing substantial repulsion between the substituents and an unfavorable conformation. Also called *eclipsed conformation*.

Clay Refractory material containing at least 12% silicon dioxide (SiO_2).

Clinkering Third stage in the processing of Portland cement in calcium silicates form.

Coarse Aggregates Aggregate particles with a diameter greater than 0.25 inches.

Coarse Pearlite Steel microstructure with thick alternating layers of cemenite and α-ferrite.

Cold-Mix Asphalt Concrete Asphalt concrete created by adding water and surfactant molecules to asphalt, resulting in the creation of asphalt concrete similar to that formed through the hot-mix asphalt concrete (HMAC) process. This serves as the primary mechanism for recycling petroleum contaminated soil.

Cold Working Deforming a material above its yield strength but below the recrystallization temperature, resulting in an increased yield strength but decreased ductility. Also known as *strain hardening*.

Collagen High-tensile-strength structural protein found in bone and skin.

Collector Largest region in a BJT transistor that surrounds the emitter and prevents the escape of all electrons or holes.

Coloring Agents Pigments or dyes that change the way light is absorbed or reflected by a polymer.

Completion Curve Line on an isothermal transformation diagram indicating when the phase transformation has been completed.

Composites Material formed by blending two materials in distinct phases causing a new material with different properties from either parent.

Compressive Testing Process of subjecting a material to a crushing load to determine the compressive strength of a material.

Concrete Most important commercial particulate composite, which consists of a blend of gravel or crushed stone and Portland cement.

Condensation Polymerization Formation of a polymer that occurs when two potentially reactive end groups on a polymer react to form a new covalent bond between the polymer chains. This reaction also forms a by-product, which is typically water. Also known as *step-growth polymerization*.

Conduction Band Band containing conductive electrons in a higher energy state.

Confidence Limit Degree of certainty in an estimate of a mean.

Configuration Spatial arrangement of substituents around the main chain carbon atom that can be altered only by the breaking of bonds.

Conformation Refers to the spatial geometry of the main chain and substituents that can be changed by rotation and flexural motion.

Constitution All issues related to bonding in polymers including primary and secondary bonding, branching, formation of networks, and end groups.

Constructive Interference Increase in amplitude resulting from two or more waves interacting in phase.

Controlled Release Agent System implanted into the body that gradually releases a pharmaceutical product at a prescribed rate.

Converse Piezoelectric Effect Change in thickness of a material in response to an applied electric field.

Coordination Number The number of anions in contact with each cation in a ceramic lattice.

Copolymer Polymer whose composition is made up of two or more different monomers covalently bonded together.

Corpuscles White blood cells, red blood cells, and platelets in the bloodstream.

Corrosion Loss of material because of a chemical reaction with the environment.

Corundum Structure HCP-lattice system with cations occupying two-thirds of the octahedral sites.

Cracking Process of breaking large organic hydrocarbons into smaller molecules.

Creel Device capable of continuously pulling filaments from multiple different rovings without stopping the process.

Creep Plastic deformation of a material exposed to a continuous stress over time.

Creep Rate Change in the slope of the strain-time plot at any given point during a creep test.

Crevice Corrosion Loss of material resulting from the trapping of stagnant solutions against the metal.

Cristobalite High-temperature polymorph of silicon dioxide ($SiO2$) that exhibits a cubic lattice.

Critical Angle of Incidence (θ_c) Angle beyond which a ray cannot pass into an adjacent material with a different index of refraction and instead is totally reflected.

Critical Resolved Shear Stress The lowest stress level at which slip will begin in a material.

Cross-Banding In plywood, the grain of the wood veneer is offset by 90 degrees from its neighbor, causing plywood to be more resistant to warping and shrinking than normal wood.

Crystallites Regions of a material in which the atoms are arranged in a regular pattern.

Crystal Mosaic Hypothetical structure accounting for irregularities in the boundaries between crystallites.

Crystal Structure Size, shape, and arrangement of atoms in of atoms in a three-dimensional lattice.

Cullet Finely ground glass powder used in ceramic recycling.

Cup-and-Cone Failure Material failure of ductile materials, in which the crack forms such that one piece of the material has a flat center with an extended rim, like a cup, while the other piece has a roughly conical tip.

Curie Temperature Temperature above which a material no longer displays ferromagnetic properties.

Curing Hardening or toughening of a polymer material through a cross-linking of polymer chains.

Current Density Density of electrical current.

D

Dahlite Hexagonal calcium phosphate mineral in dentin.

Degree of Polymerization Number of repeat units in a polymer chain.

Dental Composites Replacements for amalgams that look more like natural teeth but are more expensive and tend not to last as long.

Dental Pulp Living soft tissue in the center of a tooth.

Dentin Yellow, porous material comprised of collagen and other structural proteins blended with dahlite.

Depletion Zone Nonconductive area at the p–n junction, in which recombination takes place.

Dermis Inner layer of skin containing blood vessels, sweat glands, nerve cells, hair follicles, oil glands, and muscles.

Design for Recycling (DFR) Effort to consider the life cycle consequences when designing a material or product.

Destructive Interference Nullification caused by two waves interacting out of phase.

Developer Alkaline solution that removes exposed material when applied to a microchip.

Dialysis Membrane filtration system used in patients whose kidneys are not capable of fully managing the removal of urea, other waste products, and excess fluid from the bloodstream.

Dialysis Fluid Sterilized, highly purified water with specific mineral salts used in dialysis.

Diamond Allotropic, highly crystalline form of carbon that is the hardest known material.

Die Part of the polymer-processing apparatus through which the polymer is pushed, causing the polymer to form a simple shape, such as a rod or tube.

Dielectric Material placed between the plates of a capacitor to reduce the strength of the electric field without reducing the voltage.

Dielectric Constant Dimensionless value representing the ability of a dielectric material to oppose an electric field.

Diffraction Interaction of waves.

Diffuse Reflection Broad range of reflectance angles resulting from electromagnetic waves striking rough objects with a variety of surface angles.

Diffusion The net movement of atoms in response to a concentration gradient.

Diffusivity Temperature-dependent coefficient relating net flux to a concentration gradient.

Digital Circuit Type of integrated circuit capable of performing functions such as flip-flops, logic gates, and other more complex operations.

Diode Electronic switch that allows electrons to flow in one direction only.

Dipole Force Electrostatic interaction between molecules resulting from alignment of charges.

Dislocations Large-scale lattice defects that occur from alterations to the structure of the lattice itself.

Dislocation Climb Mechanism by which dislocations move in directions that are perpendicular to the slip plane.

Dislocation Line The line extending along the extra partial plane of atoms in an edge dislocation.

DNQ–Novolac Photoresist-light-sensitive polymer blend.

Dopant Impurity deliberately added to a material to enhance the conductivity of the material.

Dorrance Hook Split hook system for transradial amputees that offers some prehensile ability.

Drawing Process in which a metal is pulled through a die, resulting in the material forming a tube the same size as the hole in the die.

Drift Mobility Proportionality constant relating drift velocity with an applied electrical field.

Drift Velocity Average velocity of electrons due to an applied electrical field.

Ductile Materials that can plastically deform without breaking.

Ductile-to-Brittle Transition Transition of some metals in which a change in temperature causes them to transform between ductile and brittle behavior.

Ductility Ease with which a material deforms without breaking.

Dyes Additives dissolved directly into the polymer, causing the polymer to change color.

E

Eclipsed Conformation See *cis-conformation*.

Edge Dislocations Lattice defects caused by the addition of a partial plane into an existing lattice structure.

Effective Orifice Area (EOA) Estimate that measures the efficiency of a valve.

Effective Secant Modulus of Elasticity (E_c) American Concrete Institute–specified modulus of elasticity for concrete that accounts for the tendency of the modulus of elasticity for concrete varying with the stress level.

Elastic Energy Area contained under the elastic portion of a stress-strain curve, which represents how much energy the material can absorb before permanently deforming.

Elastic Modulus Slope of the stress-strain curve in the elastic region. Also called *Young's modulus* or the *tensile modulus*.

Elastic Stretching Region on a stress-strain curve in which no permanent changes to the material occur.

Elastin Structural protein that enhances the skin's strength and ability to stretch.

Elastomer Polymer that can stretch by 200% or more and still return to its original length when the stress is released.

Electrical Conductivity Ability of a material to conduct an electrical current.

Electrical Resistivity Barrier to conduction of electrons caused by collisions within the lattice.

Electrochemical Cell Device designed to create voltage and current from chemical reactions.

Electrochemistry Branch of chemistry dealing with the transfer of electrons between electrolyte and electron conductor.

Electron Acceptor Molecule that accepts electrons from another substance.

Electron Donor Molecule that donates electrons to another substance.

Electronegativity The ability of an atom in a covalent bond to attract electrons to itself.

Electroplating The electrochemical deposition of a thin layer of metal on a conductive surface.

Electronic Materials Materials that possess the ability to conduct electrons, such as semiconductors.

Enamel Hardest substance in the body, which covers the teeth and consists primarily of hydroxyapatite.

End Groups Two substituents found at both ends of a polymer chain, which have little to no effect on mechanical properties.

Endurance Limit Stress level below which there is a 50% probability that failure will never occur.

Energy Bands Split of energy levels with slight variance.

Energy Gap Gap between the conduction and valence bands.

Engineering Strain Property determined by measuring the change in the length of a sample to the initial length of that sample.

Engineering Stress Ratio of applied load to cross-sectional area.

Epidermis Outer layer of skin, which contains no blood cells but receives its nutrients from diffusion from the dermis.

Epoxy Resin Resin used as a matrix in composite materials that is more expensive than polyester resin but provides improved mechanical properties and exceptional environmental resistance.

Equivalent Property Time (EPT) Period used to force the same aging processes to occur on a sample in a shorter amount of time.

Erosion Corrosion Loss of material resulting from the mechanical abrasion of a metal by a corrosive material.

Error Bar Limit placed on the accuracy of a reported mean, based on the number of samples tested, the standard deviation, and the desired level of confidence.

Eutectic Isotherm Constant temperature line on a phase diagram that passes through the eutectic point.

Eutectic Point Point on the phase diagram at which the two solid phases melt completely to form a single-phase liquid.

Eutectoids Points in which one solid phase is in equilibrium with a mixture of two different solid phases.

Evaporation-Dehydration First stage in the processing of Portland cement in which all free water is driven off.

Extinction Condition The systematic reduction in intensity of diffraction peaks from specific lattice planes.

Extracapsular Silicone Silicone that escapes from the hard scar tissue capsule that surrounds a breast implant and is free to migrate through the body.

Extrinsic Semiconductor Created by introducing impurities called dopants into a semiconductor.

Extruder Device used in the processing of polymers that melts polymer pellets and feeds them continuously through a shaping device.

Extrusion Process in which a material is pushed through a die resulting in the material obtaining the shape of the hole in the die.

F

Face-Centered Cubic (FCC) One of the Bravais lattices that has one atom in each corner of the unit cell and one atom on each face of the unit cell.

Face Sheet Typically very strong materials used as the outer end of a sandwich composite.

Fatigue Failure because of repeated stresses below the yield strength.

Fatigue Life Number of cycles at a given stress level that a material can experience before failing.

Fermi Energy Energy level of the highest-occupied energy band.

Ferritic Stainless Steel Carbon–iron solid solutions with at least 12% chromium that contain no nickel.

Ferroelectric Materials Materials with permanent dipoles that polarize spontaneously without the application of an electric field.

Fiber Pull-Out Premature failure in a composite caused by inadequate bonding between the fiber and matrix.

Fiber-Reinforced Composite Composite in which the one material forms the outer matrix and transfers any loads applied to the stronger, more brittle fibers.

Fick's First Law Equation determining steady-state diffusion.

Fick's Second Law Equation representing the time-dependent change in diffusion.

50% Completion Curve Line on an isothermal transformation diagram indicating when half of the phase transformation has been completed.

Fillers Additives whose primary purpose is to reduce the cost of the final product.

Fine Aggregates Aggregate particles with a diameter less than 0.25 inches.

Fine Pearlite Steel microstructure with thin alternating layers of cemenite and α-ferrite.

Fireclay Clay material containing 50% to 70% silica and 25% to 45% alumina.

Flame Aerosol Process Method of producing ceramic nanoparticles in which a liquid metallic organic is brought to a flame to form a nucleus for particle growth.

Flexural Strength   Amount of flexural stress a material can withstand before breaking. Measured through the bend test.

Float Glass Process   Glass production technique in which a fine ribbon of glass is drawn from a furnace and floated on the surface of a pool of molten tin.

Foley Catheter   Tube used to replace a damaged urethra.

Forging   Mechanical reshaping of metal.

Forming Operations   Techniques to alter the shape of metals without melting.

Forward Bias   Connecting a battery with the positive terminal corresponding with a p-type site and the negative terminal corresponding with an n-type site.

Fourth Quantum Number   Number representing the spin of an electron.

Fracture Mechanics   Study of crack growth leading to material failure.

Fracture Toughness   Value that the stress concentration factor must exceed to allow a crack to propagate.

Free Radical   Molecule containing a highly reactive unpaired electron.

Free-Radical Polymerization   See *addition polymerization*.

Freeze Line   Term associated with blown-film apparatus, which indicates the point at which the molecules develop a more crystalline orientation around the bubble of air.

Frenkel Defect   Point defect found in ceramic materials that occurs when a cation diffuses onto an interstitial site on the lattice.

Fresnel Equation   Mathematical relationship describing the quantity of light reflected at the interface of two distinct media.

Friable   Forming sharp edges when broken under stress.

Fullerenes   See *Buckminster fullerenes*.

Fuller-Thompson Parameter   A parameter used in the equation used to determine the maximum packing density in concrete.

Full-Width Half Maximum (FWHM)   Standard used to measure the spread in the peak of a diffractogram, measured at the intensity value corresponding to the half highest value in the peak.

Functional Biomaterials   Materials that interact or replace biological systems with a primary function other than providing structural support.

Functional Groups   Specific arrangements of atoms that cause organic compounds to behave in predictable ways.

Functionality   Number of bonds a molecule has formed.

G

Gain Medium   Substance that passes from a higher to lower energy state and transfers the associated energy into a laser beam.

Galvanic Corrosion   Loss of material resulting from the metal that is lower on the galvanic series oxidizing in favor of the more cathodic metal.

Galvanic Series   A list ranking metals in order of their tendency to oxidize when connected with other metals in solutions with their ions.

Gamma-2 Phase Amalgam   Amalgam containing 50% mercury and 50% of an alloy powder made up of at least 65% silver, less than 29% tin, about 6% copper, and small amounts of mercury and zinc.

Gauche   Conformation that occurs when the largest substituents in a molecule are offset by 60 degrees.

Gel Pores   Spaces within the C-S-H material during the hydration of cement.

Glass Transition Temperature (T_g)   Second-order thermodynamic transition in which the onset of large-scale chain mobility occurs in polymers. Below T_g, the polymer is glasslike and brittle. Above T_g, the polymer becomes rubbery and flexible.

Glasses   Inorganic solids that exist in a rigid, but noncrystalline form.

Glaze   Blend of SiO_2 and metal oxides that are used to coat bisque material and provide color when refired.

Gradation   Process of passing aggregate through sieve trays to acquire the particle size distributions.

Graft Copolymer   Polymer in which one chain of a particular monomer is attached as a side chain to a chain of another type of monomer.

Grain Boundary   Area of a material that separates different crystallite regions.

Grain Growth   Second step in the formation of crystallites, which is dependent on temperature and can be described using the Arrhenius equation.

Grain Size Number   A numerical quantity developed by the American Society for Testing and Materials (ASTM) to characterize grain sizes in materials.

Graphene Layer Planes   Parallel planes consisting of conjugated six-member aromatic carbon rings.

Graphite   Allotropic carbon material consisting of six-member aromatic carbon rings bonded together in flat planes, allowing for the easy occurrence of slip between planes.

Green Engineering   Movement supporting an increase in the knowledge and prevention of environmental hazards caused during the production, use, and disposal of products.

Ground State   Condition in which all electrons are in their lowest energy levels.

Gummy Bear Silicone Breast Implants   Silicone breast implants that potentially eliminate or at least significantly reduce the leakage of silicone gel. Named because the texture of the pouch resembles that of the famous candy.

H

Hall-Petch Equation   Correlation used to estimate the yield strength of a given material based on grain size.

Hardenability   The ability of a material to undergo a martensitic transformation.

Hardener   Substance added to epoxy resin to cause it to cross-link; the hardener becomes incorporated in the resulting polymer.

Hardness   Resistance of the surface of a material to penetration by a hard object under a static force.

Hardness Testing   Method used to measure the resistance of the surface of a material to penetration by a hard object under a static force.

Hemodialysis   Most common form of dialysis, in which the blood of the patient is removed by a catheter and passed through a semipermeable membrane to remove toxins and excess water.

Hemoglobin   Iron-containing protein that supplies oxygen to the cells of the body.

Heterogeneous Nucleation   Clustering of atoms around an impurity that provide a template for crystal growth.

Hexagonal Close-Packed (HCP)   Most common of the noncubic Bravais lattices, having six atoms forming a hexagon on both the top and the bottom and a single atom positioned in the center, between the two hexagonal rings.

High-Copper Amalgam   Amalgam containing 50% liquid mercury and 50% of an alloy powder made of 40% silver, 32% tin, 30% copper, and small amounts of mercury and zinc.

High-Volume Thermoplastic (HVTP)   Simple polymeric material produced as a pellet in large quantities.

Hole   Positively charged vacant site left by an electron moving to a higher energy state.

Homogeneous Nucleation   Clustering that occurs when a pure material cools sufficiently to self-support the formation of stable nuclei.

Homograft   Tissues that, after being removed from a cadaver and frozen in liquid nitrogen, are thawed and installed as a direct replacement in the body.

Homopolymer   Polymer that is made up of a single repeat unit.

Honeycomb Structure   Common shaping used for the low-density material in sandwich composites to add stiffness and resistance to perpendicular stresses.

Hopper   Part of the extrusion apparatus that holds a large quantity of polymer pellets as they are fed into the barrel.

Hot-Mix Asphalt Concrete (HMAC) Process   Process used to produce most asphalt on major highways, in which asphalt is heated to 160°C before mixing in the aggregate and is laid and compacted at 140°C.

Hot Working   Process in which the forming operations are performed above the recrystallization temperature of the metal. Recrystallization occurs continuously, and the material can be plastically deformed indefinitely.

HPHT Method   Process for producing synthetic diamonds using elevated temperatures and high pressures.

Hybrid Composite   Composite materials produced with at least one phase that is a composite material itself.

Hydration Catalysts   Catalysts that alter the rate of hydration in Portland cement.

Hydraulic Cements   Binding materials that require water to form a solid.

Hydrogen Bond   Strong dipole interaction resulting between a hydrogen atom and a highly electronegative atom.

Hydroxyapatite $Ca_{10}(PO_4)_6(OH)_2$ Biomimetic material that often is used as a bone filler when osteoblasts cannot reconnect disjointed pieces of bone without aid.

Hypereutectoid Steel   Iron–carbon solid solution with more than 0.76wt% carbon.

Hypoeutectoid Steel   Iron–carbon solid solution with less than 0.76wt% carbon.

I

Immune Response   White blood cells identifying a foreign material in the body and attempting to destroy it.

Impact Energy   Amount of energy lost as the test sample is destroyed during an impact test.

Index of Refraction   Material-specific term representing the change in the relative velocity of light as it passes through a specific medium.

Initiation   First step in the process of polymerization, during which a free radical is formed.

Injection Molding   Type of polymer processing that is similar to extrusion but can be used to develop parts with complex shapes rapidly.

In-Plane Shear   Application of stresses parallel to a crack causing the top portion to be pushed forward and the bottom portion to be pulled in the opposite direction.

Insulator   Material that has no free electrons at absolute zero.

Integrated Circuits   Many transistors combined on a single microchip.

IGFET (Insulated Gate Field Effect Transistor)   Transistor made without the use of metal oxides. Term is now interchangeable with MOSFET.

Intermediate Oxides Additives used to impart special properties to glasses.

Interest Rent paid to the owner of money for the temporary use of that money.

Intergranular Corrosion Loss of material resulting from preferential attack of corroding agents at grain boundaries.

Interplanar Spacing Distance between repeated planes in a lattice.

Interstitial Defects Point defects that occur when an atom occupies a space that is normally vacant.

Interstitial Diffusion The movement of an atom from one interstitial site to another without altering the lattice.

Intrinsic Semiconductors Pure materials having a conductivity ranging between that of insulators and conductors.

Ionic Bond The donation of an electron from an electropositive material to an adjacent electronegative material.

Isostrain Condition Condition in which the quality of bonding between the fiber and matrix is sufficient that both elongate at the same rate and experience the same strain.

Isostress Condition Condition in which the fibers in a matrix offer essentially no reinforcing benefit to the matrix when a load is applied in the transverse direction, causing both to experience essentially the same strain.

Isotactic Dyad Configuration of a substituent in a polymer, in which the substituent is located on the same side of the polymer chain in all repeating units.

Isothermal Transformation Diagram Figure used to summarize the time needed to complete a specific phase transformation as a function of temperature for a given material. Also called a *T-T-T plot*.

Izod Test Impact test similar to the Charpy test in which the sample is aligned vertically with the notch facing away from the hammer.

J

Jarvik-7 Artificial Heart Artificial heart implanted into a patient named Barney Clark in 1982 that kept him alive for 112 days.

Jominy Quench Test Procedure used to determine the hardenability of a material.

K

Kaolinite Clay mineral named for the Gaolin region of China where it was discovered.

Keel Long bar that replaces the missing tibia in a transtibial prosthesis limb.

Keratin Hard structural protein contained in keratinocytes.

Keratinization Process in which after about 30 days, epidermal cells dry and fall off the body to make room for the next tier of cells.

Keratinocytes Cells that comprise 90% of the epidermis and contain a large quantity of keratin.

L

Lamellar Bone Bone that replaces woven bone, in which the collagen fibrils align along the length of the bone.

Laminar Composites Composites that are made by alternating the layering of different materials bonded to each other.

Larson-Miller Parameter Value used to characterize creep based on time, temperature, and material-specific constants.

Laser Device that produces light of a single wavelength in a well-defined beam. Acronym for light amplication by stimulated emission of radiation.

Lattice Parameters Edge lengths and angles of a unit cell.

Lead Bronzes Copper–tin alloys that also contain up to 10% lead that is added to soften the metal.

Lehr Annealing furnace used in glass manufacture.

Lever Rule Method for determining the compositions of materials in each phase using segmented tie lines representing overall weight percentages of the different materials.

Life Cycle Path taken by a material from its initial formation until its ultimate disposal.

Life Cycle Assessment Most detailed method of analyzing the life cycle of a material.

Liquidus Line Line on a phase diagram above which only liquid exists at equilibrium.

Longitudinal Direction Direction of fiber alignment.

Loose Network Molecular arrangement in glasses in which there is no long-range order, but adjacent SiO_4^{4-} tetrahedrons share a corner oxygen atom.

Low Alloy Coppers Solid solutions containing at least 95% copper.

M

Macrostructure Fourth and final level of structure in materials, describing how the microstructures fit together to form the material as a whole.

Martensite Nonequilibrium steel product formed by the diffusionless transformation of austenite.

Martensitic Stainless Steel Carbon–iron solid solutions with at least 12% to 17% chromium that can undergo the martensitic transformation.

Martensitic Transformation Diffusionless conversion of a lattice from one form to another brought about by rapid cooling.

Mask Transparent glass plate used in the photoresist process.

Matrix Material in a composite that protects, orients, and transfers load to the reinforcing material.

Matthiessen's Rule Rule that states that temperature, impurities, and plastic deformation act independently of one another in affecting the resistivity of a metal.

Maximum 28-Day Compressive Strength Compressive strength of concrete from a tested sample that was hardened at constant temperature and 100% humidity for 28 days.

Medtronic Hall Valve Most common tilting disk valve used as an artificial heart valve.

Melt Spinning Process of pushing polymers through a spinneret and winding the solidified fibers onto a tow, which imposes a shear stress on the fibers upstream as they emerge from the spinneret.

Mesophase Pitch By-product of coal or petroleum distillation formed into regions with liquid crystalline order through heat treatment.

Metallic Bonding Sharing of electrons among atoms in a metal, which gives the metal excellent conducting properties because the electrons are free to move about the electron cloud around the atoms.

Metal Matrix Composites Composites that use a metal as the matrix material in place of more common polymer matrices.

Metals Materials possessing atoms that share delocalized electrons.

Microstructure Third level of structure in materials, describing the sequencing of crystals at a level invisible to the human eye.

Miller Indices Numerical system used to represent specific planes in a lattice.

Mineralization Growth of the bone matrix onto fibrils of collagen.

Mitral Valve Heart valve separating the left ventricle from the left atrium.

Mixed Dislocations Presence of both screw and edge dislocations separated by a distance in the same lattice.

Mixed Signal Circuit Type of integrated circuit containing both analog and digital on the same chip.

Modulus of Resilience Ratio of the elastic energy to the strain at yielding, which determines how much energy will be used for deformation and how much will be translated to motion.

Modulus of Rupture (f_r) Maximum tension strength at the bottom surface of a concrete beam.

Moh Hardness Nonlinear, qualitative method used to evaluate the resistance of a material's surface to penetration by a hard object.

Monocrystals Materials in which the entire structure is a single unbroken grain.

Monomer Low-molecular-weight building block repeated in the polymer chain.

Moore's Law Empirical observation that the density of transistors doubles every 18 to 24 months.

MOSFET (Metal Oxide Semiconductor Field Effect Transistors) Transistor originally made with metal oxides, but the term is now interchangeable with IGFET.

Mullite Clay material formed by high-temperature aluminosilicates.

Mutual Termination One of the two different types of termination in the polymerization process. During this type of termination, the free radicals from two different polymer chains join to end the propagation process.

Myoelectric Arms Transradial prosthetics that use muscle impulses in the residual arm to control the function of the prosthetic.

N

Nanocrystals Crystalline materials with sizes of nanometers in length.

Necking Sudden decrease in cross-sectional area of a region of a sample under a tensile load.

Negative Climb The filling of a vacancy in the partial plane of an edge dislocation by an adjacent atom resulting in a growth of the crystal in the direction perpendicular to the partial plane.

Network Modifiers Additives used to reduce the viscosity of loose networks in glasses.

New Scrap Recycled metal from preconsumer sources.

Nitinol Nickel–titanium alloy used in stent manufacture that experiences a shape memory effect.

Nominal Stress Stress values not involving the presence of stress raisers within the material.

Nonhydraulic Cements Binding materials that do not require water to form a solid.

Nonpolar Interaction in which the electron density around adjacent atoms is symmetric.

Nordheim Coefficient Proportionality constant representing the effectiveness of an impurity in increasing resistivity.

Nordheim's Rule Method for estimating the resistivity of a binary alloy.

n-Type Semiconductor Semiconductor in which dopant donates electrons to the conduction band, causing the number of holes to be less than the number of electrons in the conduction band.

Nucleation Process of forming small aligned clusters of atoms that serve as the framework for crystal growth.

Nuclei Tiny clusters of arranged atoms that serve as the framework for subsequent crystal growth.

Number Average Molecular Weight Form of the molecular weight of a sample of polymer chains determined by dividing the mass of the specimen by the total number of moles present.

Nylon Type of polyamide in which less than 85% of the amide groups are bonded to aromatic rings.

O

Octahedral Positions   Intersitial spaces between six atoms in a lattice that form an octahedron.

Offset Yield Strength   Estimate of the transition between elastic stretching and plastic deformation for a material without a linear region on a stress-strain curve.

Old Scrap   Recycled metal from consumer products that have completed their useful life.

Oligomer   Small chain of bonded monomers whose properties would be altered by the addition of one more monomer unit.

Opening Stresses   Stresses that act perpendicularly to the direction of a crack, causing the crack ends to pull apart and opening the crack further.

Opposed Spin   Electrons with different fourth quantum numbers.

Optical Cavity   Pair of mirrors that repeatedly reflect a beam of light through the gain medium of a laser.

Optical Fibers   Thin glass or polymeric fibers that are used to transmit light waves across distances.

Optical Microscopy   Use of light to magnify objects up to 2000 times.

Osseointegration   Process in which hydroxyapatite becomes part of the growing bone matrix.

Osteoblasts   Cells located near the surface of bone that produce osteoid.

Osteoclasts   Cells that dissolve the bone matrix using acid phosphatase and other chemicals to allow the body to reabsorb the calcium in the bone.

Osteocytes   Osteoblasts trapped in the bone matrix that facilitate the transfer of nutrients and waste materials.

Osteoid   Blend of structural proteins (containing primarily collagen) and hormones that regulate the growth of bones.

Out-of-Plane Shear   Application of stress perpendicular to a crack, which pulls the top and bottom portions in opposite directions.

Oxidation   Chemical reaction in which a metal transfers electrons to another material.

Oxygen Therapeutics   Artificial blood capable of absorbing oxygen from the lungs and releasing it throughout the body.

P

Palmaz Stent   Balloon-expandable stent made from stainless steel.

Parallel Spin   Electrons with a different fourth quantum number.

Passivation   Spontaneous formation of a protective barrier that inhibits oxygen diffusion and corrosion.

Pauli Exclusion Principle   Concept that no more than two electrons can fit in any orbital and that the two electrons must have opposed spins.

Particulate Composites   Composites that contain large numbers of coarse particles, such as the cement and gravel found in concrete.

Pearlite   Mixture of cementite (Fe_3C) and α-ferrite named for its resemblance to mother-of-pearl.

Percent Cold Work (%CW)   Representation of the amount of plastic deformation experienced by a metal during strain hardening (cold working).

Performance Index (PI)   Dimensionless term that is calculated by dividing the effective orifice area by a standard.

Periclase   Refractory material containing at least 90% magnesium oxide (MgO).

Peritectics   Points at which a solid and liquid are in equilibrium with a different solid phase.

Perovskite Structure   FCC-lattice system with two species of cation, one occupying corner sites and the other occupying octahedral sites. Anion species occupies the face sites in the lattice.

Phase   Any part of a system that is physically and chemically homogeneous and possesses a defined interface with any surrounding phases.

Phase Diagram   Graphical representation of the phases present at equilibrium as a function of temperature and composition.

Phenolic Resin   Matrix material that has many voids and poor mechanical properties but does offer a level of fire resistance.

Photon   Discrete unit of light.

Photoresist   Coating that becomes soluble when exposed to ultraviolet light.

Piezoelectric Effect   Production of an electric field in response to a mechanical force.

Piezoelectric Material   Materials that convert mechanical energy to electrical energy or electrical energy to mechanical energy.

Pig Iron   Metal remaining in the steelmaking process after the slag has removed the impurities. When treated with oxygen to remove excess carbon, pig iron becomes steel.

Pigment   Coloring agent that does not dissolve into the polymer.

Pitting   Form of corrosion resulting from corrosive material collecting in small surface defects.

Pittsburgh Process   Glassmaking process developed in 1928 to reduce both cost and distortion.

Plane Strain Fracture Toughness   Fracture toughness above the critical thickness, in which the width of the material no longer impacts the fracture toughness.

Plasma   Yellow fluid that makes up 60% of the total volume of blood.

Plastic Deformation   Region on a stress-strain curve in which the material has experienced a change from which it will not completely recover.

Plasticizer Additive that causes swelling, which allows the polymer chains to slide past one another more easily, making the polymer softer and more pliable. Also used to decrease the viscosity of cement paste to make it easier to flow the concrete into its final form.

Plywood Most common type of laminar composite, consisting of thin layers of wood veneer bonded together with adhesives.

p–n Junction Area in which the p-type and n-type areas meet.

Point Defect Flaw in the structure of a material that occurs at a single site in the lattice, such as vacancies, substitutions, and interstitial defects.

Poisson's Ratio Ratio that relates the longitudinal deformation and the lateral deformation of a material under stress.

Polar Interaction in which the electron density around adjacent atoms is asymmetric.

Polyamide Polymers that contain amide (—N—) groups in the chain.

Polyester Long-chain polymers that contain at least 85% of an ester of a substituted aromatic carboxylic acid. These fibers are strong and can be dyed or made transparent.

Polyester Resin Most economical choice for a matrix material in composites in situations where the mechanical properties of the matrix are not crucial to the application.

Polyheme Improved hemoglobin-based blood substitute that the Food and Drug Administration (FDA) began clinical trials on in late 2003.

Polyimide Resins Polymeric matrix materials that are extremely expensive and used only in high-end applications, due to their ability to maintain their properties at temperatures above 250°C.

Polymers Covalently bonded chains of molecules with the small monomer units repeated from end to end.

Polymer Backbone Covalently bonded atoms, which are usually carbon, that comprise the center of the polymer chain.

Polymer Network Three-dimensional structures that result when polymer chains form significant numbers of chemical cross-links.

Polyolefins Polymers that contain only hydrogen and aliphatic carbon.

Polyurethane Broad category of polymers that includes all polymers containing urethane linkages.

Pooled Variance Value used to determine if two distinct sets of samples are statistically different.

Porcelain Whiteware with translucence caused by the formation of glass and mullite during the firing process.

Portland Cement Most common hydraulic cement made from pulverizing sintered calcium silicates.

Positive Climb Filling of a vacancy in the partial plane of an edge dislocation by an adjacent atom resulting in a shrinking of the crystal in the direction perpendicular to the partial plane.

Powder Pressing Formation of a solid material by the compacting of fine particles under pressure.

Precipitation Hardening See *age hardening*.

Precipitation Heat Treatment Second stage in age hardening in which the diffusion rate increases enough to allow one phase to form a fine precipitate.

Prepreg Fiber bundle already impregnated with matrix material, which can be converted into a composite without any additional processing.

Prepregging Process of creating prepregs by dipping fibers into a resin bath and heating them slightly to ensure that the coating sticks.

Present Worth Analysis System that enables one to form a direct comparison between the value of money spent now and the value of the same amount of money in the future.

Primary Bonding Covalent bonding of the polymer backbone and side groups.

Primary Creep First stage of *creep*, during which the dislocations in a material slip and move around obstacles.

Primary Slip System First plane and direction in a material to experience the movement of a dislocation when a tensile stress is applied.

Primary Termination Last step in the polymerization process, which occurs when the free radical of a polymer chain joins with the free radical on an end group.

Principal Quantum Number Number that describes the major shell in which an electron is located.

Projection Lithography Process of projecting ultraviolet light onto a microchip in a manner similar to a slide projector.

Propagation Second stage of the polymerization process during which the polymer chain begins to grow as monomers are added to the chain.

Prosthetic Limbs Replacement artificial limbs.

p-Type Semiconductor Semiconductor in which a dopant removes electrons from the valence band, causing there to be more holes than electrons in the valence band.

Pulmonary Valve Heart valve that controls the flow of blood into the lungs.

Pultrusion Process often used to create uniaxial, fiber-reinforced composites.

Q

Quantum Mechanics The science governing the behavior of extremely small particles such as electrons.

Quantum Numbers Four numbers used to classify individual electrons based on their energy, cloud shape, orientation of cloud, and spin.

Quench Cracking Damage caused near the surface of a metal because of the uneven temperature distribution of the rapidly cooled exterior and the hot interior.

Quenching Terminating a condensation polymerization reaction by adding a material with only one functional group.

R

Random Copolymer Polymers comprised of two or more different monomers, which attach to the polymer chain in no particular order or pattern.

Rayon Lightweight polymer that absorbs water well; the first synthetic polymer ever constructed.

Rebar Steel reinforcing bars used to reinforce the ability of concrete to handle tensile loads.

Recombination Process in which holes in a p-type semiconductor and electrons in an n-type semiconductor cancel each other out.

Recovery The first stage of annealing in which large misshapen grains form in a material and residual stresses are reduced.

Recrystallization The second stage of annealing in which the nucleation of small grains occurs at the sub-grain boundaries, resulting in a significant reduction in the number of dislocations present in the metal.

Recrystallization Temperature Temperature marking the transition between recovery and recrystallization.

Reduction Chemical reaction in which a material receives electrons transferred from a metal.

Reflection Process in which incoming photons trigger the release of identical photons such that the angle of incidence equals the angle of reflection.

Reflection Coefficient Fraction of light reflected at the interface of two media.

Refraction Process in which incoming photons have their path altered in the new media.

Refractories Materials capable of withstanding high temperatures without melting, degrading, or reacting with other materials.

Regurgitation Blood leaking back into the previous chamber of the heart when a valve fails to seal properly.

Relative Molecular Mass (RMM) Term used to represent the average molecular weight of a sample containing a wide range of polymer chain lengths. This term is used to avoid confusion between the number average molecular weight and the weight average molecular weight.

Remodeling Continuous process in which the bone is reabsorbed and replaced through the life of an organism.

Repeat Unit Smallest portion in a polymer chain that recurs over and over again. Also known as a *structural unit*.

Resin Formulation Process in which bits of chopped-up fibers are mixed or blown into the matrix material, along with any curing agents, accelerators, diluents, fillers, or pigments, in order to form a simple chopped-fiber composite.

Resin Transfer Molding Process of converting mats or weaves into composites through using a mold, in which the mat is placed and injected with resin at a high enough pressure to permeate and surround the woven mat.

Restenosis Buildup of scar tissue around a stent, leading to a restriction of blood flow.

Retarders Hydration catalysts that decrease the rate of hydration in Portland cement.

Reverse Bias Connecting a battery with the positive terminal corresponding with an n-type site and the negative terminal corresponding with a p-type site.

Rockwell Hardness Test Specific method of measuring the resistance of a material's surface to penetration by a hard object under a static force.

Rolling Thinning of a metal sheet by pressing it between two rollers, each applying a compressive force.

Roving Large number of single-fiber strands wound in parallel.

Russian Doll Model Representation of multiwalled carbon nanotubes in which outer graphene layers surround inner ones, much like nesting dolls.

Rust $Fe(OH)_3$ End product from the electrochemical corrosion of iron.

S

S–N Curve Curve plotting the results of testing multiple samples at different stress levels that is used to determine the fatigue life of a material at a given stress level.

SACH Foot Solid-ankle–cushion-heel foot. The most common transtibial prosthesis, which contains a rubber wedge in the heel and a solid keel, often wooden.

Sacrificial Anode Metal low on the galvanic series used to oxidize and transfer electrons to a more important metal.

St. Jude Bileaflet Valve Artificial heart valve featuring two leaflets that swing apart when the valve is open to create three separate regions of flow.

Saline Breast Implants Alternative to silicone breast implants that uses saline instead of silicone in the pouch.

Sandwich Composite Composite used in situations where strength is required but weight is a significant factor. Usually made up of strong face sheets on the outer ends of the composite with a low-density material, often in a honeycomb structure, which adds stiffness and resistance to perpendicular stress.

Scanning Electron Microscope (SEM) Microscope that focuses a high-energy beam of electrons at the source and collects the back-scattered beam of these electrons.

Scherrer Equation Means of relating the amount of spreading in a X-ray diffractogram to the thickness of the crystallites in the sample.

Schmid's Law Equation used to determine the critical resolved shear stress in a material.

Schottky Defect Point defect that occurs in ceramics when both a cation and an anion are missing from a lattice.

Scleroprotein High-tensile-strength structural protein.

Scrap Metal Metal available for recycling.

Screen Pack Piece of the extrusion apparatus that is used as a filter to separate unmelted particles, dirt, and other solid contaminants from the molten polymer.

Screw Dislocation Lattice defect that occurs when the lattice is cut and shifted by a row of atomic spacing.

Second Quantum Number Number that describes the shape of the electron cloud.

Secondary Bonding Highly distance-dependent bonding between adjacent polymer chains; usually includes hydrogen bonding, dipoles, and Van der Waals forces.

Secondary Creep Stage in which the rate that dislocations propagate equals the rate at which the dislocations are blocked, resulting in a fairly linear region on the strain-time plot.

Selective Leaching Preferential elimination of one constituent of a metal alloy.

Self-Expanding Stents Stents that are deployed by the use of a catheter.

Semiconductors Materials having a conductivity range between that of conductors and insulators.

Sensitized Process by which a material becomes more susceptible for corrosion when a stabilizing element precipitates out near a grain boundary.

Shape Memory Effect Effect in which the alloy does not change shape when the load is removed but does return to its initial lattice position when heated.

Shot Size Specified weight of a polymer that is injected into the mold at the end of the barrel during the injection molding process.

Side Groups Atoms attached to the polymer backbone. Also called *substituents*.

Silica Glass Noncrystalline solid formed from the cooling of molten silicon dioxide (SiO_2).

Silicone Gel Implant Breast implant that uses a silicone rubber sac filled with a silicone gel.

Simple Cubic Bravais lattice that has one atom in each corner of the unit cell.

Sintered Formed into a solid from particles by heating until individual particles stick together.

Sinus Node Cluster of cells that generate an electrical signal that controls the pacing within a normal heart.

Slip Movement of dislocations through a crystal, caused when the material is placed under shear stress.

Slip Direction Direction in which a dislocation moves during the slip process.

Slip Planes New planes formed after the material has undergone slip.

Slip System Composed of both the slip plane and the slip direction.

Smelting Process that refines metal oxides into pure metals.

Snell's Law Equation describing the change in velocity of electromagnetic waves passing between two media.

Soda–Lime Glass Most common glass composition that includes silicon dioxide (72%), soda (14%), and lime (7.9%) as primary constituents.

Sodium Chloride Structure Lattice system in which anions fill the face and corner sites in an FCC lattice while an equal number of cations occupy the interstitial regions.

Softbaking Process of removing residual solvents from the photoresist.

Sol Gel Material formed by forming a colloidal suspension of metal salts then drying them in a mold into a wet, solid gel.

Solidus Line Line on a phase diagram below which only solids exist at equilibrium.

Solubility Amount of a substance that can be dissolved in a given amount of solvent.

Solution Heat Treatment First step in age hardening that involves heating until one phase has completely dissolved in the other.

Solution Spinning Process used to make thermoset fibers by performing the polymerization reaction in a solvent as the material flows through a spinneret and into a quenching bath.

Solvus Lines Lines defining the border between the pure solid phase and a blend of two solid phases on a phase diagram.

Specular Reflection Reflection from smooth surfaces with little variation in the angle of reflection.

Spheroidite Nonequilibrium product of steel with cementite spheres suspended in a ferrite matrix.

Spin A theoretical concept that enables individual electrons within sublevels to be distinguished from each other.

Spinel Structure FCC-lattice system with two species of cation, one occupying tetrahedral sites and the other occupying octahedral sites. Anion species occupy the face and corner sites in the lattice.

Spinneret   Circular, stationary block with small holes through which molten polymer can flow to take the shape of a fiber.

Stabilizers   Additives that improve a polymer's resistance to variables that can cause bonds to rupture, such as heat and light.

Stabilization   Conversion of a carbon fiber precursor to a thermally stable form that will not melt during carbonization.

Staggered Conformation   Arrangement of the largest substituents where the substituents are offset by 120 degrees.

Stainless Steel   Carbon–iron solid solutions with at least 12% chromium.

Standard Deviation   Square root of the variance. This value provides more information about the distance from the mean a random sample is likely to be.

Starr-Edwards Valve   Only modern caged-ball artificial heart valve design approved for use.

Stenosis   Hardening of a heart valve, which prevents it from opening properly.

Step-Growth Polymerization   See *condensation polymerization*.

Strain Hardening   See *cold working*.

Stress Concentration Factor   Ratio of the maximum stress to the applied stress.

Stress Corrosion   Loss of material resulting from the combined influence of a corrosive environment and applied tensile stress.

Stress Intensity Factor   Term that accounts for the increased stress applied to an elliptical crack whose length is much greater than its width.

Stress Raisers   Cracks, voids, and other imperfections in a material that cause highly localized increases in stress.

Structural Biomaterials   Materials designed to bear load and provide support for a living organism, such as bones.

Structural Clay Products   Any ceramic materials used in building constructions, including brick and terra cotta.

Structural Unit   See *repeat unit*.

Substituent   See *side groups*.

Substitutional Defects   Point defects that result when an atom in the lattice is replaced with an atom of a different element.

Substitutional Diffusion   Movement of an atom within the lattice itself into an unoccupied site.

Sustainability   Length of time a material will remain adequate for use.

Syndiotactic Dyad   Configuration of a polymer in which the substituent is located on opposite sides of the molecule in each repeating unit.

T

Tacticity   Relative configuration of adjacent asymmetric carbons.

Temper Designation   Nomenclature that shows whether an aluminum alloy was strain hardened or heat treated.

Tensile Modulus   See *elastic modulus*.

Tensile Strength   Stress at the highest applied force on a stress-strain curve.

Tensile Test   Method used to determine the tensile strength, breaking strength, and yield strength of a sample.

Termination   Final step in the polymerization process, which causes the elongation of the polymer chain to come to an end.

Terra Cotta   Ceramic material made with an iron-oxide rich clay that is recognized by its reddish-orange color.

Tertiary Creep   Final stage of creep, during which the rate of deformation accelerates rapidly and continues until rupture.

Tetrahedral Positions   Four interstitial sites present in lattices that form a tetrahedron when lines are drawn from the center of the sites.

Theoretical Density   Density a material would have if it consisted of a single perfect lattice.

Thermoplastic   Polymer with a low melting point due to the lack of covalent bonding between adjacent chains. Such polymers can be repeatedly melted and re-formed.

Thermoset   Polymer that cannot be repeatedly melted and re-formed because of strong covalent bonding between chains.

Third Quantum Number   Number representing the orientation of the electron cloud.

Thrombosis   Blood clotting.

Tie Line   Horizontal line of constant temperature that passes through the point of interest.

Tilting Disk Valves   Artificial heart valves with circular disks that regulate the flow of blood.

Time Value of Money   Concept that future money is worth less than money in the present because of the interest it could have earned.

Tool Steels   Carbon–iron solid solutions with high carbon content that result in increased hardness and wear resistance.

TOSLINK   Common connector for optical cables.

Toughness   Property defining a material's resistance to a blow that is measured by an impact test.

Tow   Large spool that is used to wind solidified polymer fibers after they have been pushed through the spinneret.

trans-Conformation   Conformation in which the largest substituents are offset by 180 degrees. This conformation is typically the most favorable one.

Transducers   Devices that convert sound waves to electric fields.

Transistors   Amplifying or switching devices in microelectronics.

Transmission   Process in which incoming photons pass through a material without interacting.

Transmission Coefficient   Fraction of light not reflected at the interface of two media and instead enters the second media.

Transmission Electron Microscopy   Electron microscope that passes the electron beam through the sample and uses the differences in the beam scattering and diffraction to view the desired object.

Transradial   Below the elbow.

Transtibial Prosthesis   Artificial limbs beginning below the knee.

Transverse Direction   Direction perpendicular to the fibers in a composite.

Tricuspid Valve   Heart valve separating the right atrium from the right ventricle.

True Strain   Measures the ratio of the instantaneous length of the chain to the initial length of the chain.

True Stress   Ratio of the force applied to a sample and the instantaneous cross-sectional area of the sample.

Tridymite   High-temperature polymorph of silicon dioxide (SiO_2) that exhibits a hexagonal lattice.

Turbostratic   Structure in which irregularities in otherwise parallel planes cause distortion.

t-Table   Statistical table based on the degrees of freedom and the level of uncertainty in a set of reported sample values.

T-T-T Plot   See *isothermal transformation diagram.*

U

Ultrafiltration   Process in which the differential pressure causes water to pass through the membrane in dialysis, reducing the excess fluid and electrolyte concentration in blood.

Uniform Attack   Type of corrosion in which the entire surface of the metal is affected.

Unit Cell   Smallest subdivision of a lattice that still contains the characteristics of the lattice.

V

Vacancies   Point defects that result from the absence of an atom at a particular lattice site.

Vacancy Diffusion   Movement of an atom within the lattice itself into a unoccupied site.

Valence Band   Band containing covalently bonded electrons.

Variance   Statistical quantity that takes into account the random error from a variety of sources and provides information about the spread of the data.

Vascular Stent   Small metallic tubular mesh inserted into a blood vessel during angioplasty to keep the artery open after the procedure.

Ventricular Assist Device (VAD)   Device that helps a damaged heart increase its functionality and throughput.

Vinyl Ester Resin   Polymeric matrix material that combines the economic advantages of polyester resins and the exceptional properties of epoxy resins.

Vinyl Monomer   Double-bonded organic molecule used to begin addition polymerization.

Viscose   Process used to make rayon, which involves treating cellulose from wood or cotton with alkali and extruding it through a spinneret.

Vitrification   Heating process in which a glassy solid develops large-scale motion.

Vulcanization   Process by which chemical cross-linkages can form between adjacent polymer chains, strengthening the material without significantly damaging its elastic properties.

W

Warm-Mix Asphalt Concrete (WAM) Process   Process of creating asphalt concrete, by adding zeolites to lower the softening temperature by up to 25°C. This reduces the emission released and the cost, and creates more pleasant working conditions.

Weight Average Molecular Weight   One method of expressing the molecular weight of a sample of polymers with averaging based on weight. This method is more useful when large molecules in the sample dominate the behavior of the sample.

Wet-Filament Winding   Process of creating more complicated shapes of fiber-reinforced composites into a desired shape.

Wet Out   Quality of bonding between the fiber and the matrix in a composite material.

Whitewares   Fine-textured ceramics used in dinnerware, floor and wall tiles, and sculptures.

Widmanstätten Structure   Microstructure present in brass in which α-phase grains are surrounded by β-phase precipitate.

Wolf's Law   Rate of bone growth will adapt to repeated environmental stresses, becoming stronger when exposed to high stress levels and weaker when the stress is reduced.

Woven Bone   During growth and repair, bone produced in which the collagen fibrils are aligned randomly.

Wrought   Shaped by plastic deformation.

X

Xenograft   Implant of tissues from another species.

Y

Yield Strength   Stress at the point of transition between elastic stretching and plastic deformation.

Young's Modulus   See *elastic modulus*.

Y-TZP Femoral Head   Bioceramic artificial replacement femoral head using yttria-stabilized tetragonal zirconia polycrystals, which offer better wear rates and better strength than titanium or alumina femoral heads.

Z

Zener Breakdown   Rapid carrier acceleration caused by the reverse bias becoming too large, which excites other carriers in the region to cause a sudden large current in the opposite direction.

Zinc-Blend Structure   FCC-lattice system with equal number of cations and ions in which each anion is bonded to four identical cations.

INDEX

A

Abrasives, 198–200
Accelerated aging studies, 96–97
Accelerators, 238
Acetabulum socket, 277
Acrylic, 153–154
Addition polymerization, 161–163
Additives
    in glasses, 201–204
    in polymers, 176–177
Admixtures, 238
Advanced ceramics, 212–213
Age hardening, 131–132
Aggregate, 237–238
Air-entrainer, 238
Alloy steels, 125–127
Alloys 23, 110–136
Aluminum Alloys, 135–136, 142
Amalgam, 284
American Concrete Institute
    (ACI), 239
Annealing, 108–109
Aortic valve, 289
Aramid, 154
Arrhenius equation, 54
Artificial pacemaker, 292
Ashby-style chart, 9
Aspect ratio, 228
Asphalt, 241–242
ASTM standards, 68–69
Atomic arrangement, 33–34
Atomic packing factor, 37–39
Atomic structure, 33–34
Attenuation, 267
Austenite, 119
Austenizing, 121
Autograft, 291

B

Bainite, 122–124
Balloon angioplasty, 280
Bar size, 240
Bauxite, 135
Begin curve, 128
Bend test, 70, 80
Bernoulli equation, 290
Binary eutectic system, 116–118
Binary isomorphic system, 111–116
Bio-based materials, 275
Bioceramics, 212, 277

Biocompatibility, 275
Bioethics, 294–295
Biological materials, 274–275
Biomaterials
    functional, 25, 285–293
    structural, 25, 275–285
Biomimetic materials, 275
Biotechnology, 274
Bipolar Junction Transistor
    (BJT), 259
Bisque, 211
Blends, 153
Blister copper, 132
Blown-film apparatus, 179–180
Bohr Model, 10
Bonding
    covalent, 13–15
    hydrogen, 16
    ionic, 13
    metallic, 17
Bone, 275–278
Bort, 215
Boules, 60
Bradycardia, 292
Braggs' equation, 46
Branching, 156, 168
Brass, 110, 131–133
Bravais lattice, 33–35
Breaking strength, 73
Breast implants, 282–283
Brinell hardness, 81–82
Brittle material, 75
Bronze, 132–135
Buckminster Fullerenes, 219–220
Burgers vector, 55

C

Caged-ball valve
Calcination, 205
Calcium fluoride (CaF_2) structure, 195–196
Cantilever beam test, 95–96
Capacitor, 261–262
Capillary pores, 208
Capsular contracture, 282
Carbon fibers, 24, 216–218
Carbon nanotubes, 24, 220–221
Carbon steel, 118–131
Carbonization, 216
Cast alloys, 135–136
Cathodic protection, 139